Environmental Change

Environmental Change

Edited by **Rosemary Charles**

SYRAWOOD
PUBLISHING HOUSE

New York

Published by Syrawood Publishing House,
750 Third Avenue, 9th Floor,
New York, NY 10017, USA
www.syrawoodpublishinghouse.com

Environmental Change
Edited by Rosemary Charles

International Standard Book Number: 978-1-68286-174-5 (Hardback)

Printed in the United States of America.

Contents

Preface

Every book is a source of knowledge and this one is no exception. The idea that led to the conceptualization of this book was the fact that the world is advancing rapidly; which makes it crucial to document the progress in every field. I am aware that a lot of data is already available, yet, there is a lot more to learn. Hence, I accepted the responsibility of editing this book and contributing my knowledge to the community.

Human activities as well as natural ecological processes cause disturbances in the balance of the environment. Climate change is a primary result of such activities which have further negative impact on the environment. Topics such as environment quality management, ecological impacts of pollutants, waste management, climate change, etc. have been discussed in detail in this book. Scientists and students actively engaged in this field will find this text full of crucial and unexplored concepts. It will also prove beneficial for environmentalists, ecologists, climatologists, etc.

While editing this book, I had multiple visions for it. Then I finally narrowed down to make every chapter a sole standing text explaining a particular topic, so that they can be used independently. However, the umbrella subject sinews them into a common theme. This makes the book a unique platform of knowledge.

I would like to give the major credit of this book to the experts from every corner of the world, who took the time to share their expertise with us. Also, I owe the completion of this book to the never-ending support of my family, who supported me throughout the project.

Editor

PCDD/F and Dioxin-Like PCB Determinations in Mtoni Estuarine Sediments (Tanzania) Using the Chemically Activated Luciferase Gene Expression (CALUX) Bioassay

Matobola J. Mihale[1,2], Kim Croes[1], Clavery Tungaraza[3], Willy Baeyens[1] & Kersten Van Langenhove[1]

[1] Department of Analytical and Environmental Chemistry, Vrije Universiteit Brussel (VUB), Pleinlaan, Brussels, Belgium

[2] Department of Physical Sciences, Open University of Tanzania (OUT), Dar es Salaam, Tanzania

[3] Department of Physical Sciences, Sokoine University of Agriculture (SUA), Chuo Kikuu Morogoro, Tanzania

Correspondence: Kersten Van Langenhove, Department of Analytical and Environmental Chemistry, Vrije Universiteit Brussel (VUB), Pleinlaan 2, Brussels 1050, Belgium.

Abstract

Sediments from Mtoni estuary and 2 tributaries, Tanzania, were screened for polychlorinated-p-dibenzodioxins, -dibenzofurans (PCDD/Fs) and dioxin-like PCBs (dl-PCBs) using the chemically activated luciferase gene expression (CALUX) bioassay approach. PCDD/Fs expressed as bio analytical equivalence (BEQ) values ranged from 5.7 ± 1.4 to 39.9 ± 5.8 pg BEQ/g sediment in the wet season and from 14.1 ± 2.0 to 32.8 ± 4.7 pg BEQ/g sediment in the dry season, with higher levels observed in Kizinga River and stations close to the mouth of that river. Dioxin-like PCB levels ranged from 0.21 ± 0.03 to 0.53 ± 0.03 pg BEQ/g sediment in the wet season and from 0.22 ± 0.03 to 0.59 ± 0.04 pg BEQ/g sediment in the dry season. Higher PCDD/F and dl-PCB levels in sediments are probably related to open burning of plastic scraps, household burning of wood or charcoal and traffic related emissions, all of which occur in the Dar es Salaam region. The denser population and the more intense industrial activities in the Kizinga River basin may explain the enhanced PCDD/F and dl-PCB levels observed in the sediments of that river compared to the levels in the Mzinga River basin. A third sampling campaign, including also stations in the downstream estuary, confirmed the enhanced levels in the Kizinga River (maximum of 400 pg-BEQ/g) and also showed that a clear decreasing concentration gradient in the downstream direction exists. It cannot be excluded that the levels of these pollutants in the sediments of the Mtoni estuary pose a threat to the local biological community.

Keywords: PCDD/Fs, Dioxin-like PCBs, sediment, Mtoni Estuary, Tanzania, CALUX Bioassay, ecological risk

1. Introduction

Polychlorinated dibenzo-p-dioxins and polychlorinated dibenzofurans, which are collectively referred to as PCDD/Fs, have no commercial use and occur in the environment as unintended by-products of technological processes (Pan et al., 2010; Roots, Henkelmann, & Schramm, 2004). PCDD/Fs can be formed from natural combustion processes like bushfires and volcanoes (Birch, Harrington, Symons, & Hunt, 2007) or during incomplete anthropogenic combustion processes of chlorinated organic wastes such as for example incineration of polyvinyl chloride plastics (Birch et al., 2007; Terauchi, Takahashi, Lam, Min, & Tanabe, 2009). Formation of dioxins in such incinerators occurs due to the presence of both chlorine and catalytic metals (Manahan, 2008). Industrial activities such as metallurgy and manufacture of chlorinated chemicals, like wood preservatives and pesticides, can also produce dioxins (El-Kady et al., 2007; Ryoo et al., 2005).

Polychlorinated biphenyls (PCBs) were once produced commercially (Koistinen, Stenman, Haahti, Suonper, & Paasivirta, 1997; Srogi, 2007) and used in industrial and consumer products (Liu et al., 2006; Wang et al., 2007), such as anti-corrosion materials, coolants and insulators in heat transfer systems, electronic appliances and hydraulic fluids (Shen et al., 2009; Yang, Shen, Gao, Tang, & Niu, 2009) and as capacitors in electrical industries (Pan et al., 2010). Common PCB sources in the environment include the use as well as disposal of PCB-containing products and the formation of PCBs as by-products at low temperature (less than 800 °C) waste incineration (Chi, Chang, & Kao, 2007). Twelve of these PCBs have a planar structure and elicit biochemical and toxic responses similar to dioxins and are therefore known as dioxin-like PCBs or dl-PCBs (Okay et al., 2009).

Both natural and anthropogenic sources can lead to increased levels of these compounds in estuaries and coastal marine ecosystems (Kumar, Sajwan, Richardson, & Kannan, 2008). Major anthropogenic activities are linked to population growth, urbanization and industrialization (Kumar et al., 2008; Müller et al., 2002) and include effluents from municipal wastewater plants (Moon, Choi, Choi, Ok, & Kannan, 2009), combustion processes from waste incinerators or cement manufacturing, power plants and automobile exhausts (Zhang et al., 2010) and industrial processes like pulp bleaching and metallurgy (Bruckmeier et al., 1997).

Mangrove sediments can act as sinks and later as sources of PCDD/F and PCB contaminants to marine environments (Chi et al., 2007; Guzzella et al., 2005; Müller et al., 1999). Hence, sediments can be used to evaluate pollutant sources, historical trends and fate processes, since the amounts of these compounds in sediments will reflect regional discharges (Lee, Kim, Chang, & Moon, 2006; Moon et al., 2009). Contaminated sediments may therefore threaten the lives of organisms in the marine environment due to the toxicity, long time persistence, bioaccumulation and biomagnification of these lipophilic organic micro-pollutants (Kumar et al., 2008; Zhao et al., 2010).

Various analytical methods have been used to characterize PCDD/Fs and dl-PCBs in a sediment matrix. Gas chromatography-high resolution mass spectrometry (GC-HRMS) offers a possibility to chemically identify and quantify individual congeners (Besselink et al., 2004; Denison et al., 2004; Denison, Pandini, Nagy, Baldwin, & Bonati, 2002; Schecter et al., 1999) in the matrix and to enable the assessment of risks associated with the congeners (Long et al., 2006). To estimate the risks from the GC-HRMS results it is assumed that the additivity principle of a pollutant's response or effect is valid, which means the absence of agonistic and antagonistic interactions, and that these effects are produced through the same mechanism of toxicity. However, it has been shown that complex mixtures of PCDD/F and dioxin-like PCB congeners elicit synergistic and/or antagonistic interactions (Joung, Chung, & Sheen, 2007; Schroijen, Windal, Goeyens, & Baeyens, 2004). In addition, chemical analysis of individual congeners, particularly in small concentrations, can be very expensive and time consuming. Presence of compounds with aryl hydrocarbon receptor (AhR) affinity, but not commonly measured, and the absence of toxicological equivalents (TEQ) for several congeners further limit the use of this analytical method (Joung et al., 2007; Long et al., 2006). To overcome some of these drawbacks, biological assays utilizing either biomolecular techniques (e.g. immunoassays) or living materials (e.g. *in vitro* chemically activated luciferase gene expression, CALUX) have been used as rapid and cost-effective screening methods for chemicals with selective and specific biochemical interactions (Roy, Mysior, & Brzezinski, 2002). For example, the CALUX bioassay screens for chemicals with AhR potential (Denison et al., 2004, 2002; Schecter et al., 1999;) and produces a single integrated biological equivalent of the mixtures (Besselink et al., 2004). CALUX also measures a response which is a single toxicity end-point produced by AhR active compounds that cannot be measured and/or are below the detection level of chemo analysis (Joung et al., 2007). The major drawback of a bioassay such as CALUX is that there is no information about the congener pattern.

The CALUX method has been explained by various authors (Denison et al., 2004, 2002; Murk, 1996;). It uses genetically modified cells (hepatoma cells stably transfected with a luciferase reporter gene) which respond to chemicals that activate the cytosolar AhR by induction of luciferase (Denison et al., 2004). By this method, the toxicity of pollutants such as PCDD/Fs and dioxin like-PCBs is produced either as a change in gene expression mediated through the AhR or by interference with other pathways (Hurst, Balaam, Chan-Man, Thain, & Thomas, 2004). Estimation of relative potency and toxic potential can therefore be done by measuring the activation level of AhR gene expression (United States Environmental protection Agency [USEPA], 2008). However, even when a rigorous clean-up and separation procedure of the sample extract is performed, interferences by PCDD/Fs on dl-PCBs and vice versa or by other AhR ligands are still possible (Sanctorum, Elskens, & Baeyens, 2007).

The literature regarding CALUX analyses in marine sediments is very limited. Most of the PCDD/F analyses in the world have been performed with GC-HRMS. In those studies where the CALUX technique was used, the focus was more on method development, on the comparison with the GC-HRMS method and on the screening of food and feed (Hoogenboom et al., 2006; Van Overmeire, Van Loco, Roos, Carbonnelle, & Goeyens, 2004). In addition, very little data on dl-PCBs are available in literature. The fact is that, PCDD/F BEQ levels are in general by far higher than those of the dl-PCBs. Both chemo-analysis and CALUX analysis research on marine sediments in Africa are scarcely documented (Nieuwoudt et al., 2009; Pieters, 2007). Regarding Tanzania, only total PCBs in sediment (Machiwa, 1992) and PCDD/Fs and dl-PCBs in free range chicken have been reported (International POPs Elimination Network, 2005).

Although there exist no data about the presence of PCDD/Fs and dl-PCBs in the environment (more specifically in the sediment) of Tanzania, the applications of PCBs in electrical transformers and in other equipment is known (Loomis, Browning, Schenck, Gregory, & Savitz, 1997). In coastal Tanzania, there are a lot of municipal, chemical and even hospital wastes that are discharged into the Indian Ocean after incineration and open burning of mixed wastes (Machiwa, 1992). Wood burning is a common source of fuel as most households use either charcoal or firewood for cooking. In many local households, there is uncontrolled burning of plastics. Vehicle emissions are abundant due to increased traffic and importation of old, second-hand cars (Mbuligwe & Kassenga, 1997). The current major outcry of the country has been on the vandalism of electrical transformers (Maleko, 2005) in search of their coolant for unspecified domestic or commercial use. Since Dar es Salaam is by far the largest city in Tanzania, the Mtoni estuary, being the main aquatic system in that area, was selected to study PCDD/F and dl-like PCB levels in the aquatic environment for the first time. The first objective was to assess PCDD/F and dl-PCB levels in sediments of the mixing zone of the Mtoni estuary, including the Kizinga and Mzinga River mouths, during wet and dry seasons. A second objective was to eventually link the observed levels to local sources.

2. Method

2.1 Study Area

The Mtoni estuary (Figure 1) is located at approximately 3 km south of Dar es Salaam (Tanzania) and receives fresh water input from the Kizinga and Mzinga Rivers. Numerous creeks in the estuary and the river mouths host mangrove trees such as *Avicennia marina*, *Bruguiera gymnorrhiza*, *Ceriops tagal*, *Rhizophora mucronata* and *Sonneratia alba* species.

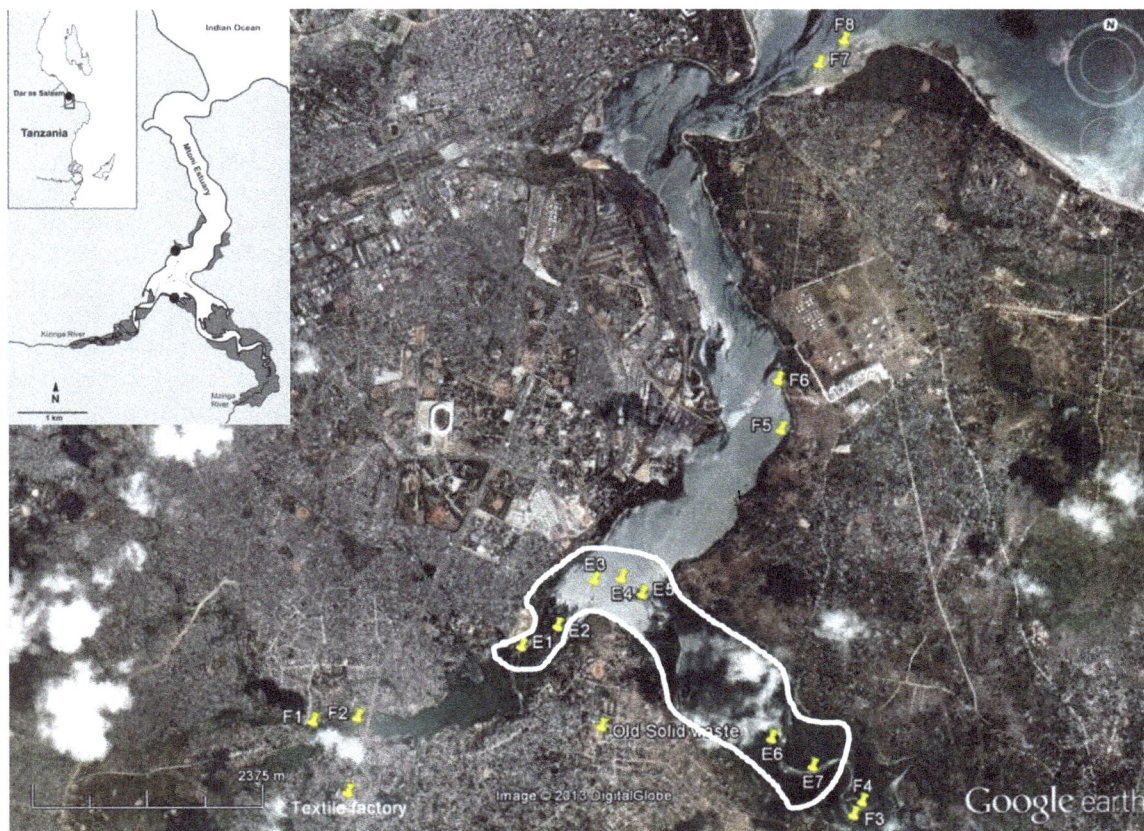

Figure 1. Sampling points in the Mtoni estuary, Dar es Salaam: E1 and E2 in the Kizinga River, E3-E5 at the confluence and E6 and E7 in the Mzinga River. Stations F1-F2 in the Kizinga River, F3-F4 in the Mzinga River, F5-F6 at the Navy shore and F7-F8 at Kigamboni Seaway are additional sampling points. White line delimits the estuarine mixing zone. The solid waste dumping site and textile factory are also indicated

The fresh water input from both rivers is low. An average base-line flow rate of 1 m^3/s is observed in Kizinga river with an increase to 8 m^3/s in the rainy season while the water-flow rate of the Mzinga River is unstable and lower than in Kizinga River (Van Camp, Mjemah, Al Farrah, & Walraevens, 2013). Hence, the effect of the river discharges on the hydrodynamics of the Mtoni estuary is very limited. The seven sampling stations (E1 to E7) are located in the mixing zone and their salinities vary from almost fresh to brackish water with somewhat higher salinities in the dry season (Kristensen et al., 2010). Downstream this mixing zone (stations F5 to F8) the water becomes rapidly sea water while the stations F1 to F4 more upstream are fresh water. This estuarine mixing zone was selected because it integrates influences of natural and anthropogenic sources in the riverine and marine systems.

The Mtoni estuary is highly impacted (Peri-urban mangrove forests as filters and potential phytoremediators of domestic sewage in east Africa [PUMPSEA], 2007) by discharges of various origin: (1) the Kizinga and Mzinga Rivers that drain the mangrove forest (Kruitwagen, Pratap, Covaci, & Wendelaar-Bonga, 2008), (2) the wastewater drainage systems from industrial and residential areas (of a population of around 500,000 inhabitants (National Bureau for Statistics [NBS], 2003)), (3) charcoal burning, (4) mangrove harvesting for residential places, (5) salt mining, (6) tourism and (7) agriculture (Taylor, Ravilious, & Green, 2003).

The Kizinga River that drains the urbanized areas of Keko, Chang'ombe, Kurasini and Temeke (approximately 400,000 inhabitants (NBS, 2003)) is suspected to carry a variety of wastes and discharges originating from agricultural, industrial as well as residential sources (Taylor et al., 2003). The Mzinga River, on the other hand, drains the rural areas of Vijibweni, Tuangoma and Mji Mwema with a population of around 90,000 (NBS, 2003). Due to rapid growth of settlements along the Mzinga creek resulting from increased human population, the river is suspected to carry agricultural and residential wastes and discharges presumed to be emptied into the creek. The estuary further receives inputs from the Dar es Salaam harbor which is located near the mouth of the estuary during diurnal tides (up to 5 m amplitude) and from the Mtoni solid waste dumping site located in between the two rivers.

2.2 Sampling

Sampling of sediments was conducted in the mangrove forests during low tides at Kizinga and Mzinga creeks (Figure 1) of Mtoni estuary. Two sampling campaigns were conducted: one during the wet season (19th-20th January, 2011) and a second during the dry season (15th-16th August, 2011). December and January have an average precipitation rate of 194 and 89 mm respectively, while these rates in July and August are much lower with 48 and 47 mm respectively. River flows in the Dar es Salaam area are mainly controlled by the precipitation rate in the previous period. The flows of Kizinga and Mzinga Rivers are highest in the wet season (the highest discharge rates can go up to 15 m^3/s for the Kizinga River and 7 m^3/s for the Mzinga River) while in the dry season, base-line flows of 1 m^3/s in the Kizinga River and even lower in the Mzinga River are observed (Van Camp et al., 2013). The impact of both rivers on the pollutant levels in the mixing zone can thus best be estimated by sampling in this mixing zone at high (wet season) and at low (dry season) river flow and comparison of these results.

Samples were collected from exactly the same locations during both campaigns. Seven sampling stations were identified using a hand-held global positioning system (GPS): two in the Kizinga River (E1 and E2), two in the Mzinga River (E6 and E7) and three at the confluence of the two rivers (E3, E4 and E5).

From the results obtained during the wet and dry seasons in the mixing zone of the estuary, it appeared that the sampling stations in Kizinga River and close to its mouth showed higher PCDD/F values than the stations in and close to Mzinga River. It was thus interesting to investigate PCDD/F levels more upstream in both rivers. In addition, the salinity gradient in the mixing zone is also small and real marine water samples were not included in the previous samplings. We were thus not able to appreciate any evolution of the PCDD/F levels from the estuarine mixing zone towards the marine environment. Therefore, an additional sampling campaign was organized at 3 end-members in October 2012: (1) one site in the Kizinga River (fresh water stations F1-F2) much more upstream than stations E1 and E2, (2) one site in Mzinga River (fresh water stations F3-F4 are slightly more upstream than stations E6 and E7, but these latter stations were yet, compared to the stations E1 and E2 in Kizinga River, much more upstream) and (3) two sites in the marine area, close to and at the mouth of the estuary (respectively marine water stations F5-F6 and F7-F8) (Figure 1).

All those samples were taken from two sub-sites within a distance of 20 m, except in the Kizinga River. The first subsample was taken at the junction of the river and the textile wastewater stream (Figure 1) and the second was taken 200 m upstream of the river very close to unauthorized human settlements.

Sediment sampling was done as described by EPA (United States Environmental protection Agency [USEPA],

2001) using a hand corer (30 cm height, 6 cm internal diameter). The corer was gently pushed in the mangrove sediments, closed at its upper end with a lid and smoothly removed by twisting and pulling. The sediments were then pushed out of the corer tube using a piston and sectioned into three segments corresponding to depth intervals of 0-3, 3-6 and 6-9 cm. All sediment samples were packed in prior labeled and zipped polyethylene bags, stored in iceboxes and later frozen to -20 °C. Sediment samples were then air-transported (frozen) to the laboratory of the Department of Analytical and Environmental Chemistry, Vrije Universiteit Brussel (VUB) in Belgium for dioxins and dioxin-like compounds analyses.

2.3 Chemical Reagents and Standards

Acetone (Pesti-S grade, minimum 99.9%), n-hexane (minimum 96% assay) and toluene (minimum 99.8% assay) both dioxins and PCB grade, were purchased from Biosolve (The Netherlands). Ethyl acetate (Pestanal, 99.8% assay) was purchased from Sigma-Aldrich (Germany). Sulphuric acid (95-97% w/w, ACS reagent) and Dimethylsulfoxide (DMSO) were obtained from Merck (Germany). Glass fibre filters were purchased from Whatman (UK). Alpha-minimal essential medium (α-MEM), fetal bovine serum (FBS) and trypsin (0.25%) were obtained from Gibco, UK. Phosphate buffered saline (PBS) was obtained from Ambion (UK). Luciferase assay substrate and buffer were purchased from Promega (The Netherlands). Anhydrous sodium sulphate was purchased from Boom (The Netherlands). The X-CARB was purchased from Xenobiotic Diagnostics Systems, XDS Inc, USA and the solution of 2,3,7,8-TCDD standard (50 µg/mL, purity 99%) was purchased from Campro Scientific (The Netherlands).

2.4 Determination of Particle Size and Total Organic Carbon (TOC)

The concentration of TOC was determined in all 50 sediment samples with a CHN analyzer (Carlo Erba) on a known amount (about 12 mg) of sediment sub-sample placed in a silver capsule and pre-treated by acidification with 5% HCl. The grain size distribution was determined by an external laboratory at 3 stations with low, medium and high organic matter (OM) content in their sediments (stations E1, E2 and E7) to test the relation between both variables. Approximately 10 g lyophilized and homogenized sediment sample was prepared by removing salts, OM and carbonates using hydrogen peroxide and hydrochloric acid respectively. A stable suspension was obtained after rinsing and adding 5 ml of a peptizing agent. The coarse fraction (above 75 µm) was separated by wet sieving on a 75 µm sieve, then dried at 105 °C, and finally dry sieved. The grain-size distribution of the fine fractions 2-75 µm and <2 µm was obtained using the Sedigraph 5100 coupled to a Mastertech 51. The precision for 10 consecutive measurements on aliquots of the same sample was around 1% for each grain-size fraction.

2.5 Analysis of Dioxin and Dioxin-Like Compounds in Sediments

2.5.1 Sample Preparation

In total, 50 sediment samples were analyzed for PCDD/Fs and dl-PCBs. Lyophilized sediment (5 g) was extracted using pressurized liquid extraction in an Accelerated Solvent Extractor, ASE®, (Dionex, USA) with a toluene:methanol (4:1 v/v) solvent system (Baston & Denison, 2011) and 33 mL extraction cells. The ASE extraction conditions were: 125 °C oven temperature; 1500 psi (100 MPa) pressure; 10 min static time; 6 min oven heating time; 60 s purge time; 60% of extraction cell volume as flush volume and 2 static cycles. The extracts were then concentrated in a vacuum centrifuge to near dryness and later re-suspended in n-hexane (5 mL).

2.5.2 Column Preparations for Clean-Up and Fractionation

The clean up and fractionation is based on the EPA Method 4435 (United States Environmental protection Agency [USEPA], 2008) from which we use the same piggybacked setup of columns. A sequential setup of columns is used to remove PAHs and break down undesired compounds (acid silica gel) and differentially elute PCDD/Fs and dl-PCBs (X-CARB affinity chromatography column). An additional, third, column was added for sulphur removal. Column preparation is described below and all columns are prepared daily.

An activated copper column (for elemental sulphur removal) was prepared by filling a Pasteur pipette from bottom to top with glass wool and 1 cm of activated (with a 20% hydrochloric acid solution) copper. The activated copper column was first rinsed with de-ionized water (Milli-Q): 3 x 1 mL and then with acetone, toluene and n-hexane (each 3 x 1 mL) in that order. The activated copper columns were stored submerged in n-hexane to avoid oxidation.

An acidified silica column was prepared by filling a 10-mL disposable column (ID 0.8mm), from bottom to top, with glass wool, sodium sulphate (0.5 cm^3), sulphuric acid-silica gel (33% H_2SO_4 on silica gel w/w; 4.3 cm^3) and sodium sulphate (0.5 cm^3). The acid silica column was then rinsed with n-hexane (3 x 10 mL).

Similarly, an X-CARB column was prepared by using an open ended tube (ID 0.8 mm), but this was filled (bottom to top) with glass wool, sodium sulphate (0.5 cm^3), 1% X-CARB (1 cm^3 packed), sodium sulphate (0.5 cm^3) and glass wool. The column was inverted and rinsed sequentially with acetone (5 ml), toluene (20 ml) and n-hexane (10 ml).

The acid silica gel column is placed on top, the copper column in the middle and the X-CARB column at the bottom end prior to sample loading. The individual columns are connected to each other and rinsed with n-hexane ensuring that the columns do not run dry.

2.5.3 PCDD/Fs and Dl-PCBs Clean up and Fractionation

The sediment extract in n-hexane was first sonicated for 5 minutes followed by vigorous vortexing. An aliquot (2 mL from the original 5 mL) was quantitatively loaded on the acid silica gel column, followed by elution of the column with n-hexane (total 21 mL). The acid silica gel and activated copper columns were removed once the solvent had passed through. The remaining X-CARB column was further rinsed with extra n-hexane (5 mL), followed by elution with a mixture of 8:1:1 of n-hexane:toluene:ethylacetate (3 x 5 mL) to collect the fraction containing coplanar PCBs (i.e. PCB fraction). The fraction containing the PCDD/Fs (dioxin fraction) was afterwards eluted (back-flushed) with toluene (3 x 5 mL) after inverting the X-CARB column. The PCB and dioxin fractions were later concentrated to dryness in a vacuum centrifuge and re-suspended in n-hexane (4 mL) for CALUX analysis.

2.5.4 CALUX Analysis

CALUX analysis was performed as described elsewhere (Van Langenhove et al., 2011). Briefly, mouse hepatoma cells (H1L7.5c1 cell line) were cultured in α-MEM with 10% FBS at 37 °C, 80% relative humidity and 5% CO_2. The cells were seeded (at an approximate density of 7.5 x 10^5 cells/mL in a clear bottom 96-well plate (Greiner Bio-One, Germany) and incubated for 24 hrs to reach a monolayer of cells.

TCDD standard solutions (from 125 nM down to 30 pM) were made in DMSO and 4 µL of this solution was transferred to 2 mL of n-hexane. Samples were serially diluted in n-hexane and 4 µL of DMSO was added to each vial as a carrier solvent. For both the TCDD and sample solutions, the remaining n-hexane was evaporated using a vacuum centrifuge, leaving behind either 4 µL DMSO with TCDD standard or 4 µL DMSO containing either PCDD/Fs or dl-PCBs from the sample extract.

Prior to dosing, these standards and sample dilutions were diluted 100 times with cell culture media (4 µL standard solution or sample in 396 µL α-MEM with 10% FBS). Final solutions were vortexed vigorously and dosed in triplicate (3 x 100 µL per well).

After 24 hrs of incubation, cells were rinsed and visually inspected for abnormalities. Then, cells were lysed and shaken for 5 min. Luciferin treatment was performed in the Glomax 96-well microplate luminometer (Promega, USA), where the light output in relative light units (RLUs) was measured. Sample responses were expressed as percentage maximum induction to 2,3,7,8-TCDD (%TCDD$_{max}$).

2.6 Statistical Analysis of Data

TCDD standards were used to generate the calibration curve. A four-variable Hill equation fitting the calibration curve (Elskens et al., 2011) was used to produce a sigmoid curve of the standard solutions. The calibration equation was then used to convert the measured RLU values of the samples into a biological equivalency (CALUX-BEQ) value (Goeyens et al., 2010) by comparing the sample response curve with the sigmoid dose-response curve (Elskens et al., 2011). Further data treatment was done using slope ratio and Box-Cox transformation methods by linearization of the non-linear Hill regression equation as described previously (Elskens et al., 2011). Effective concentrations at 50% TCDD$_{max}$ (EC$_{50}$) of standard and samples were used to determine the potency or bio analytical equivalency (BEQ) of the samples. Statistical analysis was performed using Microsoft Spreadsheet for Windows 2007 and graphical representations were performed using Spreadsheet and Sigmaplot programs (SigmaPlot 10.0). Pearson Correlation and Principal component analysis (PCA) were performed using Predictive Analytic Software (PASW, version 16.0 for Windows) with the PCDD/F and dl-PCB values and the geochemical properties (% total nitrogen (TN), % total carbon (TC) and %TOC) as variables (total 5), and using the mean concentrations of both seasons.

2.7 Quality Control (QC) and Quality Assurance (QA)

For each batch of samples, a blank sample was introduced through the complete treatment procedure (procedural blank) to monitor the activity contributed by solvents and column matrices used in the sample treatment. Moreover, DMSO and media blanks were added during dosing to detect contamination and to determine the experimental background level. All blank samples were measured in triplicate and were treated in a similar way as real samples. The results with $p < 0.05$ (Student T-test) were considered statistically significant.

The limit of detection (LOD) was calculated according to the IUPAC definition. The blank value was taken as the average background of the model fit represented in %RLU relative to the maximum TCDD-induced RLUs (Elskens et al., 2011).

QC experiments were conducted on each 96-well plate, using an in-house QC solution (0.250 pg TCDD/μL), to assess precision of the CALUX method and detect bias. Procedural blank fractions were also spiked with the same QC solution prior to dosing to detect agonists and/or potential synergetic or antagonistic effects.

3. Results and Discussion

3.1 Grain Size Distribution

Sandy particles dominated the mangrove sediments in the study area, with sand ($x > 75$ μm) contributing for more than 60% of the weight. The correlation between TOC and the fine grain size fraction (% < 2 μm) was good ($r^2 = 0.92$). This was still the case ($r^2 = 0.82$) between TOC and the mud+silt fraction (% < 75 μm). It is well-known that muddy sediments having a high TOC content but also a high amount of fine grain size fraction (< 2 μm), accumulate by far higher amounts of pollutants than sandy sediments (Baeyens et al., 1991). This means that TOC values can eventually be used to normalize the pollutant concentrations in the sediment versus the mud fraction (% < 2 μm).

In the Mtoni estuary, the high sand proportion implies that the capacity of the mangrove sediments to adsorb the dioxins and dl-PCBs is medium to low. On the other hand, a high sand fraction favors abiotic processes such as enhanced diffusion of oxygen in the sediment making pollutants more bioavailable due to faster oxidation of organic matter and simultaneous release of associated persistent organic micro-pollutants such as dioxins (Davies & Tawari, 2010; Holmer, 2003).

3.2 The CALUX H1L7.5c1 Bioassay

The CALUX bioassay integrates the responses of every AhR ligand available in the analyzed sample and because of this; it provides only an indication of the possible overall toxicity (Van Langenhove et al., 2011). The results presented in this study show the importance of full-dose curves for environmental samples because the additivity principle is not uphold, despite (1) the use of a sulphuric acid silica gel column in the sample clean up step aimed at eliminating interferences caused by poly-aromatic hydrocarbons (PAHs) and (2) separation of PCDD/Fs and dl-PCBs avoiding known antagonistic effects between these two compound groups (Van Langenhove et al., 2011).

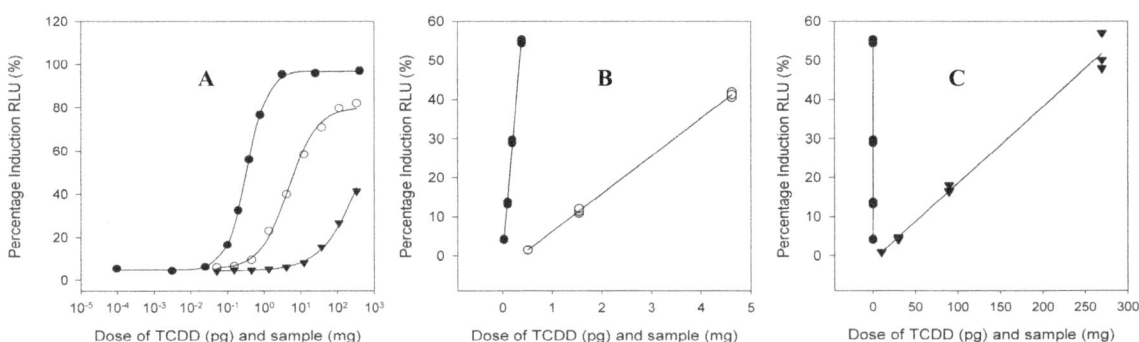

Figure 2. (A) Typical TCDD-standard curve (●), sample with full dose response (○) and sample not attaining an upper plateau (▼) using Hill regression. (B) Same data portrayed using the slope ratio method for TCDD (●) and full dose sample (○). (C) Slope ratio for TCDD (●) and sample without upper plateau (▼)

Luciferase induction was reproducible with coefficients of variation (CV) less than 15% for a given standard or sample extract measured in triplicate. Mean values of luciferase response measured in three replicate wells were used to generate the dose-response curves. The dose-response curve of TCDD standards was sigmoidal in appearance as shown in Figure 2.

3.3 Quality Control

Blank samples (n = 6) spiked with the in-house QC solution (0.250 pg TCDD/µL) ranged in recoveries from 85 to 120% for the PCB fraction and from 91 to 115% for the PCDD fraction, well in accordance with an acceptable relative standard deviation of 20% (80-120% recovery of the TCDD spike). DMSO controls showed no marked difference in response to fitted background values (p = 0.45 for a two-tailed Student T-test). Media controls were generally lower in response, but borderline not statistically significantly different (p = 0.06 for a two-tailed Student T-test).

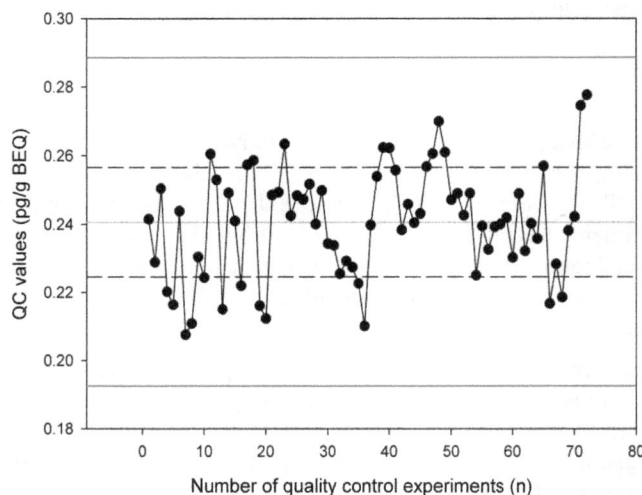

Figure 3. Scatter chart of QC analyses during the study. The dotted lines above and below the mean (0.24) equal the mean ± 1SD values. The 2 additional lines above and below the dotted lines equal the mean ± 3SD values

A total of 72 QC experiments were performed in this study. When these experiments are plotted on a control chart (Figure 3), a mean of 0.24 ± 0.02 pg/g indicated that the analysis procedures were reliable with a coefficient of variation (CV) at 7.2%. The LOD based on the averaged fitted background value and on the DMSO blank was 0.05 pg TCDD/well. Recalculating with the starting volume used in the serial dilution, an LOD of about 0.22 pg TCDD/g sediment is found.

3.4 Bio Analytical Equivalents (BEQs) in Mtoni Sediment Samples

CALUX-BEQ values of PCDD/Fs in Mtoni estuary sediments were assessed using three methods: the Hill regression equation, Box-Cox transformation (both providing an EC_{50} in pg/g BEQ) as well as the slope ratio method (providing a single BEQ value). A detailed description of the methods can be obtained elsewhere (Elskens et al., 2011). The relationships between these methods indicate that they correlate well ($r^2 > 0.8$, Figure 4), implying that either of these methods will provide reliable results in this study. Model precision for BEQ determination based on CVs ranged from 6.7-17.4% (Hill model), 5.2-14.3% (Box-Cox transformation) and 5.5-15.5% (Slope ratio method).

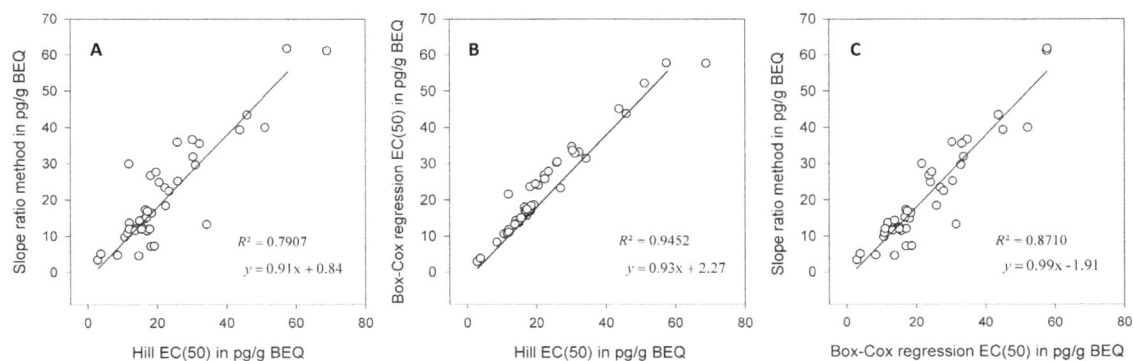

Figure 4. Relationship between PCDD/Fs concentrations (EC_{50}) in Mtoni estuary sediment samples; (A) estimated by Hill equation and slope ratio method, (B) Hill regression and Box-Cox transformation and (C) Box-Cox transformation and Slope ratio

Since the Hill equation is the most used method, and for simplicity of comparison, Hill regression BEQs will be continuously employed in this study and are presented in Table 1. To account for non-parallelism that usually exists between dose-response curves of the reference standards and the sample, the BEQ PCDD/F values in a sample were determined based on the EC_{20}, EC_{50} and EC_{80} of the maximum TCDD result ($TCDD_{max}$).

Dl-PCBs responses are lower than those of PCDD/Fs owing to their lower toxicity equivalent factors (TEFs). Their potency was therefore determined using inverse prediction, assuming a sample behaves like a diluted TCDD standard solution (Elskens et al., 2011).

In order to eliminate concentration differences emanating from variations in the sand fraction of the sediments, pollutant concentrations in the Mtoni estuary sediments were normalised to TOC. Despite the normalisation, PCDD/F and dioxin-like PCB profiles showed no obvious vertical trends. The rather steady depth profiles indicate that the estuarine sediments are fairly well-mixed making the profiles more or less uniform. It is possible that the hydrodynamics of the estuary (tidal amplitude of up to 5 m at the mouth) provide frequent sediment mixing, smoothing out possible vertical concentration gradients. Perhaps, a depth of 9-cm is not sufficient enough to observe any difference and therefore it may be interesting in the future to study pollutant profiles in deeper sediment layers.

Table 1. Mtoni estuary sediment properties and the CALUX-BEQ values for PCDD/Fs and PCBs

Code and Location	Section depth (cm)	Season	TOC (%)	CALUX-BEQ (pg TCDD/g sediment)		
				PCDD/Fs BEQ$_{50}$	PCDD/Fs BEQ$_{20}$-BEQ$_{80}$ range	Dioxin-like PCBs*
	0–3 cm	wet	0.62	31.07	44.6 – 21.6	0.35
E1:		dry	1.86	51.02	63.3 – 41.1	0.47
S 06°52.443	3–6 cm	wet	0.63	34.28	45.4 – 25.9	0.33
E 039°17.014		dry	1.77	30.37	53.0 – 17.4	1.03
	6–9 cm	wet	0.50	22.45	28.0 – 18.0	0.34
		dry	2.23	17.13	24.0 – 12.2	0.26
	0–3 cm	wet	1.02	18.20	22.3 – 14.9	0.38
E2:		dry	1.46	45.84	55.7 – 37.7	0.63
S 06°52.357	3–6 cm	wet	1.02	23.50	31.4 – 17.6	0.34
E 039°17.099		dry	0.56	11.58	15.4 – 8.7	0.26
	6–9 cm	wet	0.56	14.93	21.5 – 10.4	0.33
		dry	0.24	14.02	20.0 – 10.0	0.24
	0–3 cm	wet	2.55	68.87	85.7 – 55.4	0.47
E3:		dry	1.35	57.42	72.6 – 45.4	0.41
S06°52.058	3–6 cm	wet	3.57	30.18	38.2 – 23.9	0.15
E 39°17.355		dry	0.68	22.25	43.6 – 11.4	0.40
	6–9 cm	wet	3.83	20.54	32.2 – 13.1	0.14
		dry	0.66	16.48	19.4 – 14.0	0.11
	0–3 cm	wet	1.04	32.22	42.7 – 24.3	0.48
E4:		dry	0.98	26.93	81.2 – 8.90	0.27
S 06°52.090	3–6 cm	wet	1.15	18.10	31.5 –- 10.4	0.28
E 039°17.501		dry	1.71	43.73	49.9 – 38.4	0.55
	6–9 cm	wet	1.23	16.66	20.0 – 13.9	0.39
		dry	0.96	25.86	42.8 – 15.6	0.10
	0–3 cm	wet	2.38	10.63	12.9 – 8.75	0.25
E5:		dry	0.70	11.88	15.9 – 8.89	0.55
S 06°52.164	3–6 cm	wet	0.90	2.82	3.73 – 2.31	0.18
E 039°17.658		dry	0.80	19.66	35.3 – 11.0	0.47
	6–9 cm	wet	1.40	3.70	5.29 – 2.59	0.23
		dry	0.46	12.02	16.0 – 9.00	0.36
	0–3 cm	wet	3.12	26.00	36.9 – 18.3	0.28
E6:		dry	3.57	18.38	24.1 – 14.0	0.39
S 06°52.882	3–6 cm	wet	4.42	13.72	19.2 – 9.81	0.19
E 039°18.391		dry	3.01	17.22	25.4 – 11.7	0.26
	6–9 cm	wet	2.54	15.73	22.9 – 10.8	0.18
		dry	4.30	15.51	22.7 – 10.6	0.24
	0–3 cm	wet	7.50	16.91	24.5 – 11.7	0.54
E7:		dry	1.35	14.70	19.4 – 11.7	0.19
S 06°52.952	3–6 cm	wet	2.41	17.80	25.6 – 12.4	0.44
E 039°18.454		dry	1.72	19.17	25.4 – 14.4	0.23
	6–9 cm	wet	7.07	12.04	17.1 – 8.45	0.60
		dry	1.66	8.52	12.4 – 5.85	0.24

*Results were determined by the inverse prediction method.

The PCDD/F results from the CALUX screening for the Mtoni estuary sediment extracts are shown in Figure 5A. The PCDD/F concentrations in figure 5A were based on the Hill BEQ$_{50}$ and ranged from 5.7 ± 1.4 to 39.9 ± 5.8 pg BEQ/g sediment in wet season and were between 14.1 ± 2.0 and 32.8 ± 4.7 pg BEQ/g sediment in dry season. Higher levels of PCDD/Fs in both seasons were found in the Kizinga River and in confluence stations close to it. Wet season samples contained less PCDD/Fs than dry season samples in the Kizinga River, however at the Mzinga site it was the opposite. Higher PCDD/F levels at confluence station E3 in both seasons could be due to a local emission source in proximity to the Kizinga River. This is confirmed by the subsequent higher levels at the nearby station E4 in both seasons and lower levels at E5, a station closer to the Mzinga River. No significant difference (t = 0.55 two tailed; p = 0.59) in PCDD/F levels was observed between the two seasons (n = 42).

Response levels of dl-PCBs (Figure 5B) ranged from 0.21 ± 0.03 to 0.53 ± 0.03 pg BEQ/g sediment in wet season and from 0.22 ± 0.03 to 0.59 ± 0.04 pg BEQ/g sediment in dry season. These BEQ values are much lower (about 2 orders of magnitude) than those of the PCDD/Fs. At the 2 Kizinga stations, confluence station E5 and Mzinga station E6, lower PCBs values were observed in wet compared to dry season, while the opposite was true for confluence stations E3 and E4 as well as Mzinga station E7. High dioxin-like PCB levels were generally correlated to high PCDD/Fs values. The PCB levels detected in the estuary showed no significant variations (t = 0.27 two tailed; p = 0.79) between the seasons (n = 42).

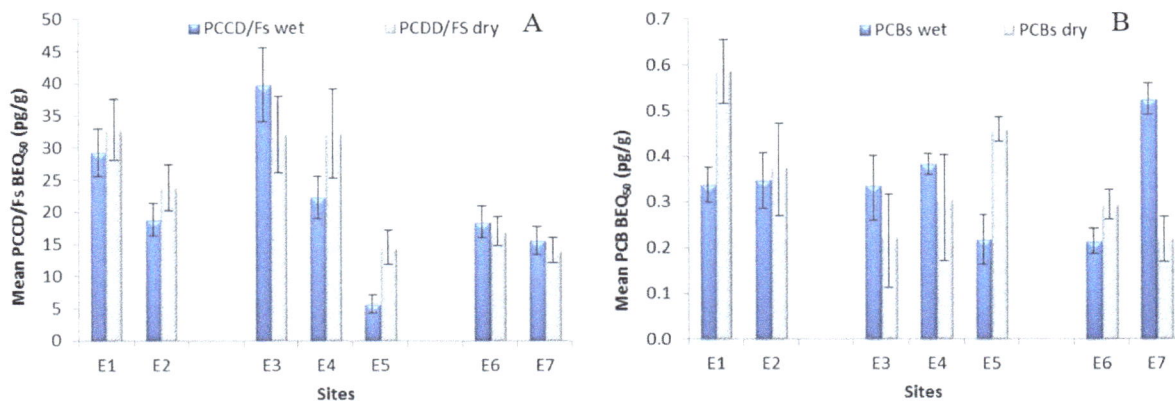

Figure 5. BEQ (with uncertainty error bars) for PCDD/Fs (A) and PCBs (B) of the Mtoni sediments presented as averaged values of the three depth layers in dry and wet seasons

The Kizinga River drains the peri-urban environments of Temeke, Mbagala, Ukonga, Charambe, Kijichi, Vituka, Keko, Kitunda, Mtoni, and Kurasini which are highly populated and have many small-scale to medium-scale industries. On the other hand, the Mzinga stream drains the rural environments of Mji Mwema, Vijibweni and Tuangoma that are less populated and have very few industries. As a result of various socio-economic activities, polymeric materials (household scraps, plastics, vehicle tires and electronic wastes) are abundantly present in most domestic and industrial wastes. Due to lack of appropriate infrastructures, most of these wastes are collected in open dumping sites and then burnt. Polymeric materials in wastes, when subjected to open burning, may lead to formation of PCDD/Fs and dl-PCBs (Estrellan & Iino, 2010). In addition, 2 other sources of PCDD/Fs and dl-PCBs have to be considered in the study area: (1) charcoal and/ or wood burning is the main source of energy for domestic purposes; (2) privately owned second hand commuter buses that need frequent services are the main means of transport for most residents. It was observed (Mkoma, 2008) that biomass burning (wood and charcoal burning) and traffic-related emissions (leaded gasoline exhausts and spills, tire wear) are the major sources of particulate matter in the Dar es Salaam atmosphere. The detected PCDD/Fs and dl-PCBs levels in the estuary can thus reasonably be associated with the open burning of polymeric waste, domestic wood/charcoal burning and with traffic-related emissions.

The presence of sources in the vicinity of the sampling stations must also be taken into account to explain the detected levels of PCDD/Fs and dl-PCBs. For example, upstream station E1 in the Kizinga River is downstream to an untreated wastewater discharge point of a textile factory. Higher levels of both PCDD/F and dl-PCBs in Kizinga stations compared to Mzinga stations may reflect the combined effects of the denser population and the more industrial activities in the Kizinga River basin because both are directly related to the various sources of PCDD/Fs and dl-PCBs mentioned above.

A third sampling campaign was organised to verify the higher PCDD/F and dl-PCB levels in the Kizinga River and to assess the change of the dioxin levels downstream the previous sampling stations in the estuary (E1 to E5). The most upstream sampling station in the Kizinga River (F1) which is 200 m upstream of a sewage factory pipe and near unauthorised human settlements, is very rich in organic matter (Table 2). Station F1 shows a high

PCDD/F level of 400 pg-BEQ/g and a dl-PCB level of 0.91 pg-BEQ/g. The second upstream Kizinga River sample (F2) is also high but not exceptional with 21 pg-BEQ/g PCDD/Fs and 0.63 pg-BEQ/g dl-PCBs. The levels in the upstream Mzinga River samples are close to those of the Kizinga River sample F2 (Table 2). A negative concentration gradient of PCDD/Fs and dl-PCBs in sediments is observed in the downstream direction. The levels at the mouth of the Mtoni estuary are very low: 1 to 2 pg-BEQ/g for PCDD/Fs and below LOD for dl-PCBs. These additional results confirm the previous conclusions that the impact of dioxins on the Kizinga River basin is higher than on that of the Mzinga River and that a dilution effect is clearly noticeable in the downstream direction.

Table 2. CALUX-BEQ values for PCDD/Fs and dl-PCBs in additional end-member samples

	Code and Location	TOC (%)	CALUX-BEQ (pg TCDD/g sediment)		
			PCDD/Fs BEQ$_{50}$	PCDD/Fs BEQ$_{20}$-BEQ$_{80}$ range	Dioxin-like PCBs*
F1	S 06°53'01.29'' E 039°15'48.33''	8.58	397	652-242	0.91
F2	S 06°52'58.07'' E 039°16'06.81''	2.08	20.7	26.7-16.0	0.68
F3	S 06°53'17.67'' E 039°18'52.06''	1.05	17.2	21.0-14.2	0.60
F4	S 06°53'13.60'' E 039°18'52.69''	0.56	12.5	15.8-9.86	0.74
F5	S 06°51'06.95'' E 039°18'22.82''	0.85	6.71	7.70-5.85	0.25
F6	S 06°50'57.46'' E 39°18'19.16"	0.72	6.82	7.47-6.23	0.23
F7	S 06°49'03.71'' E 39°18'24.11"	0.28	1.66	2.43-1.14	< LOD
F8	S 06°48'57.62'' E 39°18'34.60"	0.99	1.03	1.74-0.61	< LOD

*Results were determined by the inverse prediction method.

3.5 Correlation Between Pollutants and Sediment Geochemical Characteristics

Calculations of the Pearson correlation coefficients were performed on pollutant levels as well as sediment geochemical characteristics of all samples (n = 42) in the mixing zone in both seasons (Table 3). A significant correlation was observed between PCDD/Fs and dioxin-like PCB levels ($r^2 = 0.40$, $p \ll 0.05$), however, no significant correlations ($p > 0.05$) were found with % TOC although it is well-known that lipophilic compounds such as PCDD/Fs and dl-PCBs tend to preferably associate with organic fractions (Koh et al., 2004, 2006; Pieters, 2007).

Table 3. Correlation coefficients between pollutants and various geochemical parameters

	PCDD/Fs	PCBs	%TOC	%TN	%TC
PCDDs	1.000				
PCBs	0.395*	1.000			
%TOC	0.070	0.208	1.000		
%TN	-0.049	0.120	0.913*	1.000	
%TC	-0.065	0.172	0.823*	0.893*	1.000

* Significant at $\alpha = 0.05$.

The weak correlation between the pollutants and %TOC is contrary to the theoretical expectation explained elsewhere (Hilscherova et al., 2003), but the sources of TOC and PCDD/Fs or dl-PCBs are quite different in the Mtoni study area. TOC in the estuary and rivers is mainly originating from 2 sources: (1) mangrove degradation compounds (Marchand, Lallier-Vergès, & Baltzer, 2003) and (2) untreated domestic sewage water (Mihale, Kishimba, Baeyens, & Brion, 2013), while PCDD/Fs and dl-PCBs originate from open burning processes (Lee, Coleman, J. Jones, K. Jones, & Lohmann, 2005) and traffic related emissions (Coleman, Lee, Alcock, & Jones, 1997). A mixing of the PCDD/Fs or dl-PCBs with the organic matter in the estuary can occur in all kind of ratios. It is thus not surprising that there exists no correlation between both variables. Good and significant correlations (p < 0.05) were observed between % TOC, % TC and % TN.

3.6 Pollutant Source Analysis by PCA

Multivariate analysis can be used to identify similarities and differences between pollutants in samples as a means to detect possible sources. As we could expect, the PCA results indicated that the variables can be represented by two principal components that account for 83.7% of the total variance in the original data sets. Based on the loading distribution of the variables, % TC, % TN and % TOC constituted one related group (PC1), while the pollutants PCDD/Fs and dl-PCBs formed the other group (PC2).

3.7 Comparison with Other Studies and with Sediment Quality Guidelines

Little information is available in literature regarding CALUX analysis of PCDD/Fs and dl-PCBs in marine, estuarine and river sediments. No such research has been conducted in Tanzania as a whole. Research on dioxin-like compounds in such sediments in the world used chemo-analysis or GC-HRMS, because this method is considered as the golden standard (Baeyens et al., 2004). Furthermore, most research focused on PCDD/Fs while no data on dl-PCBs were available for comparison. Comparison with literature data indicated that the observed PCDD/F levels were covering the ranges observed in the literature (see Table 4). Only at the Belgian coast where lower values were observed (Sanctorum et al., 2007).

An attempt was made to compare the PCDD/F and dioxin-like PCB levels with the sediment quality guidelines. Since Tanzania lacks these guidelines, National Oceanic and Atmospheric Administration (NOAA), USA and Canadian Sediment quality guidelines were applied to assess the toxicity and risk of the dioxin-like pollutants in the Mtoni estuary ecosystem. Mangrove sediments from Mtoni estuary have higher PCDD/F levels than the NOAA apparent effects threshold (AET) of 3.6 pg-TEQ/g. The levels were also higher than both the threshold effect level (TEL) for Canadian sediment quality guideline (0.85 pg-TEQ/g) and the probable effect level (PEL) of 21.5 pg-TEQ/g. There were no specific guidelines for dl-PCB levels.

Table 4. Comparison of PCDD/F and PCB levels (pg-BEQ/g dw) determined by CALUX bioassay in different marine sedimentary environments

Study site	PCDD/Fs[1]	PCBs[2]	Reference
Mtoni estuary, Tanzania[3]	1.0–397	<LOD–0.91	This study
UK estuaries	1.0–88	-	(Hurst et al., 2004)
Hong Kong mudflats	3–68	-	(Wong, Giesy, Siu, & Lam, 2005)
North coast of Bohai Sea, China	3.4–28	-	(Hong et al., 2012)
Masan Bay, Korea	17–275	-	(Yoo, Khim, & Giesy, 2006)
Belgian coast	0.08–42.4	-	(Sanctorum et al., 2007)
West coast, South Korea	3.4–11	-	(Hong et al., 2012)

[1] Values presented are BEQ_{50}; [2] values are inversely predicted; [3] Values indicate the range of mean BEQ of both seasons (n = 6).

The comparisons have shown that there could currently be a risk regarding PCDD/Fs, while the risk associated with dl-PCBs is much lower. Because the toxicity of these chemicals is assumed additive, increasing levels in line with the increasing anthropogenic activities can be alarming to the local biological community (Kruitwagen, Hecht, Pratap, & Wendelaar-Bonga, 2005) such as the barred mudskippers and soft bottom mollusks that inhabit the muddy areas. Presence of these pollutants in mangrove sediments may cause impairments, like abnormal growth and malformations. The effects can reach other organisms higher in the trophic level due to their bio-accumulation, bio-concentration and persistent properties. The levels of pollutants observed in the frequently exchanged upper (0-3 cm) layer can have impacts on the distribution and fate of pollutants to mangrove ecosystems and to higher organisms that use mangrove sediment organisms as their food.

4. Conclusion

This study is the first of its kind that used *in vitro* bioassay analysis (CALUX) to determine dioxin and dioxin-like compounds in the environment of East Africa. Sediment samples were collected in the Mtoni estuary and its tributaries, the Kizinga and the Mzinga Rivers, in the vicinity of Dar es Salaam. While it is well known that anthropogenic activities as open burning of plastic scrap, household burning of wood or charcoal and traffic related emissions, which all frequently occur in the Dar es Salaam region, can lead to PCDD/F and dioxin-like PCB production, it was totally unknown if some or all of those sources resulted in the contamination of the nearby aquatic systems. Our CALUX analyses of the sediments in the Mtoni estuary and its 2 tributaries demonstrated that the range of PCDD/F values (1.0 to 400 pg-BEQ/g-sediment) covers the ranges observed in sediments from Western Europe and Eastern Asia. Neither significant seasonal variations nor vertical gradients in the sediments could be observed. For dl-PCBs we could not find sediment results obtained by CALUX in the literature. Sediments in the Kizinga River, which flows through a denser populated and more industrialised area than the Mzinga River, showed also higher PCDD/F and dioxin-like PCB levels. Finally, the overall BEQ value of dioxins (PCDD/F and dioxin-like PCB levels) observed in sediments of the Dar es Salaam region, indicate possible ecological and human risks that may emanate from these contaminants.

Acknowledgements

This work was supported by the Belgian Technical Cooperation (BTC) under the Belgian Development Agency through a scholarship offered to M. J. Mihale. The CALUX bioassay H1L7.5c1 cell line was developed with funding from the National Institute of Environmental Health Sciences Superfund Research grant (ES04699) by Prof. Michael S. Denison, UC Davis, who kindfully provided us with the cells. MJM is very grateful to Prof. Michael Kishimba, who passed away during this research, for his unlimited support.

References

Baeyens, W., Panutrakul, I. S., Elskens, M., Navez, M. L. J., & Monteny, F. (1991). Geochemical Processes in Muddy and Sandy Tidal Flat Sediments. *Geo-Marine Letters*, 188-193. http://dx.doi.org/10.1007/BF02431011

Baeyens, Willy, Verstraete, F., & Goeyens, L. (2004). Elucidation of sources, pathways and fate of dioxins, furans and PCBs requires performant analysis techniques. *Talanta, 63*(5), 1095-100. http://dx.doi.org/10.1016/j.talanta.2004.06.010

Baston, D. S., & Denison, M. S. (2011). Considerations for potency equivalent calculations in the Ah receptor-based CALUX bioassay: normalization of superinduction results for improved sample potency estimation. *Talanta, 83*(5), 1415-21. http://dx.doi.org/10.1016/j.talanta.2010.11.035

Besselink, H. T., Schipper, C., Klamer, H., Leonards, P., Verhaar, H., Felzel, E., ... Brouwer, B. (2004). Intra- and interlaboratory calibration of the DR CALUX bioassay for the analysis of dioxins and dioxin-like chemicals in sediments. *Environmental Toxicology and Chemistry/SETAC, 23*(12), 2781-9. Retrieved from http://www.ncbi.nlm.nih.gov/pubmed/15648750

Birch, G. F., Harrington, C., Symons, R. K., & Hunt, J. W. (2007). The source and distribution of polychlorinated dibenzo-p-dioxin and polychlorinated dibenzofurans in sediments of Port Jackson, Australia. *Marine Pollution Bulletin, 54*(3), 295-308. http://dx.doi.org/10.1016/j.marpolbul.2006.10.009

Bruckmeier, B. F. A., Jüttner, I., Schramm, K. W., Winkler, R., Steinberg, C. E. W., & Kettrup, A. (1997). PCBs and PCDD/Fs in lake sediments of Großer Arbersee, Bavarian Forest, South Germany. *Environmental Pollution, 95*(1), 19-25. http://dx.doi.org/10.1016/S0269-7491(96)00118-2

Chi, K. H., Chang, M. B., & Kao, S. J. (2007). Historical trends of PCDD/Fs and dioxin-like PCBs in sediments buried in a reservoir in Northern Taiwan. *Chemosphere, 68*(9), 1733-40. http://dx.doi.org/10.1016/j.chemosphere.2007.03.043

Coleman, P. J., Lee, R. G. M., Alcock, R. E., & Jones, K. C. (1997). Observations on PAH, PCB, and PCDD/F Trends in U.K. Urban Air, 1991−1995. *Environmental Science & Technology, 31*(7), 2120-2124. http://dx.doi.org/10.1021/es960953q

Croes, K., Van Langenhove, K., Elskens, M., Desmedt, M., Roekens, E., Kotz, A., ... Baeyens, W. (2011). Analysis of PCDD/Fs and dioxin-like PCBs in atmospheric deposition samples from the Flemish measurement network: Optimization and validation of a new CALUX bioassay method. *Chemosphere, 82*(5), 718-24. http://dx.doi.org/10.1016/j.chemosphere.2010.10.092

Davies, O. A., & Tawari, C. C. (2010). Season and tide effects on sediment characteristics of trans-okpoka creek , upper bonny Estuary , Nigeria. *Agriculture and Biology Journal of North America, 1*(2), 89-96.

De Wolf, H., & Rashid, R. (2008). Heavy metal accumulation in Littoraria scabra along polluted and pristine mangrove areas of Tanzania. *Environmental Pollution, 152*(3), 636-43. http://dx.doi.org/10.1016/j.envpol.2007.06.064

Denison, M. S., Pandini, A., Nagy, S. R., Baldwin, E. P., & Bonati, L. (2002). Ligand binding and activation of the Ah receptor. *Chemico-Biological Interactions, 141*(1-2), 3-24. http://dx.doi.org/10.1016/S0009-2797(02)00063-7

Denison, M. S., Zhao, B., Baston, D. S., Clark, G. C., Murata, H., & Han, D. (2004). Recombinant cell bioassay systems for the detection and relative quantitation of halogenated dioxins and related chemicals. *Talanta, 63*(5), 1123-33. http://dx.doi.org/10.1016/j.talanta.2004.05.032

El-Kady, A. A., Abdel-Wahhab, M. A., Henkelmann, B., Belal, M. H., Morsi, M. K. S., Galal, S. M., & Schramm, K. W. (2007). Polychlorinated biphenyl, polychlorinated dibenzo-p-dioxin and polychlorinated dibenzofuran residues in sediments and fish of the River Nile in the Cairo region. *Chemosphere, 68*(9), 1660-8. http://dx.doi.org/10.1016/j.chemosphere.2007.03.066

Elskens, M., Baston, D. S., Stumpf, C., Haedrich, J., Keupers, I., Croes, K., ... Goeyens, L. (2011). CALUX measurements: statistical inferences for the dose-response curve. *Talanta, 85*(4), 1966-73. http://dx.doi.org/10.1016/j.talanta.2011.07.014

Estrellan, C. R., & Iino, F. (2010). Toxic emissions from open burning. *Chemosphere, 80*(3), 193-207. http://dx.doi.org/10.1016/j.chemosphere.2010.03.057

Goeyens, L., Hoogenboom, R., Eppe, G., Malagocki, P., Vanderperren, H., Scippo, M. L., ... Haedrich, J. (2010). Discrepancies between Bio-analytical and Chemo-analytical results have a non-negligible message. *Organohalogen Compounds, 72*, 964-967.

Guzzella, L., Roscioli, C., Viganò, L., Saha, M., Sarkar, S. K., & Bhattacharya, A. (2005). Evaluation of the concentration of HCH, DDT, HCB, PCB and PAH in the sediments along the lower stretch of Hugli estuary, West Bengal, northeast India. *Environment International, 31*(4), 523-34. http://dx.doi.org/10.1016/j.envint.2004.10.014

Hilscherova, K., Kannan, K., Nakata, H., Hanari, N., Yamashita, N., Bradley, P. W., ... Giesy, J. P. (2003). Polychlorinated Dibenzo- p -dioxin and Dibenzofuran Concentration Profiles in Sediments and Flood-Plain Soils of the Tittabawassee River, Michigan. *Environmental Science & Technology, 37*(3), 468-474. http://dx.doi.org/10.1021/es020920c

Holmer, M. (2003). *Biogeochemistry of Marine Systems*. In K. D. Black & G. B. Shimmield (Eds.), *Journal of Marine Systems* (Vol. 50, pp. 1-34). Blackwell Publishing Ltd. http://dx.doi.org/10.1016/j.jmarsys.2004.05.004

Hong, S., Khim, J. S., Naile, J. E., Park, J., Kwon, B. O., Wang, T., ... Giesy, J. P. (2012). AhR-mediated potency of sediments and soils in estuarine and coastal areas of the Yellow Sea region: a comparison between Korea and China. *Environmental Pollution, 171*, 216-25. http://dx.doi.org/10.1016/j.envpol.2012.08.001

Hoogenboom, L., Traag, W., Bovee, T., Goeyens, L., Carbonnelle, S., Vanloco, J., ... Baeyens, W. (2006). The CALUX bioassay: Current status of its application to screening food and feed. *TrAC Trends in Analytical Chemistry, 25*(4), 410-420. http://dx.doi.org/10.1016/j.trac.2006.02.012

Hurst, M. R., Balaam, J., Chan-Man, Y. L., Thain, J. E., & Thomas, K. V. (2004). Determination of dioxin and dioxin-like compounds in sediments from UK estuaries using a bio-analytical approach: chemical-activated luciferase expression (CALUX) assay. *Marine Pollution Bulletin, 49*(7-8), 648-58. http://dx.doi.org/10.1016/j.marpolbul.2004.04.012

International POPs Elimination Network. (2005). *Contamination of chicken eggs near the Vikuge obsolete pesticides stockpile in Tanzania by dioxins , PCBs and hexachlorobenzene Contamination of chicken eggs near the Vikuge obsolete pesticides stockpile in Tanzania by dioxins* (pp. 0-22).

Joung, K. E., Chung, Y. H., & Sheen, Y. Y. (2007). DRE-CALUX bioassay in comparison with HRGC/MS for measurement of toxic equivalence in environmental samples. *Science of The Total Environment, 372*(2-3), 657-67. http://dx.doi.org/10.1016/j.scitotenv.2006.10.036

Koh, C-H, Khim, J. S., Kannan, K., Villeneuve, D. L., Senthilkumar, K., & Giesy, J. P. (2004). Polychlorinated dibenzo-p-dioxins (PCDDs), dibenzofurans (PCDFs), biphenyls (PCBs), and polycyclic aromatic hydrocarbons (PAHs) and 2,3,7,8-TCDD equivalents (TEQs) in sediment from the Hyeongsan River, Korea. *Environmental Pollution, 132*(3), 489-501. http://dx.doi.org/10.1016/j.envpol.2004.05.001

Koh, C. H., Khim, J. S., Villeneuve, D. L., Kannan, K., & Giesy, J. P. (2006). Characterization of trace organic contaminants in marine sediment from Yeongil Bay, Korea: Dioxin-like and estrogenic activities. *Environmental Pollution, 142*(1), 48-57. http://dx.doi.org/10.1016/j.envpol.2005.09.006

Koistinen, J., Stenman, O., Haahti, H., Suonperä, M., & Paasivirta, J. (1997). Polychlorinated diphenyl ethers, dibenzo-p-dioxins, dibenzofurans and biphenyls in seals and sediment from the gulf of finland. *Chemosphere, 35*(6), 1249-1269. http://dx.doi.org/10.1016/S0045-6535(97)00212-9

Kristensen, E., Mangion, P., Tang, M., Flindt, M. R., Holmer, M., & Ulomi, S. (2010). Microbial carbon oxidation rates and pathways in sediments of two Tanzanian mangrove forests. *Biogeochemistry, 103*(1-3), 143-158. http://dx.doi.org/10.1007/s10533-010-9453-2

Kruitwagen, G., Hecht, T., Pratap, H. B., & Wendelaar-Bonga, S. E. (2005). Changes in morphology and growth of the mudskipper (Periophthalmus argentilineatus) associated with coastal pollution. *Marine Biology, 149*(2), 201-211. http://dx.doi.org/10.1007/s00227-005-0178-z

Kruitwagen, G., Pratap, H. B., Covaci, A., & Wendelaar-Bonga, S. E. (2008). Status of pollution in mangrove ecosystems along the coast of Tanzania. *Marine Pollution Bulletin, 56*(5), 1022-31. http://dx.doi.org/10.1016/j.marpolbul.2008.02.018

Kumar, K. S., Sajwan, K. S., Richardson, J. P., & Kannan, K. (2008). Contamination profiles of heavy metals, organochlorine pesticides, polycyclic aromatic hydrocarbons and alkylphenols in sediment and oyster collected from marsh/estuarine Savannah GA, USA. *Marine Pollution Bulletin, 56*(1), 136-49. http://dx.doi.org/10.1016/j.marpolbul.2007.08.011

Lee, Coleman, P., Jones, J. L., Jones, K. C., & Lohmann, R. (2005). Emission Factors and Importance of PCDD/Fs, PCBs, PCNs, PAHs and PM 10 from the Domestic Burning of Coal and Wood in the U.K. *Environmental Science & Technology, 39*(6), 1436-1447. http://dx.doi.org/10.1021/es048745i

Lee, S. J., Kim, J. H., Chang, Y. S., & Moon, M. H. (2006). Characterization of polychlorinated dibenzo-p-dioxins and dibenzofurans in different particle size fractions of marine sediments. *Environmental Pollution, 144*(2), 554-61. http://dx.doi.org/10.1016/j.envpol.2006.01.040

Liu, H., Zhang, Q., Cai, Z., Li, A., Wang, Y., & Jiang, G. (2006). Separation of polybrominated diphenyl ethers, polychlorinated biphenyls, polychlorinated dibenzo-p-dioxins and dibenzo-furans in environmental samples using silica gel and florisil fractionation chromatography. *Analytica Chimica Acta, 557*(1-2), 314-320. http://dx.doi.org/10.1016/j.aca.2005.10.001

Long, M., Andersen, B. S., Lindh, C. H., Hagmar, L., Giwercman, A., Manicardi, G. C., ... Bonefeld-Jorgensen, E. C. (2006). Dioxin-like activities in serum across European and Inuit populations. *Environmental Health, 5*(1), 14. http://dx.doi.org/10.1186/1476-069X-5-14

Loomis, D., Browning, S. R., Schenck, A. P., Gregory, E., & Savitz, D. A. (1997). Cancer mortality among electric utility workers exposed to polychlorinated biphenyls. *Occupational and Environmental Medicine, 54*(10), 720-728. http://dx.doi.org/10.1136/oem.54.10.720

Machiwa, J. F. (1992). Anthropogenic pollution in the Dar es Salaam harbour area, Tanzania. *Marine Pollution Bulletin, 24*(11), 562-567. http://dx.doi.org/10.1016/0025-326X(92)90709-F

Maleko, G. C. (2005). *Impact of Electricity Services on Microenterprise in Rural Areas in Tanzania Development University of Twente , Enschede.*

Manahan, S. A. (2008). *Fundamentals of Environmental Chemistry* (3rd ed., p. 1264). CRC Press.

Marchand, C., Lallier-Vergès, E., & Baltzer, F. (2003). The composition of sedimentary organic matter in relation to the dynamic features of a mangrove-fringed coast in French Guiana. *Estuarine, Coastal and Shelf Science, 56*(1), 119-130. http://dx.doi.org/10.1016/S0272-7714(02)00134-8

Mbuligwe, S. E., & Kassenga, G. R. (1997). Automobile air pollution in Dar es Salaam City, Tanzania. *Science of The Total Environment, 199*(3), 227-235. http://dx.doi.org/10.1016/S0048-9697(97)05461-2

Mihale, M. J., Kishimba, M. A., Baeyens, W., & Brion, N. (2013). Carbon and Nitrogen dynamics in tropical estuarine mangrove sediments of Mtoni, Tanzania. Unpublished manuscript.

Mkoma, S. (2008). *Physico-Chemical Characterisation of Atmospheric Aerosols in Tanzania, with Emphasis on the Carbonaceous Aerosol Components and on Chemical Mass Closure.* Retrieved from https://biblio.ugent.be/publication/470145/file/1880861.pdf

Moon, H. B., Choi, M., Choi, H. G., Ok, G., & Kannan, K. (2009). Historical trends of PCDDs, PCDFs, dioxin-like PCBs and nonylphenols in dated sediment cores from a semi-enclosed bay in Korea: tracking the sources. *Chemosphere, 75*(5), 565-71. http://dx.doi.org/10.1016/j.chemosphere.2009.01.064

Müller, J. F., Haynes, D., McLachlan, M., Böhme, F., Will, S., Shaw, G. R., ... Connell, D. W. (1999). PCDDs, PCDFs, PCBs and HCB in marine and estuarine sediments from Queensland, Australia. *Chemosphere, 39*(10), 1707-21. Retrieved from http://www.ncbi.nlm.nih.gov/pubmed/10520488

Müller, Jochen, F., Gaus, C., Prange, J. A., Päpke, O., Poon, K. F., Lam, M. H. W., & Lam, P. K. (2002). Polychlorinated dibenzo-p-dioxins and polychlorinated dibenzofurans in sediments from Hong Kong. *Marine Pollution Bulletin, 45*(1-12), 372-8. Retrieved from http://www.ncbi.nlm.nih.gov/pubmed/12398408

Murk, A. (1996). Chemical-Activated Luciferase Gene Expression (CALUX): A Novel In Vitro Bioassay for Ah Receptor Active Compounds in Sediments and Pore Water. *Fundamental and Applied Toxicology, 33*(1), 149-160. http://dx.doi.org/10.1006/faat.1996.0152

National Bureau for Statistics (NBS). (2003). Tanzania populations and housing census–General report. Retrieved from http://web.archive.org/web/20040313070101/http://www.tanzania.go.tz/census/census/tables.htm

Nieuwoudt, C., Quinn, L. P., Pieters, R., Jordaan, I., Visser, M., Kylin, H., ... Bouwman, H. (2009). Dioxin-like chemicals in soil and sediment from residential and industrial areas in central South Africa. *Chemosphere, 76*(6), 774-83. http://dx.doi.org/10.1016/j.chemosphere.2009.04.064

Okay, O. S., Karacik, B., Başak, S., Henkelmann, B., Bernhöft, S., & Schramm, K. W. (2009). PCB and PCDD/F in sediments and mussels of the Istanbul strait (Turkey). *Chemosphere, 76*(2), 159-66. http://dx.doi.org/10.1016/j.chemosphere.2009.03.051

Pan, J., Yang, Y., Geng, C., Yeung, L. W. Y., Cao, X., & Dai, T. (2010). Polychlorinated biphenyls, polychlorinated dibenzo-p-dioxins and dibenzofurans in marine and lacustrine sediments from the Shandong Peninsula, China. *Journal of Hazardous Materials, 176*(1-3), 274-9. http://dx.doi.org/10.1016/j.jhazmat.2009.11.024

Peri-urban mangrove forests as filters and potential phytoremediators of domestic sewage in east Africa (PUMPSEA). (2007). *Peri-urban mangrove forests as filters and potential phytoremediators of domestic sewage in East Africa (2007). Distribution and fate of heavy metals and indicator pathogens in sewage exposed mangroves, Project No. INCO – CT2004 – 510863.*

Pieters, R. (2007). *An assessment of dioxins, dibenzofurans and PCBs in the sediments of selected freshwater bodies and estuaries in South Africa.* North-West University, Potchefstroom Campus.

Roots, O., Henkelmann, B., & Schramm, K. W. (2004). Concentrations of polychlorinated dibenzo-p-dioxins and polychlorinated dibenzofurans in soil in the vicinity of a landfill. *Chemosphere, 57*(5), 337-42. http://dx.doi.org/10.1016/j.chemosphere.2004.06.012

Roy, S., Mysior, P., & Brzezinski, R. (2002). Comparison of dioxin and furan TEQ determination in contaminated soil using chemical, micro-EROD, and immunoassay analysis. *Chemosphere, 48*(8), 833-842. http://dx.doi.org/10.1016/S0045-6535(02)00129-7

Ryoo, K. S., Ko, S.-O., Hong, Y. P., Choi, J. H., Cho, S., Kim, Y., & Bae, Y. J. (2005). Levels of PCDDs and PCDFs in Korean river sediments and their detection by biomarkers. *Chemosphere, 61*(3), 323-31. http://dx.doi.org/10.1016/j.chemosphere.2005.02.093

Sanctorum, H., Elskens, M., & Baeyens, W. (2007). Bioassay (CALUX) measurements of 2,3,7,8-TCDD and PCB 126: interference effects. *Talanta, 73*(1), 185-8. http://dx.doi.org/10.1016/j.talanta.2007.03.009

Sanctorum, H., Windal, I., Hanot, V., Goeyens, L., & Baeyens, W. (2007). Dioxin and dioxin-like activity in sediments of the Belgian coastal area (Southern North Sea). *Archives of Environmental Contamination and Toxicology, 52*(3), 317-25. http://dx.doi.org/10.1007/s00244-006-0063-x

Schecter, A. J., Sheu, S. U., Birnbaum, L. S., Devito, M. J., Denison, M. S., & Päpke, O. (1999). A comparison and discussion of two differing methods of measuring dioxin-like compounds: gas chromatography-mass spectrometry and the CALUX bioassay–implications for health studies. *Organohalogen Compounds, 40*, 247-250.

Schroijen, C., Windal, I., Goeyens, L., & Baeyens, W. (2004). Study of the interference problems of dioxin-like chemicals with the bio-analytical method CALUX. *Talanta, 63*(5), 1261-8. http://dx.doi.org/10.1016/j.talanta.2004.05.036

Shen, H., Han, J., Tie, X., Xu, W., Ren, Y., & Ye, C. (2009). Polychlorinated dibenzo-p-dioxins/furans and polychlorinated biphenyls in human adipose tissue from Zhejiang Province, China. *Chemosphere, 74*(3), 384-8. http://dx.doi.org/10.1016/j.chemosphere.2008.09.094

Smith, A. H., & Lopipero, P. (2001). *Evaluation of the toxicity of dioxins and dioxin-like PCBs: A health risk appraisal for the New Zealand population: A report to the New Zealand Ministry for the Environment* (pp. 17-101).

Song, M., Jiang, Q., Xu, Y., Liu, H., Lam, P. K. S., O'Toole, D. K., ... Jiang, G. B. (2006). AhR-active compounds in sediments of the Haihe and Dagu Rivers, China. *Chemosphere, 63*(7), 1222-30. http://dx.doi.org/10.1016/j.chemosphere.2005.08.065

Srogi, K. (2007). Levels and congener distributions of PCDDs, PCDFs and dioxin-like PCBs in environmental and human samples: a review. *Environmental Chemistry Letters, 6*(1), 1-28. http://dx.doi.org/10.1007/s10311-007-0105-2

Taylor, M., Ravilious, C., & Green, E. P. (2003). *Mangroves of East Africa–UNEP* (p. 28).

Terauchi, H., Takahashi, S., Lam, P. K. S., Min, B. Y., & Tanabe, S. (2009). Polybrominated, polychlorinated and monobromo-polychlorinated dibenzo-p-dioxins/dibenzofurans and dioxin-like polychlorinated biphenyls in marine surface sediments from Hong Kong and Korea. *Environmental Pollution, 157*(3), 724-30. http://dx.doi.org/10.1016/j.envpol.2008.11.028

United States Environmental protection Agency (USEPA). (2001). *Methods for Collection, Storage and Manipulation of Sediments for Chemical and Toxicological Analyses–Technical Manual*, (EPA-823-B-01-002 October 2001), 208.

United States Environmental protection Agency (USEPA). (2008). Method 4435 for Toxic Equivalents (TEQs) determinations for dioxin-like chemical activity with the CALUX bioassay. Retrieved from http://www.epa.gov/osw/hazard/testmethods/pdfs/4435.pdf

Van Camp, M., Mjemah, I. C., Al Farrah, N., & Walraevens, K. (2013). Modeling approaches and strategies for data-scarce aquifers: example of the Dar es Salaam aquifer in Tanzania. *Hydrogeology Journal, 21*(2), 341-356. http://dx.doi.org/10.1007/s10040-012-0908-5

Van Langenhove, K., Croes, K., Denison, M. S., Elskens, M., & Baeyens, W. (2011). The CALUX bio-assay: analytical comparison between mouse hepatoma cell lines with a low (H1L6.1c3) and high (H1L7.5c1) number of dioxin response elements. *Talanta, 85*(4), 2039-46. http://dx.doi.org/10.1016/j.talanta.2011.07.042

Van Overmeire, I., Van Loco, J., Roos, P., Carbonnelle, S., & Goeyens, L. (2004). Interpretation of CALUX results in view of the EU maximal TEQ level in milk. *Talanta, 63*(5), 1241-7. http://dx.doi.org/10.1016/j.talanta.2004.05.034

Wang, H., He, M., Lin, C., Quan, X., Guo, W., & Yang, Z. (2007). Monitoring and assessment of persistent organochlorine residues in sediments from the Daliaohe River watershed, northeast of China. *Environmental Monitoring and Assessment, 133*(1-3), 231-42. http://dx.doi.org/10.1007/s10661-006-9576-z

Wong, H. L., Giesy, J. P., Siu, W. H. L., & Lam, P. K. S. (2005). Estrogenic and dioxin-like activities and cytotoxicity of sediments and biota from Hong Kong mudflats. *Archives of Environmental Contamination and Toxicology, 48*(4), 575-86. http://dx.doi.org/10.1007/s00244-004-0166-1

Yang, Z., Shen, Z., Gao, F., Tang, Z., & Niu, J. (2009). Occurrence and possible sources of polychlorinated biphenyls in surface sediments from the Wuhan reach of the Yangtze River, China. *Chemosphere, 74*(11), 1522-30. http://dx.doi.org/10.1016/j.chemosphere.2008.11.024

Yoo, H., Khim, J. S., & Giesy, J. P. (2006). Receptor-mediated in vitro bioassay for characterization of Ah-R-active compounds and activities in sediment from Korea. *Chemosphere, 62*(8), 1261-71. http://dx.doi.org/10.1016/j.chemosphere.2005.07.007

Zhang, H., Zhao, X., Ni, Y., Lu, X., Chen, J., Su, F., ... Zhang, X. (2010). PCDD/Fs and PCBs in sediments of the Liaohe River, China: levels, distribution, and possible sources. *Chemosphere, 79*(7), 754-62. http://dx.doi.org/10.1016/j.chemosphere.2010.02.039

Zhao, L., Hou, H., Zhou, Y., Xue, N., Li, H., & Li, F. (2010). Distribution and ecological risk of polychlorinated biphenyls and organochlorine pesticides in surficial sediments from Haihe River and Haihe Estuary Area, China. *Chemosphere, 78*(10), 1285-93. http://dx.doi.org/10.1016/j.chemosphere.2009.12.007

2

Energy and Environmental Impacts of Urban Buses and Passenger Cars–Comparative Analysis of Sensitivity to Driving Conditions

Leonid Tartakovsky[1], Marcel Gutman[1], Doron Popescu[1] & Michael Shapiro[1]

[1] Faculty of Mechanical Engineering, Technion–Israel Institute of Technology, Haifa, Israel

Correspondence: Leonid Tartakovsky, Faculty of Mechanical Engineering, Technion–Israel Institute of Technology, Haifa 32000, Israel. E-mail: tartak@technion.ac.il

Abstract

A methodology is suggested for a comparative analysis of energy and environmental impacts of various urban transport modes. A total emission indicator is used as a tool for integral assessment of vehicle emissions. An environmental impact factor is suggested in order to compare between the various transport modes that use different energy sources. Vehicle occupancy values, yielding equality of the specific environmental impact factors and specific energy consumption of the compared transport modes are used for the analysis purposes. This methodology is applied for a comparison between the buses and the passenger cars at various levels of service, road gradients and urban road types. The comparison results reveal that the environmental impact of the bus for driving at an urban access road falls below the one of the passenger car when the bus occupancy is 14–18 persons. Urban buses turn out to be energetically beneficial over passenger cars at occupancy values substantially lower compared with those providing a similar environmental impact.

Keywords: energy and environmental impacts, total emission indicator, environmental impact factor, urban bus, passenger car

1. Introduction

Urban air pollution induced by road vehicles has become a deep concern all over the world. Vehicle emissions that include particles and toxic gases adversely affect cardio-vascular, respiratory and immune systems, thus by increasing the risk of stroke and cancer development (Pope & Dockery, 2006; Sram et al., 2011). Nano-particles penetrate through the blood cells into the human brain, liver etc. with concomitant negative health effects (Wang et al., 2012).

Various studies ensuring urban sustainable mobility are being explored worldwide to resolve problems of land use, traffic congestion, energy consumption and air pollution caused by road transportation. The possible solutions to circumvent this is to promote public transport as an alternative to individual transportation; tax incentives to bring on the roads to become environmentally friendly for vehicles implementation of Intelligent Transportation Systems (ITS), such as road tolling, congestion pricing, cybernetic transportation, etc.; sustainable urban planning; technological advancements aimed at a reduction of vehicle energy and environmental impacts(Silva et al., 2007; Parent, 2007; Radovic, 2009). Typical examples of such a progress include: engine downsizing and shut-off during idling, developing more efficient vehicle propulsion systems (plug-in hybrids, fuel cells, battery-electric vehicles, etc.), increasing availability of alternative fuels (Ahman, 2001; Dincer et al., 2010; Tartakovsky et al., 2013).

Public transportation in cities can be an effective tool for mobility improvement, traffic congestion reduction and mitigation of urban air pollution. Energy and environmental impacts of public transport depend on the type of vehicles used, driving pattern, road conditions, passengers load and other factors. These impacts should be carefully assessed and compared with those of other transportation modes, first of all–the individual transportation fleet. If the compared transportation modes use different energy carriers, the primary energy consumption and emissions are normally used as a basis for the comparison–well-to-wheel approach (Ahman, 2001).

Emissions from buses or passenger cars of various technologies and at different driving conditions are described quite thoroughly in the literature. Most of these studies are focused on a comparison of a few specific motor vehicles at some type of driving conditions. Such a comparison is very important because it provides specific and detailed information on the benefits and drawbacks of the studied vehicle types in the considered usage conditions. Nylund et al. (2004, 2007) compared emissions from several preselected diesel and natural gas buses of the emission certification ranged from Euro 3 to the enhanced environmentally friendly vehicle. Fox and Eweka (2009) investigated the performance of hybrid buses under a variety of operating conditions and duty cycles. De Vlieger et al. (2000) studied the influence of driving behavior and traffic conditions on fuel consumption and emissions of nine passenger cars. Tsang et al. (2011) examined the effects of road gradient and traffic conditions on the emissions and fuel consumption of a Euro 4 gasoline car. No comparison has been reported between the energy and the environmental impacts of different urban transportation modes in these works. Information on the emissions comparison between urban buses and passenger cars is quite limited and mainly focused on carbon dioxide emissions (Bradley & Associates, 2007). Lenaers and De Vlieger (1997) performed on-board measurements of CO, HC and NO_x emissions from six gasoline-driven cars and 5 diesel urban buses on regular city lines. Their research performed in 1997 dealt with vehicles of Euro 0 to Euro 2 technology generations. Silva et al. (2007) investigated the effects of vehicle occupancy on specific fuel consumption and emissions (per person·km) for two different driving patterns of low and congested traffic. Two models of diesel urban buses and two models (diesel- and gasoline-driven) of light-duty vehicles were considered. The comparison was performed separately for each pollutant type (CO_2, CO, HC, NO_x, and PM). The results vary significantly for different pollutants, which makes it difficult to compare between the various transportation modes.

Most of the published works dealt with a comparison of emissions only (Dincer et al., 2010; Granovskii et al., 2006). If road transportation modes based on electric and motor vehicles are to be compared, their environmental impact depends not only on the emissions level, but also on the number of people exposed to the polluted air and therefore subjected to health damage. This requires a development of an overall comparison approach. To resolve this problem we suggest a methodology for comparative analysis of energy and environmental impacts of various urban transport modes. It is based on a total emission indicator as a tool for integral assessment of vehicle emissions. An environmental impact factor is suggested in order to compare between transport modes using different energy sources. The methodology is applied for a comparison of the fleet averaged impacts of buses and passenger cars at various levels of service, road gradients and urban road types. The analysis uses the real world data collected for vehicle fleet composition.

2. Method

2.1 Calculation of Fleet Averaged Emissions and Energy Consumption

Fleet averaged emission factors and energy consumption of various road transport modes can be calculated by using one of the validated complex Road Emission Models, such as: ARTEMIS, COPERT and others. If electric vehicles (EVs) are presented in the fleet, an appropriate calculation model should be used. Emissions produced due to EVs operation largely depend on technology of electricity generation (Dincer et al., 2010). Fuel mix used for electricity production varies in composition depending on a country or region considered. In our work for the calculation of energy consumption and emission factors of motor vehicles we use the ARTEMIS model (André et al., 2009). The model allows a prediction of emissions and energy consumption for each vehicle and fuel type, technology generation, engine volume (for passenger cars) and driving cycle. It relies on the detailed classification of the vehicles into several categories (cars, buses, etc.). Vehicle categories are further divided into segments according to technology, fuel type and size. For example, buses of standard-, midi- and articulated-configuration may be considered. Passenger cars (PCs) are classified by fuel type, engine size, etc. (Andre & Rapone, 2009). The segments are broken down by the emission concepts (pre-Euro, Euro1 to 5 plus several other cases). This model is based on the largest in Europe data base containing experimentally measured emission factors. It consists of: emission data set including all the emission data and functions needed for the computation of different emissions, a fleet module enabling a calculation of the detailed fleet composition, an emission factor processor enabling a computation of all the relevant emissions factors, a traffic data set module combining traffic scenarios with applications and an emission computation module providing the total emissions for a given case study. The ARTEMIS model is accepted Europe-wide for the assessment of real-world emissions by various transportation modes. A detailed description of the ARTEMIS model can be found in the work of André et al. (2009).

2.2 Vehicle Fleet Data

Information on the fleet composition of the considered transportation mode can be collected from the vehicle operators or published statistical sources. The data on average number of cold starts by various vehicle types is usually less accessible. For an urban bus this data was obtained from the bus operators. There are two cold starts per day in average. The number of PC cold starts was assessed with the aid of a questionnaire distributed among 160 drivers. The cars in the sample were selected to represent the local PC fleet composition. A distribution of PC cold starts derived on the basis of the drivers' survey is shown in Figure 1. The mean number of PC cold starts that was used in simulations is 4.7 per day.

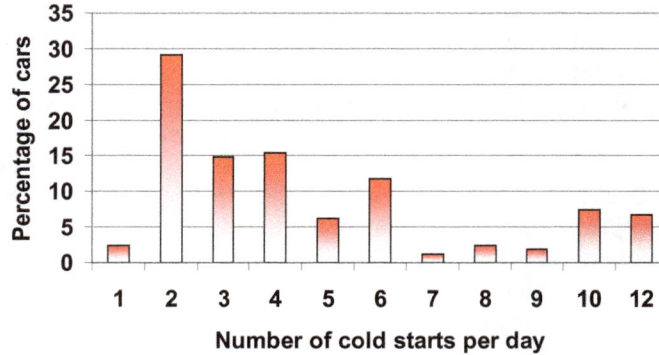

Figure 1. Distribution of PC cold starts

2.3 Environmental Impact Assessment

2.3.1 Total Emission Indicator

The values of weighted fleet-average emissions by each considered transportation mode should be further processed, to provide a total emission indicator (*TEI*). The latter is a useful tool for an integral quantitative assessment of vehicle emissions and also for a comparison between different transportation modes. For each vehicle type *TEI* is defined as the sum of normalized emission values of different pollutants:

$$TEI = c_{cor} \sum (EM_i / TLV_i) \ \ (g/km) \tag{1}$$

where: EM_i–the emission of pollutant i (g/km); TLV_i–threshold limit value for pollutant i, (mg/m^3); c_{cor} –dimension correction coefficient (c_{cor} = 1 mg/ m^3). The TLV_i values were taken from the list of Threshold Limit Values and Biological Exposure Indices of the American Conference of Governmental Industrial Hygienists (2010). Some relevant values of TLV_i are presented in Table 1.

Table 1. Threshold limit values for selected pollutants (American Conference of Governmental Industrial Hygienists, 2010)

Pollutant	TLV_i (mg/m^3)
Nitrogen dioxide NO$_2$ (for normalization of NO$_x$ emissions)	5.6
Carbon monoxide CO	28.5
1,3-Butadiene (for normalization of HC emissions)	4
Particulate matter PM2.5 (for normalization of PM emissions)	3

2.3.2 Environmental Impact Factor

The environmental impact of a road transportation mode is a function not only of the emissions level, but also of the number of people that are exposed to the polluted air and therefore exposed to health damage. Thus,major differences are possible between electric vehicles and motor vehicles. To account for this fact, we propose to introduce an environmental impact factor (*EIF*) to allow a comparison between different transportation modes. The dimensionless *EIF* value is calculated as:

$$EIF = TEI \cdot D_s \tag{2}$$

where D_s is the receptor density in the site of consideration. Receptor density is calculated as the population per

km^2 of the site area. Since it is usually impossible to distinct between various pollution sources that were involved in the additional electricity production due to electric vehicle activity, the uniform background approach was suggested by Curtiss and Rabl (1996). This approach is applied to calculate EIF values for EVs based transportation modes. Had the world been homogeneous, the receptor density would have been uniform D_u. The receptor density D_s depends on the site. The relative receptor density f can be defined as:

$$f = D_s/D_u \tag{3}$$

The dependence of EIF on f can be used to analyze the environmental impact of different transportation modes.

2.3.3 Environmental Impact Equality Value

We will further aim at characterizing the environmental impact of a vehicle group per person moved. Towards this goal for each vehicle group we define the specific environmental impact factor ($SEIF$) calculated as:

$$SEIF = EIF/Vehicle\ Occupancy \tag{4}$$

Environmental impacts of different transportation modes, if they emit hazardous pollutants at the same site with the same receptor density, can be determined by the level of their emissions. In this case the specific total emission indicator ($STEI$) calculated per person·km will be enough for comparison purposes:

$$STEI = TEI/Vehicle\ Occupancy(\text{g/person·km}) \tag{5}$$

$SEIF$ or $STEI$ values can be used to compare the environmental impacts of the considered transportation modes. For example, if urban buses are compared with passenger cars, we can calculate the bus occupancy (BO) for any given PC occupancy, yielding equality of the Bus and PC $SEIF$ or $STEI$. We will call this quantity $SEIF$ ($STEI$) equality BO. In the considered example, for any specified PC occupancy $STEI$ equality BO represents the minimal number of bus passengers required to ensure that the buses $STEI$ remains below $STEI$ of the passenger cars.

2.4 Energy Impact Assessment

Similar to $STEI$ equality BO, bus occupancy values that provide the equality of specific fuel consumption (SFC) between a bus and PC (called hereafter SFC equality BO) can be computed and used for energy impact comparison. SFC is calculated as:

$$SFC = FC/Vehicle\ Occupancy(\text{g/person·km}) \tag{6}$$

where FC is vehicle fuel consumption in (g/km).

If the compared transportation modes include vehicles using different energy carriers, the primary energy consumption will be used as a basis for the comparison. A vehicle occupancy value that provides the equality of the specific energy consumption should be computed and used in this case.

2.5 Considered Case: Buses and Passenger Cars

The methodology described in sections 2.1–2.4 was applied for the case of urban buses and passenger cars comparison at various levels of service, road gradients and urban road types. In all calculations we presumed the use of diesel fuel and gasoline with sulfur content lower than 10 ppm that meets the European Directive 2009/30/EC. Calculations were carried out for different values of the bus or car passenger occupancy. Weighted fleet-average values of emission factors were calculated for buses and PCs using the fleet composition data presented in Tables 2 and 3, respectively.

Table 2. Composition of the considered urban buses fleet (%)

Bus type	Technology generation						Total percentage
	Euro 0	Euro 1	Euro 2	Euro 3	Euro 4	Euro 5	
Articulated	4.2	0.9	0.8	9.7	5.4	0	21
Standard	0.1	6.8	21	35	13	2.1	78
Midi	0	0	1	0	0	0	1

A comparison between energy and environmental impacts of buses and PCs was performed for different traffic and road conditions. The following levels of service, commonly considered in the traffic situation scheme, were analyzed (André et al., 2006):

- Free flow–free flowing conditions with low and steady traffic flow. Constant and quite high speed. On roads with a speed limit of 50 km/h the speed is 45–60 km/h.

- Stop & Go–heavily congested flow, stop and go or gridlock. Variable and low speed and stops typical for city center driving with traffic jams. Indicative speed 5–15 km/h on roads with a speed limit of 50 km/h.

At each level of service calculations were performed for the average road gradients of 0, 2, 4 and 6%. The road types considered in these simulations were the urban access road with a speed limit of 50 km/h and the urban distributor type road with a speed limit of 80 km/h. A detailed description of these road types appears in (André et al., 2006).

Table 3. Composition of the considered PC fleet (%)

PC by engine volume (l) & fuel type	Technology generation						Total percentage
	Euro 0	Euro 1	Euro 2	Euro 3	Euro 4	Euro 5	
<1.4, gasoline	1.3	0.8	1.7	2.4	2.1	0.9	9.2
<1.4, diesel	0	0	0	0	0	0	0
1.4-2.0, gasoline	12.2	6.8	14.8	21.3	18.5	8.2	81.8
1.4-2.0, diesel	0	0.1	0.6	0.8	0.8	0.2	2.5
>2.0, gasoline	0.9	0.5	1.2	1.7	1.5	0.7	6.5
>2.0, diesel	0	0	0	0	0	0	0

3. Results and Discussion

3.1 Comparison of Environmental Impacts

Figure2 presents the dependence of *STEI equality BO* values on PC occupancy for urban access roads. At the "free flow" conditions and average PC occupancy of 1.2 (typical European PC occupancy for commuting trips to/from work) the fleet averaged emissions by an urban bus, expressed in specific *TEI* values, become lower compared to a passenger car, if the bus passenger load exceeds 18. For the "stop & go" traffic and the same PC occupancy of 1.2, an urban bus is starting to be environmentally beneficial, if its passenger occupancy is higher than 14. In other words, (for any given PC occupancy) the *STEI equality BO* value becomes lower, if the traffic is more congested. Therefore, the growth of emissions per passenger with an increasing traffic load is lower for buses than for passenger cars. One can see that with respect to the increasing traffic load buses are a more environmentally beneficial vehicle group than PC.

One of the probable reasons of this trend is a difference between the diesel and SI engine's mechanical efficiencies due to throttling losses. This difference is higher at lower loads typical for the "stop & go" traffic. In the case considered in this paper the fleet of PCs almost entirely consists of cars with spark ignition engines (see Table 3) as opposed to buses equipped by diesel engines that have no throttle. The difference between *STEI equality BO* values for the free flow and "stop & go" levels of service is growing with an increase in PC occupancy. The results obtained for the average road gradient of 6% are similar and indicate the same tendency.

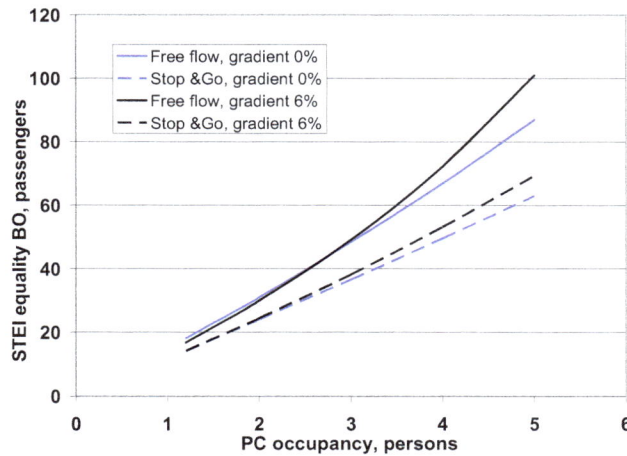

Figure 2. *STEI equality BO* versus PC occupancy. Free flow and stop & go levels of service, road gradients 0 and 6%, urban access road

Calculation results show that the influence of road gradient on the bus occupancy level providing equality of the bus *STEI* with that of PC is quite weak (few percent). However, it becomes sensible at road gradients exceeding 4% and PC occupancy values above 3 persons per vehicle. It is quite remarkable that for a high PC occupancy a reverse trend is observed, namely *STEI equality BO* grows with an increasing road gradient by up to 16%. In other words, as expressed in terms of *STEI equality BO*, at high PC occupancies the increasing road steepness is a factor less environmentally disadvantageous for PCs. These tendencies prevail for the free flow traffic and become weaker with increasing traffic congestion. The latter observation is explained by very low engine loads at "stop & go" level of service and a dominant influence of transient regimes on emissions formation, rather than load increase due to road gradient raise.

Figure 3 shows a comparison of the *STEI equality BO* for various road types and PC occupancy of 1.2 persons.As was already mentioned, the dependence of the *STEI equality BO* on the road gradient is quite weak and almost unaffected by the road type. No differences were observed between *STEI equality BO* values for various road types at stop & go level of service. This is because of a very weak dependence of the average speed on the speed limit in the congested traffic conditions (Table 4). There are also no significant differences between the average speeds of buses and a PCs in the "stop & go" traffic for both studied road types.

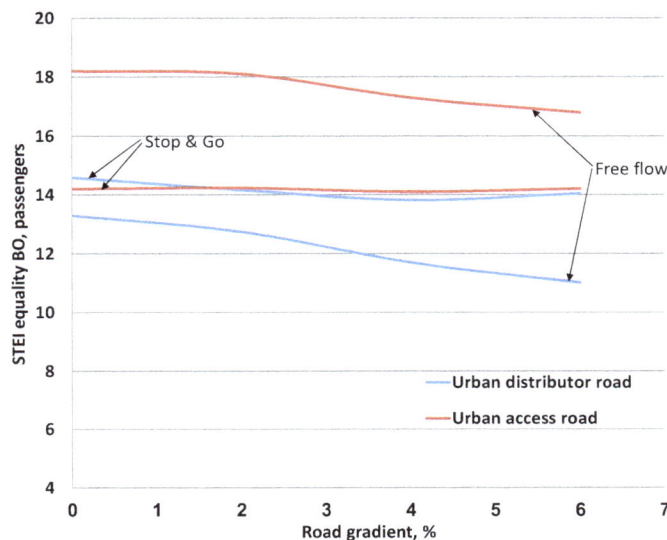

Figure 3. Comparison of *STEI equality BO* for different road types, PC occupancy–1.2 persons

Table 4. Average speeds of buses and PCs at different road types and levels of service

Level of service	Free flow		Stop & Go	
Road type	Access road	Distributor road	Access road	Distributor road
Speed limit, km/h	50	80	50	80
Urban bus	25.7	53.7	11.8	13.5
Passenger car	45.8	70.1	12.7	16

The data shown in Table 4 suggest that at free flow traffic conditions the average PC speed is close to the speed limit at both road types. The average bus speed at free flow traffic conditions is about 50% of the speed limit at urban access road and increases up to 67% of the speed limit at urban distributor road. The average bus speed at the urban distributor road is closer to the speed limit, thus indicating that a number of bus acceleration/deceleration events is reduced compared with a free flow driving at the urban access road. Normally this stems from a reduction in the number of bus stops per km. The number of transient events for buses is lower on urban distributor roads than on access roads and this explains the concomitantly lower vehicle emissions and fuel consumption. The PC average speed is close to the speed limit for both road types and this explains why the bus becomes environmentally and energetically beneficial over the PC at lower occupancy values if it is driven on a distributor road with higher average speed compared to the access road. The results shown in Figure 3 support this explanation.

3.2 Comparison of Energy Impacts

Figure 4 shows the *SFC equality BO* calculated for the various PC occupancies at road gradients 0 and 6% and urban access road. The data presented in Figure 4 show that buses become to be energetically beneficial with respect to the passenger cars at bus occupancies substantially lower compared with those providing the *STEI* equality. For example, at typical PC occupancy of 1.2 and a flat road *SFC* equality is achieved at a bus occupancy of 7 and 4.8 passengers per bus (for free flow and stop & go conditions, respectively) compared with 18 and 14 passengers required to achieve the *STEI* equality. The reported trend is attributed to the PCs fleet composition, which almost entirely consists of cars with spark ignition (SI) engines (Table 3) as opposed to buses equipped by diesel engines. It is known that diesel engines are featured by better efficiency together with comparable or even worse pollutant emissions compared to SI engines. Therefore, if the share of diesel cars in the PC fleet is significant (a situation typical for many European countries), higher bus occupancy values will be required to achieve the *SFC* equality. On the other hand, this PC fleet change will lead to a reduction of bus occupancy values yielding the environmental impact equality.

The tendency of changing *SFC equality BO* values as a function of PC occupancy and level of service is similar to that one described earlier for *STEI equality BO*. For a given PC occupancy, a reduction of bus fuel consumption with less congested traffic is lower compared with that of passenger cars. As was noted earlier, this is mainly the result of a different contribution of throttling losses to bus and PC fuel consumption.

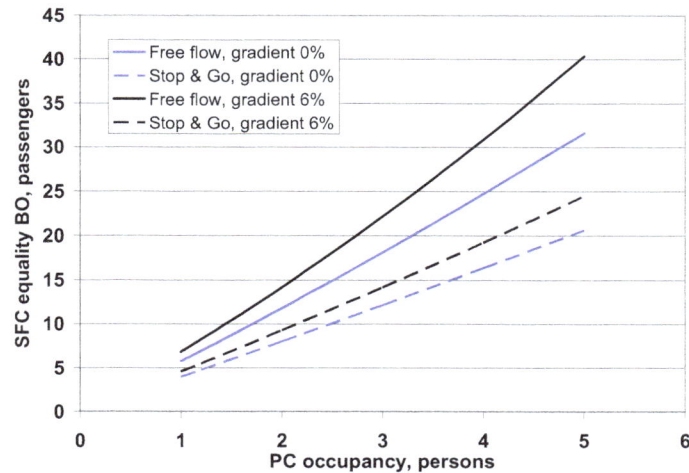

Figure 4. Dependence of *SFC equality BO* on PC occupancy. Free flow and stop & go levels of service, road gradients 0 and 6%, urban access road

Figures 5, 6 show the influence of the road gradient on *SFC equality BO* for the urban access road and the different PC occupancies at free flow and stop & go levels of service, respectively. In contrast to the conclusions drawn for emissions comparison, the road gradient effect on *SFC equality BO* does not change its character depending on PC occupancy. For all studied occupancy values the *SFC equality BO* increases (up to 25%) with the road gradient growth. The dependence of *SFC equality BO* on road gradient becomes weaker at lower PC occupancies and more congested traffic. These results show that the fuel consumption of buses rises with increase of the road gradient or vehicle occupancy more sharply than that of PCs. The observed effect may be a combined result of the mentioned above differences in throttling losses and differences in driving style of bus and car drivers. The former is mainly a bimodal style very roughly called "full gas–full brakes" driving (Tartakovsky et al., 2003) and aimed at keeping a constant driving speed. The latter is aimed at the achievement of the best possible fuel economy. As a result, at steep road or high occupancies the bus fuel consumption increases at a higher extent compared to that of a PC. This hypothesis is supported by the results obtained for the "stop & go" level of service. The very low traffic speeds in the latter case quite limit the realization of a bimodal bus driving style, thus leading to a reduction of differences in the fuel consumption change as a function of the road gradient or vehicle occupancy. It should be noted that the applied ARTEMIS model is based on measured driving cycles, thus the differences in the driving style are reflected in the modeling results.

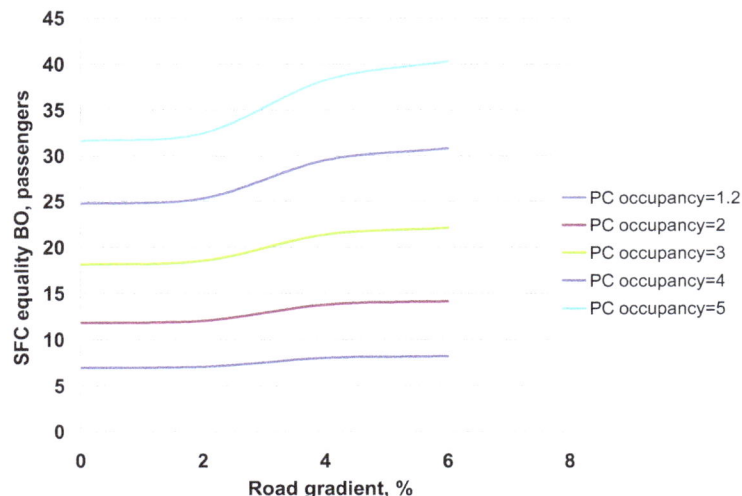

Figure 5. Dependence of *SFC equality BO* values on road gradient, free flow level of service, urban access road

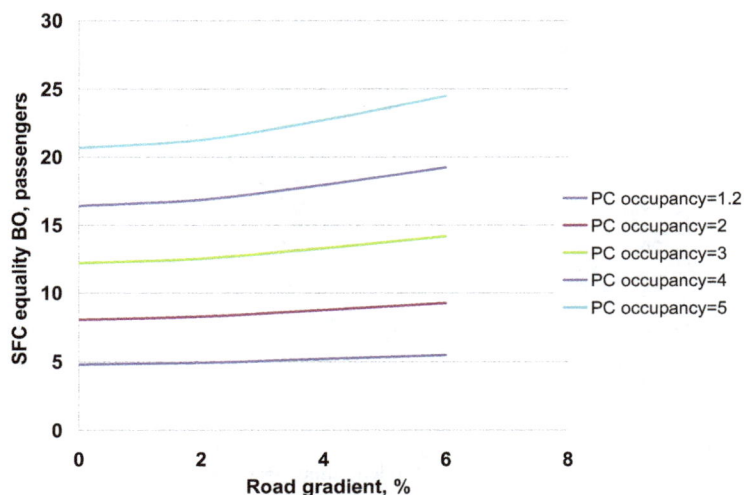

Figure 6. Dependence of *SFC equality BO* values on road gradient, stop & go level of service, urban access road

4. Conclusions

A methodology is developed for a comparative analysis of energy and environmental impacts of various urban transport modes. A total emission indicator is used as a tool for integral assessment of vehicle emissions. The environmental impact factor is suggested for a comparison between transport modes using different energy sources. Vehicle occupancy values, yielding equality of the specific environmental impact factors and specific energy consumption of the compared transport modes are used for the analysis purposes.

The developed methodology takes into account both the emissions level and the number of people exposed to the polluted air. This allows projecting health impacts caused by air pollution from vehicles that use different energy sources.

The methodology is applied for a comparison between buses and passenger cars at various levels of service, road gradients and urban road types. The results of the comparative analysis show that for driving on urban access roads with a speed limit of 50 km/h and PC occupancy of 1.2 persons the fleet averaged specific emissions (*STEI*) by urban buses become lower compared with PC, if the buses passenger load is higher than 14–18 persons. Buses become to be energetically beneficial over passenger cars at occupancy values substantially lower compared with those providing the emissions equality—4.8–7 persons depending on the level of service. This is explained by the fact that in the considered case almost the whole PC fleet has SI engines as opposed to buses equipped by more efficient diesel engines. If the share of diesel cars in the PC fleet is significant (situation typical for many European countries) higher bus occupancy values will be required to achieve the *SFC equality*.

The influence of the road gradient on *STEI equality bus occupancy* is quite weak and becomes sensible at steeper roads. The tendency of the *STEI equality BO* dependence on road gradient remains quite similar for different road types. At free flow level of service, when the average PC speed is close to the speed limit, urban buses become to be environmentally and energetically beneficial over PCs at lower occupancy values on roads with higher average speed. No significant differences were observed between *STEI* and *SFC equality BOs* for various road types at stop & go level of service.

Acknowledgements

The research leading to these results has received funding from the European Community's Seventh Framework Program under the grant agreements 218636/Conduits and 231341/CATS. This financial support is gratefully acknowledged.

References

Ahman, M. (2001). Primary energy efficiency of alternative powertrains in vehicles. *Energy, 26*, 973-989. http://dx.doi.org/10.1016/S0360-5442

American Conference of Governmental Industrial Hygienists. (2010). *Threshold Limit Values and Biological Exposure Indices*. Cincinnati, OH: American Conference of Governmental Industrial Hygienists.

André, M., & Rapone, M. (2009). Analysis and modeling of the pollutant emissions from European cars regarding the driving characteristics and test cycles. *Atmospheric Environment, 43*(5), 986-995. http://dx.doi.org/10.1016/j.atmosenv.2008.03.013

André, M., Fantozzi, C., & Adra, N. (2006). Development of an approach for the estimation of the road transport pollutant emissions at a street level. *ARTEMIS WP1000 report LTE06*, Bron: INRETS.

André, M., Keller, M., Sjödin, Å.,Gadrat, M., McCrae, I.,& Dilara, P. (2009). The Artemis European Tools for Estimating the Transport Pollutant Emissions. *Proc. 18th International Emission Inventories Conference* (pp. 1-10).

Bradley, M. J., & Associates. (2007). Comparison of energy use and CO_2emissions from different transportation modes.*Report to American Bus Association*, 17. Retrieved from http://www.buses.org/files/ComparativeEnergy.pdf

Curtiss, P. S., & Rabl, A. (1996).Impacts of air pollution: general relationships and site dependence. *Atmospheric Environment, 30*(19), 3331-3347. http://dx.doi.org/10.1016/1352-2310

De Vlieger, I., De Keukeleere, D., & Kretzschmar, J. G. (2000). Environmental effects of driving behavior and congestion related to passenger cars. *Atmospheric Environment, 34*, 4649-4655. http://dx.doi.org/10.1016/S1352-2310(00)00217-X

Dincer, I., Rosen, M., & Zamfirescu, C. (2010). Economic and Environmental Comparison of Conventional and Alternative Vehicle Options. In G. Pistoia (Ed.), *Electric and Hybrid Vehicles* (pp. 1-17). Amsterdam: Elsevier.

Fox, H., & Eweka, E. (2009). Simulation of hybrid buses: a study of fuel economy and emissions. *WIT Transactions on the Built Environment, 107*, 129-141. http://dx.doi.org/10.2495/UT090131

Granovskii, M., Dincer, I., & Rosen, M. (2006). Economic and environmental comparison of conventional, hybrid, electric and hydrogen fuel cell vehicles. *Journal of Power Sources, 159*, 1186-1193. http://dx.doi.org/10.1016/j.jpowsour.2005.11.086

Lenaers, G., & De Vlieger, I. (1997). On-board emission measurements on petrol-driven cars and diesel city buses. *International Journal of Vehicle Design, 18*(3-4), 368-378.

Nylund, N. O., Erkkilä, K., Lappi, M., & Ikonen, M. (2004). Transitbus emissions study: comparison of emissions from diesel and natural gas buses. *VTT research report PRO3/P5150/04*, 63. Retrieved from http://www.vtt.fi/inf/pdf/jurelinkit/VTTNylund.pdf

Nylund, N.O., Erkkilä, K., & Hartikka, T. (2007). Fuel consumption and exhaust emissions ofurban buses. Performance of the new diesel technology. *VTT Tiedotteita, Research Notes 2373*, 48. Retrieved from http://www.vtt.fi/inf/pdf/tiedotteet/2007/T2373.pdf

Parent, M. (2007). Advanced urban transport: automation is on the way. *IEEE Intelligent Systems, 22*(2), 9-11. http://dx.doi.org/10.1109/MIS.2007.20

Pope, C. A., & Dockery, D. W. (2006). Health effects of fine particulate air pollution. *Journal of the Air & Waste Management Association, 56*, 709-742. http://dx.doi.org/10.1080/10473289.2006.10464485

Radovic, D. (2009). *Eco-Urbanity: towards well-mannered built environments*. London and New York: Routledge.

Silva, C., Bravo, J., Gonçalves, G., Farias, T., & Mendes-Lopes, J. (2007). Bus public transport energy consumption and emissions versus individual transportation. *Proc. Transportation Land Use Planning and Air Quality Congress*, Copyright ASCE, 147-160. http://dx.doi.org/10.1061/40960(320)15

Sram, R. J., Binkova, B., Beskid, O., Milcova, A., Rossner, P., Rossner, P. Jr., ... Topinka, J. (2011). Biomarkers of exposure and effect interpretation in human risk assessments. *Air Quality Atmosphere and Health 4*, 161-167. http://dx.doi.org/10.1007/s11869-011-0133-8

Tartakovsky, L., Gutman, M., & Mosyak, A. (2012). Energy efficiency of road vehicles – trends and challenges. In E. F. S. Cavalcanti, & M. R. Barbosa (Eds.), *Energy Efficiency: Methods, Limitations and Challenges* (pp. 63-90). New York: Nova Science Publishers.

Tartakovsky, L., Mosyak, A., & Zvirin, Y. (2013). Energy analysis of ethanol steam reforming for internal combustion engine. *International Journal of Energy Research, 37*, 259-267. http://dx.doi.org/10.1002/er.1908

Tartakovsky, L., Zvirin, Y., Motzkau, M., Van Poppel, M., Riemersma, I., Veinblat, M., ... Gutman, M. (2003). Measurements and analysis of real-world driving behavior of urban buses. *Proc. the 12th International Scientific Symposium on Transport and Air Pollution*, Avignon (France).

Tsang, K. S., Hung, W. T., & Cheung, C. S. (2011). Emissions and fuel consumption of a Euro 4 car operating along different routes in Hong Kong. *Transportation Research Part D, 16*, 415-422. http://dx.doi.org/10.1016/j.trd.2011.02.004

Wang, T., Bai, J., Jiang, X., & Nienhaus, G. U. (2012). Cellular uptake of nanoparticles by membrane penetration: a study combining confocal microscopy with FTIR spectroelectrochemistry. *ACS Nano, 6*(2), 1251-1259. http://dx.doi.org/10.1021/nn203892h

3

An Empirical Study of the Environmental Kuznets Curve in Sichuan Province, China

Chuanqi Fan[1] & Xiaojun Zheng[2]

[1] College of Economics & Management, Sichuan Agricultural University, Chengdu, China

[2] School of International Law, Southwest University of Political Science and Law, Chongqing, China

Correspondence: Chuanqi Fan, College of Economics & Management, Sichuan Agricultural University, Chengdu, China. E-mail: 523141835@qq.com

Abstract

The empirical Environmental Kuznets Curve (EKC) literature is colorful but far from conclusive. The environmental Kuznets hypothesis (EKC) confirms an inverse U-shaped relationship between environmental pollution and per capita income. Many authors have analyzed the existence of an EKC for various pollutants. Others have used the EKC framework to identify country characteristics that help to explain the income–environment relationship. But for a local area, such as a province, studies are rare indeed. In this framework, based on the GDP per capita and emissions of industrial waste from 1985 to 2010 in Sichuan Province, China, the relationship is analyzed using regression between economic development and environment in Sichuan Province. Our evidence suggests that there exists a U-shaped or an inverted N-shaped relationship between environmental pollution and economic development in Sichuan Province, that is to say, the environmental Kuznets hypothesis is invalid in Sichuan Province. There are two possible reasons for this conclusion: firstly, KEC curve will not appear at any level of the economic development in Sichuan Province; secondly, the Environmental Kuznets Curve in Sichuan Province exists objectively, but the economic development in Sichuan Province at current stage is not sufficient enough to promote the appearance of KEC curve. However, more attention must be paid to the relation between environmental pollution and per capita income and appropriate environmental policies are required.

Keywords: environmental pollution, economic development, Environmental Kuznets Curve

1. Introduction

Through the analysis of economic growth and income gap in 18 countries, American economist Kuznets drew the following conclusions in the study of the relationship between economic growth and income distribution inequality: the inequality of income distribution is rapidly widening at early stages of economic growth, followed by the short-term stability, then it gradually declines at the later stages of growth, while the income inequality in developing countries at the early stages of development is more serious than developed countries, which is called the income Kuznets Curve (Researched by Kuznets, 1995). Suppose that the horizontal axis representing the economic growth and the vertical axis representing the inequality of income distribution, then KC curve is a parabola bent down after upwardly curved, usually referred as the "inverted U-shaped" curve. Based on the income Kuznets Curve theory, scholars Grossman and Krueger empirically analyzed the relationship between environmental quality and per capita income of the North American Free Trade Area for the first time in 1991. They concluded: pollution would increase with the rise in per capita GDP at low income levels, and decline with the growth of per capita GDP at the high-income level (Grossman & Krueger, 1991) .Panayotou first defined the relationship between environment-quality and per capita income as the Environmental Kuznets curve (EKC) using the inverted U-shaped curve in 1993. EKC reveals that the environmental quality would deteriorate with the increase in revenue at first, and then it would improve when income rises to a certain level, namely environmental quality and income show an inverted U-shaped relationship (Panayotou, 1993).

Different scholars hold different opinions about the environmental Kuznets curve hypothesis; some support the inverted U-shaped curve, while other studies show that it is U-shaped, N-shaped, monotonously rising and monotonously decreasing. Verbeke and Managi used empirical analysis to test the environmental Kuznets curve, and results showed that the EKC changing trajectory exists in most countries (Verbeke, 2006; Managi, 2006),

However, Richmond and Galeotti found that the EKC was not widespread by studying countries at different income levels, member countries in the Organization for Economic Cooperation and Development (OECD) generally show EKC relationship, while non-member countries do not show it (Richmond & Kaufmann, 2006; Galeotti et al., 2006). Khanna and Maddison doubted EKC, they believed that the increase of income is not the main factor of the improvement of the environment quality, economic growth and environment are mutual promoted and mutual influenced (Khanna & Plassmann, 2004; Maddison, 2006). Does the quality of the environment will be improved in the course of economic growth in a specific area, such as a province?

Given this issue, this paper adopts an econometric strategy to address this question. We propose a linear regression model which we will use in the fourth section to check the income-environment relationship with GDP Per capita and industrial three wastes, where industrial wastes, including industrial emissions, industrial wastewater and industrial solid waste. Whether the inverted U-shaped curve exists in Sichuan Province, the last section will conclude.

2. Economic Development and Environmental Pollution in Sichuan Province

2.1 Economic Development from 1985 to 2010 in Sichuan

GDP Per capita in Sichuan Province has almost shown exponential growth trend in the last 26 years from 1985 to 2010: it has increased to 21,361 Yuan, almost 36 times compared with 570 Yuan in 1985, and the annual average growth rate is 15.47%. GDP Per capita has grown rapidly since 2000, and the situation can be found in details in Figure 1 below. In the figure1 the horizontal axis represents the year.

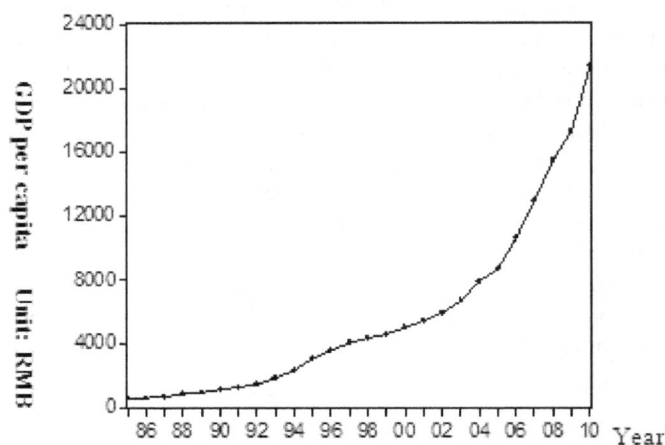

Figure 1. GDP per capita in Sichuan Province in 1985-2010

As shown in Figure 1, Sichuan province has experienced a period of rapid economic growth from 1985 to 2010.

2.2 Industrial Emissions in Sichuan

Industrial emissions in Sichuan generally showed an upward trend, which started from 378.7 billion standard m³ in 1985 and reached 2.0107 trillion standard m³ in 2010. Before 2002, industrial emissions were under 700 billion standard m³, and then it increased rapidly, and reached the historical high of 2.297 trillion standard m³ in 2007. It temporarily declined in 2008, pulled up again after 2009 and finally ended in 2.0107 trillion m³ in 2010, as shown in Figure 2 below.

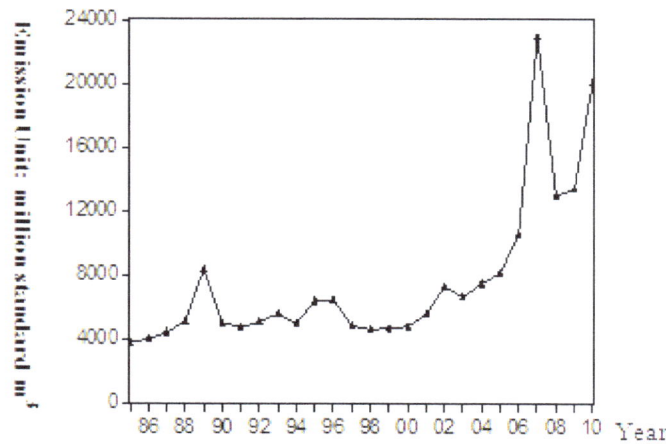

Figure 2. Industrial emissions in Sichuan Province

2.3 Industrial Wastewater Emissions in Sichuan

Sichuan industrial wastewater emissions in 1985 to 2010 roughly presented a "Λ+M+⌢" curve trend, where "Λ" presents the trend which first increased and then decreased , "M" means the trend of tow Λ, and "⌢" indicates the trend of convex shape, in the past 26 years Sichuan has made tremendous achievements in diminishing industrial wastewater emissions. In 1985, industrial wastewater emissions in Sichuan Province was 236830 ten thousand tons and its emissions in 2010 was 934.44 million tons, a net reduction of 1.43386 billion tons. In the meantime, the lowest emissions of 1.09701 billion tons appeared in 1989, in 1997 Sichuan industrial wastewater emissions were stabilized at low level and kept to the present, as shown in Figure 3.

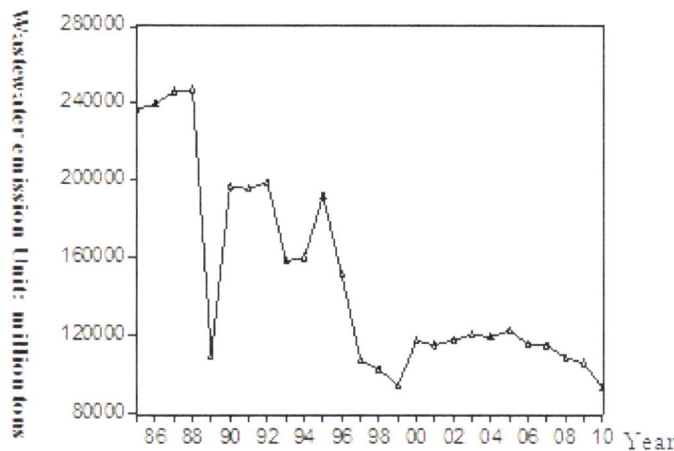

Figure 3. Industrial wastewater emissions in Sichuan Province

2.4 The Emissions of Industrial Solid Waste S in Sichuan

The emissions of industrial solid waste in Sichuan Province in the last 26 years roughly showed a downward trend, and this decreasing trend is very distinct, presenting a "M + ⌢" curve trend, where "⌢" signifies a trend of an decrease, it was 523 million tons in 1985, the highest in history was 9.98 million tons in 1998, later the emissions dropped steeply to 3.2 million tons in 2010, which was the lowest emission in history, as shown in Figure 4.

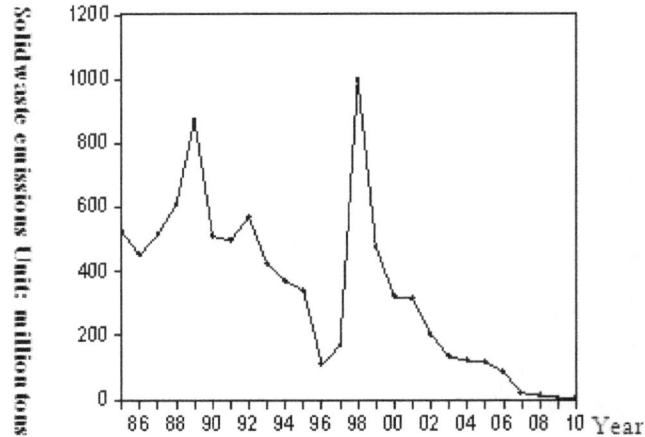

Figure 4. Industrial solid waste emissions in Sichuan Province

In general, great progress has been made in Sichuan's economic development from 1985 to 2010. Although environmental remediation has made some achievements, the environment problems are still very prominent, mainly caused by the ineffective control of industrial emissions.

3. Index Selections, Data Sources and Processing

3.1 Index Selections

Per capita GDP in Sichuan Province is selected as the indicator to measure the economic development, industrial emissions, industrial wastewater emissions, as well as industrial solid waste emissions are selected as indicators to measure the degree of environmental pollution in Sichuan Province.

3.2 Data Sources and Processing

This article selects the annual data of per capita GDP, industrial emissions, emissions of industrial waste water as well as the emissions of industrial solid waste in Sichuan Province from 1985 to 2010; they are derived from *Sichuan Statistical Yearbook, China Environment Statistical Yearbook, China City Statistical Yearbook* and *China's Regional Economic Statistical Yearbook.*

The two indexes used for environmental Kuznets empirical analysis are: per capita GDP index and environmental pollution index, which are obtained by processing the raw data. The per capita GDP Index is calculated on the basis per capita GDP in 1985, assume the per capita GDP is 1 in 1985, and divide each year's number by the per capita GDP in 1985 to get the yearly per capita GDP index. Index of environmental pollution is weighted and calculated from industrial waste gas emission index, industrial wastewater emission index and industrial solid waste emission index. Firstly, convert the three sequences of industrial emissions, industrial waste water emissions, emissions of industrial waste water, and industrial solid waste emissions in Sichuan to index. Specifically, the three sequences are all based on 1985, assume the indicators in 1985 as 100, divide emissions in each year by emissions in 1985 and multiply it by 100 to get the index. For example, industrial waste gas emission was 378.7 billion m³ in 1985 and 397.9 billion standard m³ in 1986, industrial emissions in 1985 is 100, then the indicator of industrial emissions in 1986 is (3979/3787) * 100 = 105. In this paper, the weight is determined using the principal component analysis method, the weights of the following three indicators: industrial emissions, industrial waste water, and industrial solid waste were 0.35, 0.31, and 0.34. So the Environmental Pollution Index in 1985 = 0.35×100+0.31×100+0.34×100 = 100. Likewise the environmental pollution index in subsequent years is calculated in the same way.

4. Empirical Analysis

Quadratic curve and cubic curve are used to fit in studying the relationship between economic development and environmental pollution in Sichuan Province, the quadratic curve fitting relationship is:

$$Y_t = \alpha + \beta_1 AGDP_t + \beta_2 AGDP^2_t + \varepsilon_t \tag{1}$$

The cubic curve fitting relationship is:

$$Y_t = \alpha + \beta_1 AGDP_t + \beta_2 AGDP^2_t + \beta_3 AGDP^3_t + \varepsilon_t \tag{2}$$

where α is constant, β_1 is the time coefficient, β_2 is the quadratic coefficient, β_3 is the cubic coefficient, ε_t is the regression error term, AGDP is per capita GDP index in Sichuan Province. For Equations (1) and (2), we assume that:

(a) $E(\varepsilon_t) = 0, (t=1,2,3\ldots,T)$.

(b) $Var(\varepsilon_t) = \sigma^2, (t=1,2,3\ldots,T)$.

(c) $Cov(\varepsilon_i, \varepsilon_j) = 0, (i \neq j)$.

(d) $Cov(\varepsilon_i, AGDP_t) = 0$.

For quadratic curves and cubic curves, different combinations of coefficient symbols have different curve forms.

Table 1. Curve shape of the relationship between environment and income

Model	Value of β_i	Forms of the curve
Model	$\beta_1 = \beta_2 = \beta_3 = 0$	no
Model 2 (linear)	$\beta_1 > 0, \beta_2 = \beta_3 = 0$	Linear monotonically increasing
Model 3 (linear)	$\beta_1 < 0, \beta_2 = \beta_3 = 0$	linear monotonically decreasing
Model 4 (quadratic)	$\beta_1 < 0, \beta_2 > 0, \beta_3 = 0$	U-shaped relationship
Model 5 (quadratic)	$\beta_1 > 0, \beta_2 < 0, \beta_3 = 0$	inverted U-shaped relationship
Model 6 (cubic)	$\beta_1 > 0, \beta_2 < 0, \beta_3 > 0$	N-type relationship
Model 7 (cubic)	$\beta_1 < 0, \beta_2 > 0, \beta_3 < 0$	inverted N-type relationship

As shown in Table 1, there are seven models for the curve, and the meanings vary from model to model, broadly speaking, linear monotonically increasing means that the environment quality deteriorates as income increases, linear monotonically decreasing means that the environmental quality improves with income increases. U-shaped relationship means that when income levels are in the lower stages, the environment quality improves as income rises, when the income level is at a high stage; the environment quality deteriorates as incomes rise. What's more, inverted U-shaped relationship means that when income levels are in the lower stages, the environment quality deteriorates as incomes rise and when income levels are in the high stages, the environment quality improves as income rises. Besides, N-type relationship is a kind of curve that as income levels rise gradually, the environment quality deteriorates before further improvement, and finally it falls into deterioration. On the contrary, inverted N-type relationship is totally opposite that as income levels raise gradually, the environment quality first improves before deterioration and at last improves.

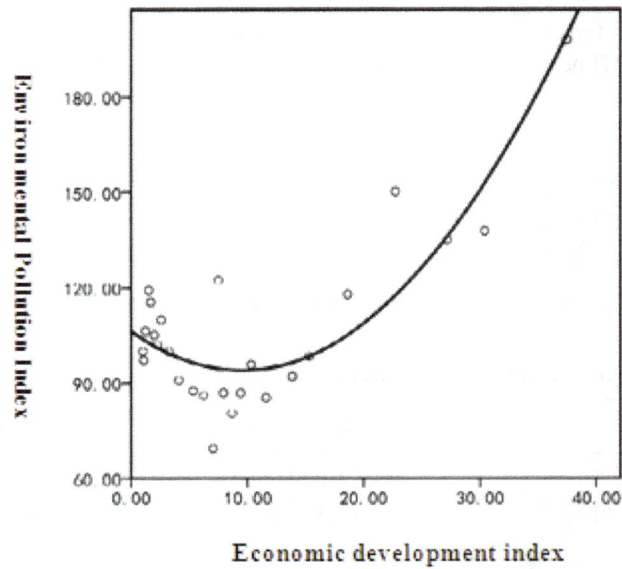

Figure 5. Quadratic curve fitting for the environmental pollution and AGDP

Figure 5 shows the quadratic curve fitting diagram for the environmental pollution and the per capita GDP in Sichuan Province.

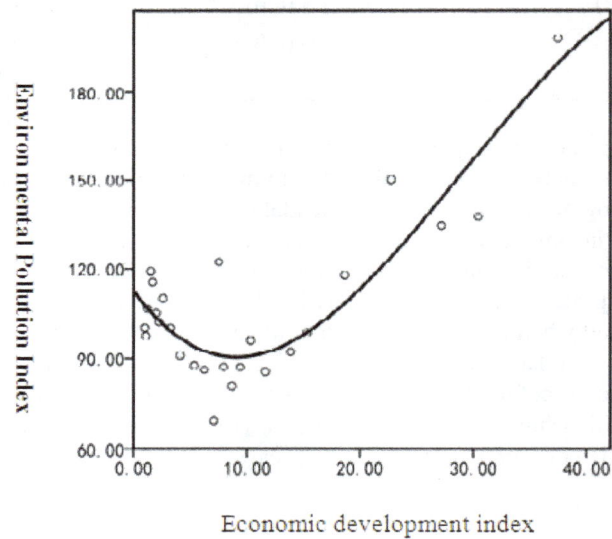

Figure 6. Cubic curve fitting for the environmental pollution and AGDP

Figure 6 shows the cubic curve fitting diagram for the environmental pollution and per capita GDP in Sichuan Province, the horizontal axis in this figure represents economic development index in Sichuan Province, and the vertical axis indicates the environmental pollution index.

It can be seen from the Figure 5 and Figure 6 that both quadratic curve and cubic curve fit well, the signs of the fitting coefficients are shown in Table 2 below:

Table 2. Fitting results for environmental pollution and per capita GDP

Model Form	Dependent variable	Model factors	Coefficient	T statistics	Coefficient of determination
Quadratic	environmental pollution degree	constant	108.77	24.2	0.803
		AGDP	-2.971	-3.8	
		AGDP^2	0.142	6.3	
Cubic	environmental pollution degree	constant	112.02	19.54	0.81
		AGDP	-5.202	-2.56	
		AGDP^2	0.34	3.08	
		AGDP^3	-0.04	-2.92	

As shown in Table 2, the t statistics of coefficients in the quadratic curve fitting are significant, coefficient of determination in the model is 0.803, the F-statistic is 37.51, indicating that the overall effect of the model is very good; t statistics of coefficients in the cubic curve fitting are also significant, the coefficient of determination in the model is 0.81, F statistics is 27.36, indicating the construction of the model is reasonable, so the quadratic formula for environment and income in Sichuan Province is:

$$Y = 0.142AGDP2-2.971AGDP+108.77 \tag{3}$$

The cubic relationship is:

$$Y = -0.04AGDP3+0.34AGDP2-5.202AGDP+112.02 \tag{4}$$

As the quadratic curve coefficients $\beta_1 < 0$, $\beta_2 > 0$, it determined that the environmental quality and income levels in Sichuan Province present "U-shaped" curve in the quadratic curve fitting, as the coefficients of its cubic curves $\beta_1 < 0$, $\beta_2 > 0$, $\beta_3 < 0$, it determined that the environmental quality and income levels in Sichuan Province present "И" type curve in the cubic curve fitting, that is the inverted "N" type curve, as shown in Figure 7 and Figure 8.

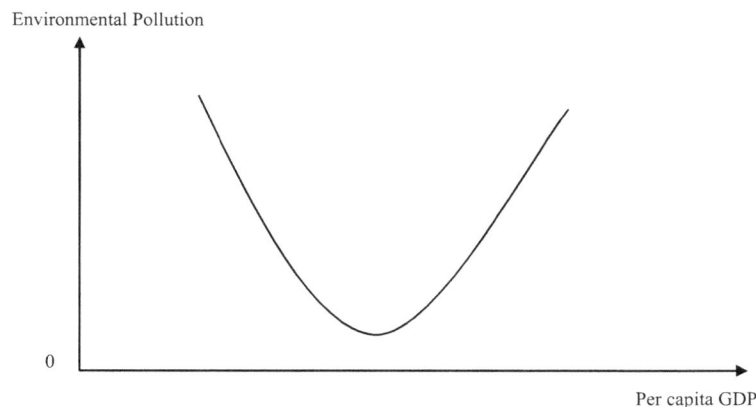

Figure 7. Curve for environment and income in Sichuan province

Environmental Pollution

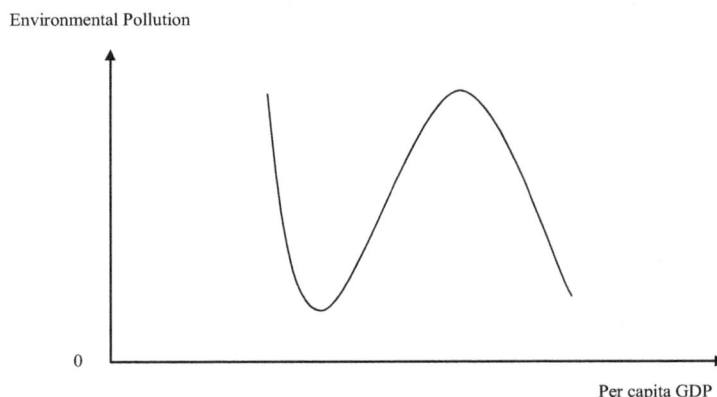

0

Per capita GDP

Figure 8. Curve for environment and income in Sichuan province

As shown in Figure 7 and Figure 8, neither quadratic curve fitting nor cubic curve fitting in Sichuan Province at current stage do not support the environmental Kuznets curve hypothesis, namely the KEC curve does not exist in Sichuan Province.

5. Conclusions and Solutions

Through empirical analysis, the relationship between environmental pollution and economic development in Sichuan Province support the environmental Kuznets curve hypothesis, which means the KEC curve does not exist in Sichuan Province. There are two possible reasons for this phenomenon: firstly, KEC curve will not appear at any level of the economic development in Sichuan Province objectively; secondly, the Environmental Kuznets Curve in Sichuan Province establishes objectively, but the economic development in Sichuan Province at current stage is not sufficient enough to promote the inflection point on KEC curve, that is to say the "И" shaped and "U" shaped relationship between environmental pollution and per capita income in Sichuan Province in this study are part of the KEC curve in Sichuan Province in the future.

Economic development provides more financial support for environmental remediation, but the increase in income is not the main incentive of the improvement in the environment quality. The environmental improvement is closely related to the environmental protection awareness of the people in Sichuan, the ideological quality of local people and the relevant policies of the government. Therefore, improvement of the environment should not only be built on the economic development. Sichuan Province must control the industrial emissions, so far industrial emissions is the biggest factor affecting the environment quality in Sichuan, Sichuan needs to phase out some industrial enterprises which are small-scale, high energy consumption, low-income or low production, especially those petrochemical enterprises. Most importantly, we need to promote the transformation of economic structure, shift from the domination of the secondary industry to the domination of the third industry, develop service industry vigorously, and improve people's awareness of environmental protection continuously.

Admittedly, this article has its own limitations and leaves more space for further researches in the future. Due to the lack of research data, it can't demonstrate the existence of Kuznets Curve in Sichuan province accurately. In addition, researchers need to explore a more scientific and effective method when confirm the weight in research. As for the study of Kuznets Curve in Sichuan province, latter researchers should collect more comprehensive and accurate data, amend the index of environment pollution, and make further discussion about the research methods.

Acknowledgments

During the process of writing this article, I got a lot of supports form many people. Firstly, the authors would like to thank two anonymous referee of this journal for useful comments and suggestion; any remaining errors are exclusively ours. I'm grateful for my tutor, Professor Jing Tan, and Professor Shishun Xiao for their valuable advice on this essay and careful revision. Then, I'd like to show many thanks to Miss Ting Zhou, my senior sister apprentice for abundant data she supplied. All in all, I extend my thanks to them from the bottom of my heart.

References

Galeotti, M., Lanza, A., & Pauli, F. (2006). Reassessing the environmental Kuznets curve for CO_2 emissions: a robustness exercise. *Ecological Economics, 57*, 152-163. http://dx.doi.org/10.1016/j.ecolecon.2005.03.031

Grossman, G., & Krueger, A. (1991). Environmental impacts of the North American Free Trade Agreement. *NBER*, working paper, No. 3914.

Khanna, N., & Plassmann, F. (2004). The demand for environmental quality and the environmental Kuznets curve hypothesis. *Ecological Economics, 51*, 225-236. http://dx.doi.org/10.1016/j.ecolecon.2004.06.005

Kuznets, S. (1955). Economic growth and income inequality. *American Economic Review, 45*, 1-28.

Maddison, D. (2006). Environmental Kuznets curves: a spatial econometric approach. *Journal of Environmental Economics and Management, 51*, 218-230. http://dx.doi.org/10.1016/j.jeem.2005.07.002

Managi, S. (2006). Are there increasing returns to pollution abatement? Empirical analytics of the environmental Kuznets curve in pesticides. *Ecological Economics, 58*, 617-636. http://dx.doi.org/10.1016/j.ecolecon.2005.08.011

Panayotou, T. (1993). Empirical Tests and Policy Analysis of Environmental Degradation at Different Stages of Economic Development, Working Paper WP238. *Technology and Employment Program*, International Labor Office, Geneva.

Richmond, A. K., & Kaufmann, R. K. (2006). Is there a turning point in the relationship between income and energy use and/or carbon emission? *Ecological Economics, 56*, 176-189. http://dx.doi.org/10.1016/j.ecolecon.2005.01.011

Verbeke, T., & Clercq, M. D. (2006). The income-environment relationship: evidence from a binary response model. *Ecological Economics, 59*, 419-428. http://dx.doi.org/10.1016/j.ecolecon.2005.11.011

4

Accumulation of Lead (Pb) in the Talus Lichenes Contained in Mahogany Tree Stands of Roadside of Medan City

Ashar Hasairin[1], Nursahara Pasaribu[2], Lisdar I. Sudirman[3] & Retno Widhiastuti[2]

[1] Biology Education Department, Faculty of Mathematic and Science, State University of Medan, Indonesia

[2] North Sumatra University, Medan, Indonesia

[3] Bogor Agricultural Institute, Indonesia

Correspondence: Ashar Hasairin, Biology Education Department, Faculty of Mathematic and Science, State University of Medan, Jl. William Iskandar Psr V Medan Estate, Medan, Indonesia. E-mail: asharunimed2014@gmail.com

Abstract

This study reports on the accumulation of lead (Pb) in *the talus Lichenes* found on roadside stands of mahogany trees in the city of Medan, North Sumatra, Indonesia. Samples were taken by purposive, ie location based on the level of traffic density with different air pollution. Pb analysis was performed using Atomic Absorption Spectroscopy (AAS). Identified as many as 8 kinds of Lichens with 2 types, namely *talus Crustose* and foliose. Type of *Lepraria incana* and *Pertusaria amara*, which is found in the three study sites belonging to the cosmopolitan types. Pb accumulated in the *talus Pertusaria amara* ranged from 5.23 to 15.07 ppm. Being on *Lepraria incana* ranged 1.19 to 4.88 ppm. *Pertusaria amara* much larger than the *Lepralia incana*, have potential as bio-indicators of resistance. *Lichenes Pb* correlation with traffic density showed *Pertusaria amara* has a very high level and significant correlation compared with other types.

Keyword: Accumulation of Pb, *talus*, *Lichenes*, tree stands

1. Introduction

Environmental problems, especially in urban areas from day to day getting out of control. This is due to the increasing industrial development and transportation. The growth of industry and transport increases in Medan, impact on environmental degradation. One of the sources of pollutants are very harmful to living beings is lead (Pb). These metals enter the human body through the respiratory and digestive systems. Pb biggest polluters in the air is the transport sector. Motor vehicles are a major source of *Pb* that pollute the air in urban areas. An estimated 60 to 70% *Pb* particles in urban air comes from motor vehicles, and roughly 75% of *Pb* is added to the fuel will be emitted back into the atmosphere (Dahlan, 1992; O'neil, 1993). *Lead* (Pb) contained in the air, which is the result of motor vehicle exhaust emissions, can accumulate in the body tissues of living things, especially on *talus Lichens*. This is a good indicator of air pollution (Bargagli et al., 1987).

In the area of Tuscany-Italy, the concentration of *Pb* in the *talus Lichenes* there are 13.2 µgg-1 dry weight. Highest concentration of *Pb* was found in the area around the vehicle parking area and near the highway. Accumulation of *Pb* in *Parmelia physodes* decreased proportionally to the distance away from the highway (Kovacs, 1992). Deruelle research results (1981) also showed that at a distance of 15 m from the highway accumulation of *Pb* was found in 1002 as µgg-1 dry weight, while at a distance of 600 m from the highway only 65 µgg accumulation of *Pb-1* dry weight.

The level of air pollution in urban areas can be tested with a bio indicator to determine the air quality. Bioindicator is an organism whose presence can be used to detect, identify and qualify environmental pollution (Conti & Cecchetti 2000; Sudirman, 2009). Research the use of bio-indicators in monitoring air pollution is still limited, so it needs to be studied more in depth lichens that can be used as bio-indicators of air pollution. Information about the environmental quality of air in the city of Medan - North Sumatra - Indonesia is illustrated by the findings of accumulation of *Lead (Pb)* in *Lichens Talus* at roadside stands of mahogany trees.

2. Method

Location is the object of this study were taken by purposive based on the level of traffic congestion and air pollution are different in the city of Medan. The location is divided by three categories, namely (1) The area of high traffic density, (2) Regional traffic levels are, (3) Regional traffic levels are low. Traffic density measurements is done by calculating the total motor vehicles passing through the counting station using a hand tally counter. Then proceed with sampling *Lichens* on bark surface of the mahogany tree. Samples identified in Plant Taxonomy Laboratory, Department of Biological Science, State University of Medan and *Pb* analysis conducted at the Laboratory of Pharmaceutical North Sumatra University using Atomic Absorption Spectroscopy (AAS). Data were analyzed descriptively to determine and compare the accumulation of Pb between the location of the traffic levels are different.

3. Results and Discussion

Lichenes contained in the Standing Mahogany

Corticolous is kind of *Lichenes* that live on the bark of trees. This species is very limited to the tropics and subtropics, the most humid environmental conditions. Existing *Lichens* on trees generally grow on the trunk or lower part of the stem (Fink, 1961). Exploration results *Lichens* on tree stands of mahogany three sampling sites, found as many as 897 samples of *Lichens* which includes 8 species and 5 families. Lichens are classified into 2 types talus, namely: the type of *Foliose* (leaf resembles the structure of the talus, often found green to grayish green) of 3 types. Type of *Crustose*, crust layer structure of the Talus as firmly attached to the substrate with the talus colors vary by 5 types. Medium type of *Squamulose* and fructicose is not found. Type of *Lichens* are found throughout the study sites shown in Table 1.

Table 1. Number of type of *Lichenes* found in all locations observations

No	Species Name	Family (suku)	Type Talus	Location / Total of *Talus*			
				I	II	III	Total
1	*Lepraria sp.*	*Leprariaceae*	*Crustose*	-	32	-	32
2	*Parmelia sp.*	*Parmeliaceae*	*Foliose*	-	3	9	12
3	*Parmelia glabratula*	*Parmeliaceae*	*Foliose*	-	8	19	27
4	*Parmelia saxatilis*	*Parmeliaceae*	*Foliose*	60	10	-	70
5	*Lepraria incana*	*Leprariaceae*	*Crustose*	64	45	155	264
6	*Graphis scripta*	*Graphidaceae*	*Crustose*	-	87	154	241
7	*Opegrapha atra*	*Opegraphaceae*	*Crustose*	-	7	13	20
8	*Pertusaria amara*	*Pertusariaceae*	*Crustose*	3	52	176	230
	Total of *Colonies* per location			127	244	526	897
	Mean			15,87	30,50	65,75	112,12
	Deviation Standard			28,50	29,52	79,94	
	Percentage presence of *Lichenes* (%)			5,87	11,27	24.29	100

Description: Location I. Jl. Yos Sudarso Medan; Location II. Jl Sudirman Medan; Location III. Jl. Cik Ditiro Medan

There are 6 genera and 8 species of *Lichens* were obtained. Type of *Lepraria incana* and *Pertusaria amara* is found in the three study sites. The types of these *Lichens* belonging to the cosmopolitan and tolerant type because it can be found throughout the observation location. The number and types of *Lichens* are very varied. Each type of *Lichens* were found to have characteristics that are so diverse between one species and another. It can be noted from the start of the *Talus* type, shape, color, surface and other characteristics. *Lichens* have morphological characteristics and properties are different from one another.

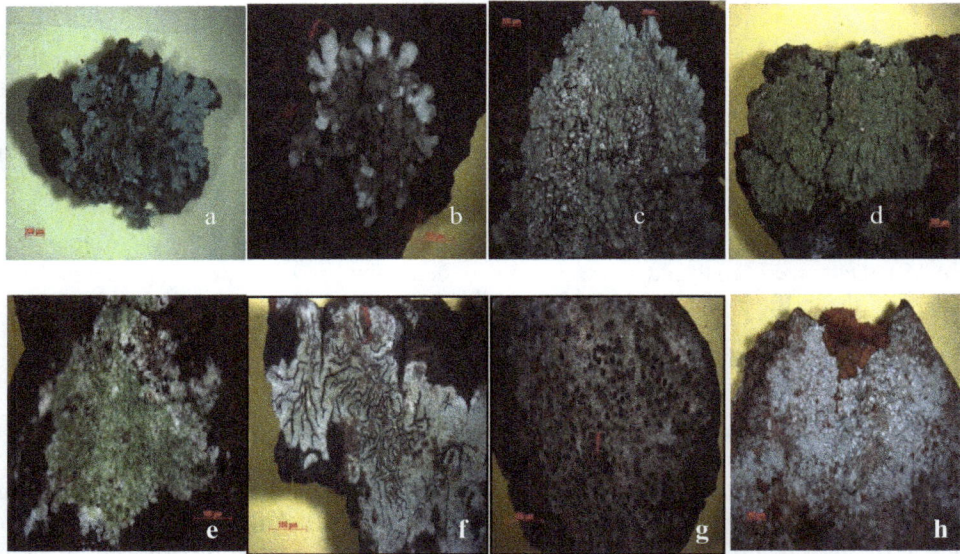

Figure 1. Type of *Lichens* are found throughout the study site, namely: a. *Parmelia grabatula*; b. *Parmelia saxatilis*; c. *Parmelia sp.*; d. *Lepraria incana*; e. *Lepraria sp.*; f. *Graphis scripta*; g. *Opegrapha atra*; h. *Pertusaria amara*

Lichenes diversity of different proportions in each study site, due to contaminated areas with a density of motor vehicles. Types of *Lichens* are found in *Swietenia* mahogany tree on Jl. T. Cik Ditiro Medan, Jl. Sudirman Medan and Jl. Yos Sudarso Medan, is as many as eight types. It appears that there are variations in the number and types of *Lichens* for each study site. It shows that there are differences in the tolerance level of *Lichens* on the level of air pollution. That difference can simply be shown in the following diagram.

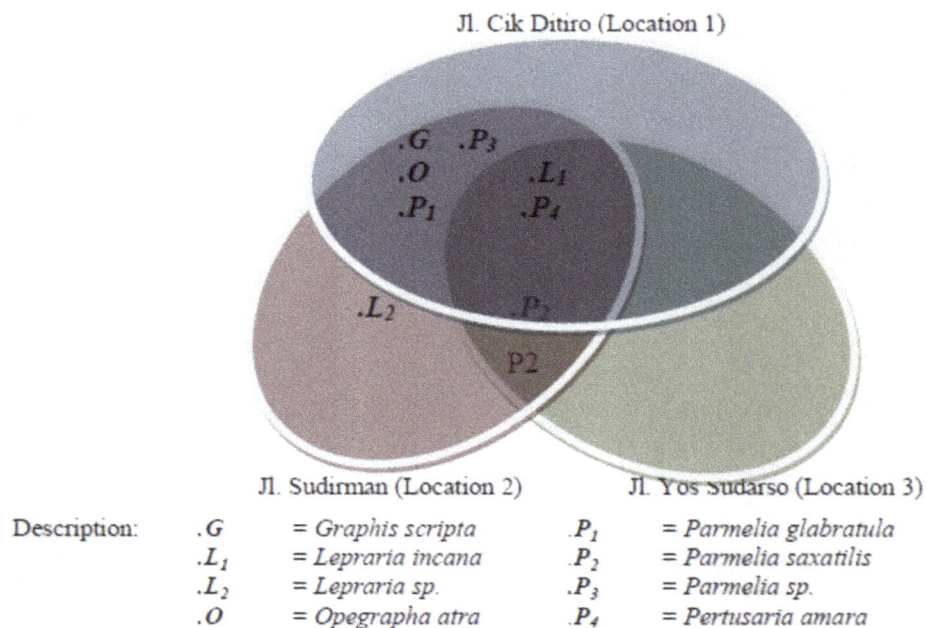

Description:	.G	= *Graphis scripta*	.P_1	= *Parmelia glabratula*
	.L_1	= *Lepraria incana*	.P_2	= *Parmelia saxatilis*
	.L_2	= *Lepraria sp.*	.P_3	= *Parmelia sp.*
	.O	= *Opegrapha atra*	.P_4	= *Pertusaria amara*

Metal accumulation *Pb* at *Tallus Lichens*

Lichens exploration results at the three sampling sites, indicating that there are differences in the level of tolerance of *Lichens* to air pollution levels (Table 1). It is characterized by differences in the type and amount of

Lichens are found in each of the sampling sites, with the accumulation of pollutants *Lead (Pb)*.

Table 2. Pb metal accumulation in Tallus Lichens in Medan

No	Type of *Talus*	Species Name	Timbal content (mcg/gram)		
			Location 1	Loction 2	Location 3
1.	*Crustose*	1. *Lepraria sp.*	-	46,39	-
	Crustose	2. *Lepraria incana*	4,88	4,88	1,19
	Crustose	3. *Grafis scripta*	-	20,47	0,95
	Crustose	4. *Opegrapha atra*	-	0,47	<0,02
	Crustose	5. *Pertusaria amara*	14,43	15,07	5,23
2.	*Foliose*	6. *Parmelia sp.*	-	44,99	19,99
	Foliose	7. *Parmelia glabratula*	-	29,82	38,04
	Foliose	8. *Parmelia saxatilis*	21,37	33,82	-

Description: Location 1. Jl. Yos Sudarso, Medan; Location 2. Jl.. Sudirman, Medan; Location 3. Jl. Cik Ditiro, Medan

Pb accumulated in the *talus Pertusaria amara* ranged from 5.23 to 15.07 ppm. In *Lepraria incana* is ranged from 1.19 to 4.88 ppm. Type of *Lichens* belonging to the cosmopolitan and tolerant type because it can be found throughout the observation location *Lepraria incana* and *Pertusaria amara*. This type classified as the most resistant type of presence both in the percentage of clean air and polluted air. According Panjaitan et al (2010) Family of *Leprariaceae* characterized by talus characteristics resembling flour, spread unevenly, with margins that form small lobes and pale green to whitish yellow.

Figure 2. (a) *Lepraria incana* dan (b) *Pertusaria amara* and arrows sign indicate location of accumulates particles lead (Pb)

Particle accumulation of *Lead (Pb)* in *Lepraria incana* and *Pertusaria amara* shown with blackish brown stain under a layer of algae seen in the upper part of the cortex. Type of *Lepraria incana* were found in all study sites, including the type that easily adapt to poor air quality conditions. Use of *Lepraria sp.* as bioindicator of air pollution ever undertaken in the city of Bandung (Taufikurahman et al., 2010). *Lichens* sensitivity to air pollution can be seen through changes in its diversity and accumulation of pollutants in its' talus. Meanwhile, according to Ohmura et. al (2009) type of lichen is very sensitive to sulfur dioxide (SO_2) and can only live in areas with clean air quality.

Further research of Lichenes crustose types is found as many as 5 types are more than the *Foliose* types. *Crustose Lichens* survive stronger than other types. This is due to a smaller flat *Crustose*, thin strongly attached to the *Corticoleus* (bark). *Crustose* has been used in Japan as a bioindicator of air pollution. McCune (2006) said Lichens crustose considered more tolerant of air pollution as it has the structure of the talus is relatively simple compared to other types of *Lichen Talus*. Type of *Tolerant Lichens* can survive in areas with environmental conditions where the air is polluted. Meanwhile, Lichens are sensitive types can not usually be found in areas with poor air quality. Lichens difference sensitivity to air pollution related to the ability to accumulate pollutants (Conti & Ceccheti, 2000).

Lichens are found to have a type derived from the *Talus Foliose Parmeliaceae* family. *Parmeliaceae* family is the largest group of *Foliose Lichens* have a specific form of the *Talus* and easily recognizable. Its' *Talus* has upper and lower cortex, there is often rizin to help adhesion to the substrate. *Lichens* found are the type of *Parmeliaceae* family, namely *Parmelia caperata*; *Parmelia glabratula*; *Parmelia saxatilis*; and *Parmelia sp.* Also note that there are different levels of *Pb* content in the *Talus Lichens Lepralia incana* and *Pertusaria amara*. *Amara Pertusaria Pb* concentration is much greater than the *Lepralia incana*, means *Pertusaria amara* which have potential as bio-indicators of resistance that can be found in various areas with different levels of air pollution.

Several types of *Lichenes* are sensitive to pollutants in the air so rarely found in polluted areas. The types are more tolerant of pollutants can accumulate a certain amount to the extent of concentration that can be tolerated. The types that are tolerant can be used as an indicator to detect the levels of accumulation of pollutants contained in the air especially. Metals are absorbed by *Lichens* accumulate in tissues its' *Talus*. *Lichens talus* structure is one of the factors that affect the efficiency of absorption of the metal. According Kinaliouglu et. al (2010) that the efficiency level of accumulation of pollutants at successive *Talus* is *Foliose > Crustose > Fruticose*. Furthermore Scerbo et. al (2002) said that causes extensive talus surface *Foliose Lichens* have greater contact with pollutants so that the accumulation of pollutants is more efficient than other types of *Talus*. *Foliose Lichen* id also called leafy *Lichens*, has extensive talus structure and can be easily removed from the substrate. *Foliose* types found in this study are in some kind of *Parmelia sp.*

In addition to traffic levels, abiotic conditions were also measured to determine the specific environmental conditions of the habitat Lichens, including air temperature, air humidity and light intensity of the wind speed. Measurement of environmental factors in the three study sites shown in Table 3.

Table 3. Environmental parameters measurement result data of each location

No	Parameter	Location			Total	Mean
		I	II	III		
1.	Traffic density Vehicles / hour	15909	8893	7765	32567	10855
2.	Humidity (%)	77	79	81	237	79
3.	Temperatures (°C)	27,6	27,8	27,9	83,3	27,76
4.	The intensity of light (Lux)	500	371	241	1112	370,66
5.	The wind speed (m/s)	2,9	1,8	1,6	6,3	2,1

Description: Location1. Jl. Yos Sudarso, Medan; Location 2. Jl.. Sudirman, Medan; Location 3. Jl. Cik Ditiro, Medan

Traffic levels and environmental factors were measured at the three sampling locations vary from each other. The high level of traffic congestion caused because it is situated. The position or location of the road are different causes of traffic levels at each observation location. Location 1 is the protocol that is located on various streets and lanes every day is always filled by a current motor vehicle. Therefore, this location has traffic levels are high compared to most other locations.

Location 2 is located in the city center adjacent to the location 3 with lane specific, the number of vehicles crossing this location is less than 1, while locations 3 locations located in schools and offices that are low traffic access. This led to the location 3 has traffic levels are low compared to most other observation location. These considerations were taken to determine the quality of the environment that may be affected by each type of vehicle exhaust gases passing at that location. According Nursal, et al (2005) the high density of traffic is one of the sources of Pb pollution in the air.

When linked with the accumulation of *Pb* were measured on the talus Lichenes obtained from each location turns on the *Talus Lichens Pb* accumulation correlates with the level of traffic congestion and other environmental factors were measured. Power accumulation of Pb in each type Lichens are not the same, even certain types have a high accumulation of power. Pertusaria amara has a very high correlation levels than other types.

Environmental factors including humidity, air temperature, light intensity and wind speed had greater influence on the accumulation of *Pb* in the *Talus Lichens* compared to other measurable factors. This relates to the nature of life and growth of the *Talus Lichens* are better suited to the more humid air conditions. Humidity and air temperature at the three study sites are not much different, compared with the light intensity and wind speed. According Nursal, et al (2005) in a more humid environment *Lichens* can live better and fertile, so that the absorption of water, minerals and accumulation of pollutants become more effective and much more.

In Table 3 it appears that the air temperature is relatively the same, while the intensity of light at each study site showed varying numbers, the highest light intensity is at location 1 of 500 Joules, decreased at location 2 is 371 Joules; location 3 is 241 Joules. Air humidity showed numbers vary at each study site, the most humid area is the location of 1 is 81%, decreased in 3 locations namely 79%, location 2 is 77%. Meanwhile, according to Noer (2004) in Pratiwi (2006) which states lichens like dry place with air humidity 40% - 69%. Average air humidity above that range. For wind speed, location 1 showed a higher rate than the two other studies tage location 2, and 3 are relatively similar locations respectively are 1.8 m /s and 1.6 m /s.

Lichenes habitat characteristics for each study site when compared with conditions during the last five years (2009-2013) are relatively constant. Lichens krakteristik habitat conditions in detail attached at the end of this paper. The following section describes the characteristics of the habitat conditions lichens in July between the years 2009-2013. For location 3, the air temperature for the month of July during the year 2009-2013 are among $27,70^0$ C - $28,20^0$ C, with humidity between 79% -83%, the intensity of light in the month of July 2013, declined to 34% compared to the previous four years (2009-2012) is between 50% - 67%, the wind speed is between 1,05m/s - 1,80 m/s. For location 2, the air temperature is in the range $27,30^0$ C - $27,90^0$ C, with humidity between 77% -79%, the intensity of light reaching 60% in 2011 and further decreased to 41% in 2013, indicating an increase in wind speed is significantly from the year 2009 that 1,40 m/s to 3,26 m/s in 2012, then decreased to 3,18 m/s in 2013 to location 1, the air temperature is in the range of between $27,30^0$ C - $27,70^0$ C, with humidity between 80% - 84%, light intensity 47% - 69%, the wind speed in 2009 is 0,90 m/s, decreased to 0,10 m/s in 2010, then continued to increase until 2013, that is 1.40 m/s.

Ambient Air Quality

Content of ambient air measurements at each study site is intended to describe the pollution level three regions. Parameters were observed, namely CO, CO_2, NO_2, and SO_2. These measurements were performed by Environmental Health and Engineering Center for Disease Controling (EHECDC) Class I Medan. Content of the ambient air measurement results can be seen in Table 4 below.

Table 4. Content of ambient air in location research

No	Parameter	Ambien air quality			Absolut quality
		Lokasi 1	Lokasi 2	Lokasi 3	
1	*Carbon monoksida (CO)*	14,00	10,00	9,00	29 ppm
2	*Nitrogen Oksida (NO₂)*	9,55	32,17	31,44	400 µg/m³/jam
3	*Sulfur dioksida (SO₂)*	11,55	32,38	6,10	900 µg/m³
4	*Carbon dioksida (CO₂)*	3.300,00	2.150,00	1.800,00	-

Description:

Location 1. Jl. Yos Sudarso (high traffic density)

Location 2. Jl.Sudirman (medium traffic density)

Location 3. Jl. Cit Ditiro (low traffic density)

Ambient air sampling was conducted at three locations during the day at 09:30 pm to 12:00 pm on September 12, 2013, which the transport activity is expected to contribute to the ambient air in the surrounding environment. Measurement of value content of the ambient air samples with parameters of carbon monoxide (CO), nitrogen dioxide (NO_2) and sulfur dioxide (SO_2) and carbon dioxide (CO_2) is still far below the air quality standard threshold according to Government Regulation No. 41 of 1999.

Based on the measurement results, the location of the ambient air contains carbon monoxide (CO) and carbon dioxide (CO_2) in a row from the top is the location area 1 (high traffic density) followed by location 2, (medium traffic congestion) and location 3 (low traffic density). CO and CO_2 measurement results demonstrate significant value in the three study sites and shows a striking comparison of values in a row. This parameter indicates the level of pollution that contrast.

Overall, the results of these measurements indicate that the relative location of the three regions with the lowest level of pollution and the location 1 is the area with the highest contamination levels among the three study sites. While the location of which 2 are medium or moderate level of contamination. The level of pollution is also linked to the volume of vehicles passing in each study site.

Lichenes Correlation with Physical Properties of Growing Media and Ambient Air

Correlation analysis performed lichens namely: 1) *Lichens* with ecological characteristics (physical factors of environmental chemistry); 2) *Lichens* with ambient air. Spearmans Correlation Analysis with SPSS computerized method 18 shown in Table 5 below.

Table 5. Correlation analysis in location research Lichenes

Location	No	Species name	Lichenes Correlation with Physical Properties of Growing Media			Lichenes Correlation with Ambient Air		
			r	A	B	r	A	B
	1	*Parmelia saxatilis*	-0,320	VL	NSig	0,553	S	Sig
1	2	*Lepraria incana*	-0,437	VL	NSig	0,568	S	Sig
	3	*Pertusaria amara*	-0,470	VL	NSig	0,555	S	Sig
	1	*Lepraria sp.*	0,500	S	Sig	0,302	L	Sig
	2	*Parmelia sp.*	0,626	H	Sig	0,213	L	Sig
	3	*Parmelia glabratula*	0,614	H	Sig	0,441	S	Sig
2	4	*Parmelia saxatilis*	0,294	L	Sig	0,388	S	Sig
	5	*Lepraria incana*	0,394	L	Sig	0,234	L	Sig
	6	*Grafis scripta*	0,560	S	Sig	0,537	S	Sig
	7	*Opegrapha atra*	0,550	S	Sig	0,486	S	Sig
	8	*Pertusaria amara*	0,437	S	Sig	0,459	S	Sig
	1	*Parmelia sp.*	0,634	H	Sig	0,515	S	Sig
	2	*Parmelia glabratula*	0,765	H	Sig	0,130	VL	Sig
3	3	*Lepraria incana*	0,386	L	Sig	0,668	H	Sig
	4	*Grafis scripta*	0,627	H	Sig	0,281	L	Sig
	5	*Opegrapha atra*	0,187	VL	Sig	0,236	L	Sig
	6	*Pertusaria amara*	0,291	L	Sig	0,544	S	Sig

Specification:

r = correlation; A = level of correlation; H = Height; S = Suficient; L = Low; VL = Very Low; B = Significant Level; NSig = Not Significant; Sig = Significant

Zone 1 locations are classified as high-traffic areas. *Lichens* type found is only three types. Correlation (r) with the physical properties of the media *Lichens* grow extremely low level of correlation (-) not significant and proportional. Negative correlation indicates that the lower colony Lichens, the lower the Physical Properties of Growing Media (air temperature, air humidity and light intensity of wind speed). Moderate correlations (r) Lichens with ambient air showed a positive correlation with the level of correlation is quite significant and proportional. The positive correlation indicates that the higher colony Lichens, the higher the ambient air in the form of carbon monoxide (CO), nitrogen dioxide (NO_2) and sulfur dioxide (SO_2) and carbon dioxide (CO_2).

Correlation of *Lead (Pb)* Lichenes with traffic density

Lichenes has the ability to absorb lead from the air. The rate of accumulation can lead increases with the density of traffic flow and decreases when more distant from the edge of the highway. According to Dahlan (1989) lead content in plants that are on the edge of the road may reach 50 ppm, but that number will decrease 2-3 ppm at a distance of 150 meters from the highway. Furthermore, to determine the correlation of Pb Lichens with traffic density were calculated using a computerized program SPSS 18 shown in Table 6.

Table 6. Correlation analysis Pb Lichenes with traffic density

No	Species name	Location 1			Location 2			Location 3		
		R	A	B	R	A	B	R	A	B
1	*Grafis scripta*	-	-	-	0,951	VH	Sig	0.999	VH	Sig
2	*Lepraria sp.*	-	-	-	0,778	H	Sig	-	-	-
3	*Lepraria incana*	0,537	S	Sig	0,937	VH	Sig	0,853	VH	Sig
4	*Opegrapha atra*	-	-	-	0,955	VH	Sig	0,509	VH	Sig
5	*Parmelia sp.*	-	-	-	0,869	VH	Sig	0,918	VH	Sig
6	*Parmelia glabratula*	-	-	-	0,984	VH	Sig	0,672	VH	Sig
7	*Parmelia saxatilis*	-0,009	VL	NSig	0.958	VH	Sig	-	-	-
8	*Pertusaria amara*	0,945	VH	Sig	0,955	VH	Sig	0,891	VH	Sig

Description: Location 1. Jl. Yos Sudarso, Medan; Location 2. Jl .. Sudirman, Medan; Location 3. Jl. Cik Ditiro, Field. r = Correlation. A = Level of Correlation (VH = Very High, H = Hight; S = sufficient; L = Low; VL = Very Low). B = Significant Level (NSig = Not Significant S = Significant)

The results of the analysis of the content of *Pb* in 3 different locations very real with *Pb* at two other locations. This is due to the location 3 has traffic density is much lower than the other sites. 3 locations located in schools and offices that are low traffic access. Value content of *Pb* in *Lichen Talus* derived from location 1 and location 2 does not have a significant difference, caused ecological almost simultaneously.

Location 1 *Pb Lichens* which correlation with traffic density correlation *Parmelia saxatilis* is very low and not significant. Negative correlation (-) indicates that the correlation in the opposite direction or inversely proportional, which means if the volume of traffic is high then the less extensive *Talus Lichens* and vice versa. Type Pertusaria amara has a very high level and significant correlation compared with other types.

Being on Location 2 and Location 3 Pb Lichens correlation with density has a positive and significant correlation with the degree of correlation is very high and directly proportional. Jamhari research results (2010); Walterbeek, et.al (2003) the morphology and physiology of Lichens are considered relevant to metal accumulation. Lichens show tolerance to metals. Determination of metal concentrations commonly used approach by utilizing bioindikasi Lichens metal pollution (Garty, 2001). Some common metals measured include black lead (Pb), Cadmium (Cd), Chromium (Cr), Zinc (Zn) and copper (Cu).

4. Conclusion

There are 8 types of *Lichenes* the mahogany tree stands in the city of Medan, which the highest types number is in location 2, as many as 8 species, and the lowest is at location 1, as much as 3 types. *Lichens* found are divided in 2 type of talus, namely *Crustose* and *Foliose* type, medium type, and the *squamulose fructicose Liken* type *Lepraria unidentified sp.*, and *Pamelia sp* are not found.

Pb accumulated in the *Talus Pertusaria amara* ranged from 5.23 to 15.07 ppm. Being on *Lepraria incana* ranged 1.19 to 4.88 ppm. *Pertusaria amara* is much larger than the *Lepralia incana*, which have potential as bio-indicators of resistance is found in a variety of areas with different levels of air pollution. *Pertusaria amara* and *Lepraria incana Lichen* is belonging to the tolerant cosmopolitan type, because it can be found throughout the observation location types which are classified as the most resistant to the percentage of attendance, either clean air or polluted air. Type of *Pertusaria amara* has a very high level and a significant correlation compared with other types. The correlation of *Pb* accumulated in *Parmelia saxatilis talus* with correlation traffic density is very low and not significantly negative. It shows that the correlation in the opposite direction reversed or which means if the traffic volume is high, then the less extensive *Talus Lichens* and vice versa. Pb Lichens with traffic density has a very high degree of correlation with the very real significance. Pb accumulation of high power, there is the type of *Graphics scripta* (r = 0.999), which has a highest degree of correlation than other types.

Ecological characteristics (chemical physics environmental factors) of each study site when compared with conditions during the last five years (2009-2013) are relatively constant, while the ambient air with parameters (CO), (NO$_2$), (SO$_2$) and (CO$_2$) is still much to be below the threshold applicable air quality standards in Indonesia. Lichenes Correlation with Physical Properties of Growing Media is significant, except at the location of Jl. KL Yos Sudarso Medan, a very low level of correlation (negative) was not significant and proportional. While Correlation with *Ambient Air Lichenes* of all of locations is positive correlation significantly and proportionally. The higher colony *Lichenes* is the higher the ambient air.

Acknowledgments

Authors are grateful to Hamonangan Tambunan, Lecturer at the Department of Electrical Engineering Education, Faculty of Engineering, State University of Medan for helping the author.

References

Bargagli, R., D'Amato, & Iosco, F. P. (1987). Lichen Biomonitoring of Metals in the San Rossore Park: Contrast With Previous Pine Deedle Data. *J. Environmental Monitoring and Assessment, 9*(3), 285-294. http://dx.doi.org/10.1007/bf00419901

Conti, M. E., & Cecchetti, G. (2000). Biological monitoring: Lichens as bioindicators of air pollution assessment – a review. *Environmentall Pollution, 114*, 47-492. http://dx.doi.org/10.1016/s0269-7491(00)00224-4

Dahlan, E. N. (1989). *Studi Kemampuan Tanaman Dalam Menyerap Timbal Emisi Dari Kenderaan Bermotor.* Tesis Pascasarjana IPB. Bogor.

Dahlan, E. N. (1992). *Hutan Kota; Untuk pengelolaan dan Peningkatan Kualitas Lingkungan Hidup.* PT. Enka Parahayangan. Jakarta.

Deruccle. (1981). The reability of lichens as biomonitors of lead polution. *Ecologie (Brunoi), 27*(4), 285-290.

Fink, B. (1961). *The Lichen Flora of The United States.* Michigan The University of Michigan Press.

Galun, M., & Ronen, R. (2000). Interaction of Lichens and Pollutants. In Galun (Ed.), *Handbook of Lichenology* (Vol III). CRC Press, Florida.

Jamhari, M. (2009). *Lichen sebagai Bioindikator Pencemaran Udara di Malang dan Upaya Pengembangan Produk Pembelajarannya.* Tesis. Universitas Negeri Malang.

Kinaliglou, K., Ozbucak, T. B., Kutbay, H. G., Huseyinova, R., Bilgin, A., & Demirayak, A. (2010). Biomonitoring of Trace Elements with Lichens in Samsun, Turkey. *Ekoloji, 19*(75), 64-70. http://dx.doi.org/10.5053/ekoloji.2010.759

Kovacs, M. (1992). *Biological Indicators in Environmental Protection.* Ellis Horwood. New York.

McCune, B., Grenon, J., & Martin, E. (2006). *Lichens in Relation to Management Issues in the Sierra Nevada National Parks.* Department of Botany and Plant Pathology, Oregon State University.

Nursal, F., & Basori. (2005). Akumulasi Timbal (Pb) Pada Talus Lichens di Kota Pekanbaru. *Jurnal Biogenesis, 1*(2), 47-50.

O'Neill, P. (1993). *Environmental Chemistry* (2nd ed.). Chapman & Hall, London.

Ohmura, Y., Kawachi, M., Kasai, F., Sugiura, H., Ohtara, K., Kon, Y., & Hamada, N. (2009). Morphology and Chemistry of *Parmotrema tinctorum* (Parmeliaceae, Lichenized Ascomycota) Transplanted into sites with different Air Pollution Levels. *Buletin National Museum of Nature and Science, 35*(2), 91-98.

Panjaitan Desi Maria, Fitmawati dan Atria Martina. (2010). Keanekaragaman Lichen Sebagai Bioindikator Pencemaran Udara di Kota Pekanbaru Provinsi Riau.

Pratiwi, M. E. (2006). *Kajian Lichens Sebagai Bioindikator Kualitas Udara (Studi Kasus Kawasan Industri Pulo Gadung, Arboretum Cibubur dan Tegakan Mahoni Cikabayan).* Institut Pertanian Bogor, Bogor.

Scerbo, R., Ristori, R., Possenti, L., Lampugnani, L., Barale, L., & Barghigiani, C. (2002). Lichen (*Xanthoria parietina*) Biomonitoring of Trace Element Contamination and Air Quality Assessment in Livorno Province (Tuscany, Italy). *The Science of the Total Environment, 241*, 91-106. http://dx.doi.org/10.1016/s0048-9697(99)00333-2

Stamenkovic, S., Cvijan, M., & Arandjelovic, M. (2010). Lichens As Bioindicators of Air Quality in Dimitrovgrad (South-Eastern Serbia). *Arc. Biology Science Belgrade, 62*(3), 643-648. http://dx.doi.org/10.2298/abs1003643s

Taufikurahman, F. M., & Sari, R. M. (2010). Using Lichen as Bioindicator for Detecting Level of Environmental Pollution. *Proceedings of the Third International Conference on Mathematics and Natural Sciences.*

Walther, G. R., Post, E., Convey, P., Menzel, A., Parmesan, C., Beebee, T. J. C., Fromentin, J. M., Hoegh-Guldberg, O., & Bairlein, F. (2003). Ecological responses to recent climate change. *Nature, 416,* 389-395. http://dx.doi.org/10.1038/416389a

5

^{133}Cesium Uptake by 10 Ornamental Plant Species Cultivated under Hydroponic Conditions

Hiromi Ikeura[1], Nanako Narishima[2] & Masahiko Tamaki[2]

[1] Organization for the Strategic Coordination of Research and Intellectual Properties, Meiji University, Kanagawa, Japan

[2] School of Agriculture, Meiji University, Kanagawa, Japan

Correspondence: Masahiko Tamaki, School of Agriculture, Meiji University, Kanagawa, 214-8571, Japan. E-mail: mtamaki@meiji.ac.jp

Abstract

We focused on the Cs uptake capacities of ornamental flowers. Ornamentals have the advantage of beautifying contaminated environments, and this may have therapeutic effects for individuals, especially in disaster areas. Furthermore, the use of ornamental plants will reduce the risk of pollutants entering the food chain. We hypothesized a strong correlation between high aboveground biomass and high Cs uptake in plants. We assessed the potential of 10 ornamental plant species for remediation of ^{133}Cesium in hydroponic solutions. Sunflower, rapeseed, and cosmos took up larger amounts of ^{133}Cs and showed better growth rates than the other 7 species. When these 3 species were exposed to 3 different concentrations of ^{133}Cs (0.5, 2, and 5 mg/L CsCl), more than 48% of the ^{133}Cs was remediated after 7 days in each case. The highest remediation rate was 67%, by sunflowers grown in 5 mg/L CsCl. Among the 3 species, shoot and root dry weights were highest in sunflower and lowest in cosmos. The rate of ^{133}Cs uptake was strongly correlated with aboveground plant biomass. The ^{133}Cs concentration did not affect plant growth rates in any of the three species.

Keywords: ^{133}Cesium, remediation, sunflower, cosmos, rapeseed

1. Introduction

Due to the nuclear accident at Fukushima, Japan, in 2011, a radioisotope of cesium (^{137}Cs) was spread over an extensive area. ^{137}Cs is moved from soil and water to plants easily, arriving humans directly and indirectly through the food chain. Hence, in order to reduce the risk of radiation for humans it will be essential to remove the ^{137}Cs from contaminated soils and soil solutions (Singh et al., 2009). However, the removal of contaminated surface soils or the immobilization of radionuclides in the soil are physically difficult, costly, and impractical (Zhu & Shaw, 2000).

Phytoremediation which is low-cost and environmentally friendly removal process using plants and a promising technique for the cleaning up of radionuclides such as ^{137}Cs and heavy metal has noted considerably (Kelly & Pinder, 1996; Singh et al., 2009). This possibility has stimulated interest in the study of Cs uptake by plants, since this may be a low-cost alternative for remediating Cs-contaminated sites (Lasat et al., 1997). The land plants have been used to clear toxic ions such as Pb^{2+} from solutions and their many roots cultivated hydroponically (McCutcheon & Schnoor, 2003). Specific transporters for nonessential metal ions do not exist in plants, and the transport systems for essential ions mediated the transport of Cs across a membrane. Additionally, most of the chemical features of Cs are analogous to those of potassium (Pinder et al., 2006). Thus, Cs is absorbed easily by plants (Dabbagh et al., 2008). It is commonly assumed that Cs is taken up by plants via mechanisms participated in the uptake of K^+ and Ca^{2+}. Many transport proteins (low affinity inward-rectifying K^+ channels; nonspecific, voltage insensitive cation channels; high affinity K^+–H^+ symporters; voltage-dependent Ca^{2+} channels; and outward-rectifying cation channels) promote the permeation of Cs^+ ions across root cell membranes in plants (Bystrzejewska-Piotrowska & Bazala, 2008).

Effective remediation will depend on the ability of the phytoremediation crop to accumulate Cs in the aerial parts. There are many studies about efficacious methods for Cs removal from contaminated soils or solutions by many plant species (Broadley & Willey, 1997; Broadley et al., 1999; Tang & Willey, 2003; Singh et al., 2009;

Moogouei et al., 2011; Borghei et al., 2011). Soudek et al. (2004, 2006) showed that there is no significant difference the uptake of Cs between radioactive (^{137}Cs) and stable cesium (^{133}Cs) isotopes by sunflower (*Helianthus annuus* L). White & Broadley (2000) also found that plants did not differentiate between ^{133}Cs and ^{137}Cs isotopes, and they concluded that plants' responses to ^{133}Cs are exemplary of their responses to ^{137}Cs isotopes. Thus, the uptake patterns of ^{137}Cs and ^{133}Cs in plants are similar.

In this study we focused on the Cs uptake capacities of ornamental flowers. Ornamentals have the advantage of beautifying contaminated environments, and this may have therapeutic effects for individuals, especially in disaster areas. Furthermore, the use of ornamental plants will reduce the risk of pollutants entering the food chain. Sunflower (Soudek et al., 2004) and rapeseed (Chou et al., 2005) have high Cs uptake capacities and large aerial biomasses. Therefore, we postulated a strong correlation between many aerial biomass and high Cs uptake in plants. To address the urgent need for remediation of Cs-contaminated land in Fukushima, we examined the ability of remediation of Cs using various flowering species including sunflower and rapeseed. Cs uptake by plants is affected by soil texture, particularly the clay and humus contents, because these components affect the strength of Cs adhesion in the soil (Kang et al., 2012). Therefore, the present study was conducted under hydroponic conditions in order to clarify the potential abilities for Cs uptake by various plants.

2. Materials and Method

2.1 Experiment 1. Differences in ^{133}Cs Uptake and Growth Rates among 10 Species

The following ornamental species are popular and easy to purchase and cultivate in Japan: sensitive plant (*Mimoza pudica*), gazania (*Gazania splendens*), cosmos (*Cosmos bipinnatus*), zinnia (*Zinnia hybrida*), California poppy (*Eschescholzia californica*), saffron thistle (*Carthamus tinctorius*), basket flower (*Centaurea americana*), and cypress vine (*Quamoslit pennata*). Therefore we tested these species and compared their performances with sunflower (*Helianthus annuus*) and rapeseed (*Brassica rapa*), which were reported to have high Cs uptake capacities (Soudek et al., 2004; Chou et al., 2005). Healthy seeds of each species were germinated in cell trays (3.0 cm diameter × 4.5 cm depth) containing commercial horticultural soil, and grown for 14 days in a chamber (MLR-351, Sanyo Electonic Co. Ltd., Osaka, Japan) at 20 °C with relative humidity 80% and a day length of 12 h. After their roots were washed thoroughly in distilled water, the seedlings were moved to containers (310 mm × 235 mm × 960 mm) containing equal volumes of the following nutrient solutions: Farm Ace No.1 (N: 10%, P: 8%, K: 26%, Mg: 0.1%, B: 0.1%, Fe: 0.15%, Mn: 0.1%, Cu: 0.002%, Zn: 0.006 %) and No. 2 (N: 11%, Ca: 16.4%) (Kaneko Seed Co., Ltd., Gunma, Japan). Each container contained 6 plants. Plants were grown in a green house at Meiji University for 7 days. Then the plants were transferred to fresh nutrient solution (as described above) that also contained ^{133}CsCl (5 mg/L, with the concentration of the ^{133}Cs$^+$ ion at 4.47 mg/L). The plants were grown in the ^{133}CsCl solution for 7 days. All solutions continuously aerated with a pump. The experiment was performed in triplicate.

The shoot height, the longest root length, and the shoot and root dry weights for each plant were measured at the end of the cultivation period. Samples of the ^{133}CsCl solutions were analyzed for ^{133}Cs concentrations. In all experiments the ^{133}Cs concentrations were determined using atomic absorption spectrophotometry (AA-6200, Shimadzu Co., Kyoto, Japan). Each sample was analyzed in triplicate. The percentage of metal uptake was calculated using the equation:

$$\% \ uptake = (C0\text{-}C1/C0) \times 100 \tag{1}$$

where C0 and C1 are the initial and remaining concentrations of the metal, respectively, in the solutions (mg/L) (Moogonei et al., 2011).

2.2 Experiment 2. Differences in 133Cs Uptake and Growth among Sunflower, Cosmos, and Rapeseed in Solutions with 3 Different 133Cs Concentrations

Sunflower, cosmos, and rapeseed showed the highest 133Cs uptake levels and highest growth rates in experiment 1. Therefore, these 3 species were used in experiment 2. The cultivation method was as described for experiment 1. Based on the experiments of Borghei et al. (2011) and Moogonei et al. (2011) we used 3 CsCl solutions of 0.5, 2, and 5 mg/L, with Cs$^+$ ion concentrations of 0.47, 1.58 and 3.95 mg/L, respectively. The experiment was performed in triplicate.

2.3 Statistical Analysis

Statistical analyses were performed using Excel statistics software (Excel Statistics 2008 for Windows, Social Survey Research Information Co., Ltd., Tokyo, Japan). All data were subjected to analyses of variance to identify significant differences, and means comparisons were obtained using the Fisher LSD test ($P < 0.05$).

3. Results

3.1 Experiment 1. Differences in ^{133}Cs Uptake and Growth Rates among 10 Species

The percentages of ^{133}Cs taken up by the 10 species after 7 days are shown in Table 1. All plants remained healthy after being transferred to the ^{133}Cs solutions. Sunflower, rapeseed, and cosmos showed the highest rates of ^{133}Cs uptake, with percentages of 69.24, 63.22, and 57.58, respectively.

The mean shoot height, longest root length, and dry weights of shoots and roots are shown for each species in Table 2. Cypress vine (a climbing plant) had the greatest shoot height (37.7 cm) and it was followed by cosmos (23.23 cm) and sunflower (22.0 cm). Cosmos, sunflower and rapeseed had the longest root lengths (24.8, 23.05, and 20.2 cm, respectively). Sunflower yielded the highest dry weights in both shoots and roots, and it was followed by rapeseed with the second-highest weights, and then cosmos.

3.2 Experiment 2. Differences in ^{133}Cs Uptake and Growth among Sunflower, Cosmos, and Rapeseed in Solutions with 3 Different ^{133}Cs Concentrations

The percentages of ^{133}Cs taken up by rapeseed, cosmos, and sunflower over a period of 7 days are shown in Figure 1. In general, the 3 species were able to remove at least 50% of the Cs that was present in the growth solution at the beginning of the experiment, regardless of the starting concentration. The only exception was cosmos grown in 2 mg/ml CsCl, which removed approximately 48% of the ^{133}Cs from the solution. The highest rate of uptake (67%) was by sunflower grown in 5 mg/ml CsCl. Relatively high rates (62-63%) were also shown by sunflower grown in 2 mg/ml CsCl and rapeseed grown in 5 mg/ml CsCl. The rate of uptake by sunflower increased significantly with increasing starting concentrations of CsCl. However, no clear correlations were observed between rate of uptake and starting concentration for either rapeseed or cosmos. Among the three species there were no significant differences in the uptake rates for plants grown in 0.5 mg/L CsCl.

The mean shoot height, length of longest root, and dry weights of shoots and roots are shown for the 3 species in Table 3. None of these parameters were affected by the ^{133}Cs concentrations of the solutions.

Table 1. Uptake of Cs from a hydroponic solution by 10 plant species

Plants	Initial concentration (mg/L)	After 7 days (mg/L $\pm$ SD)		Uptake (%)	
Rapeseed		1.65	$\pm$ 0.13	63.22	a[z]
Sensitive Plant		3.67	$\pm$ 0.16	17.99	b
Gazania		3.88	$\pm$ 0.33	13.24	bc
Cosmos		1.90	$\pm$ 0.28	57.58	d
Zinnia	4.47	2.13	$\pm$ 0.21	52.44	cde
California poppy		3.76	$\pm$ 0.38	15.87	bf
Common sunflower		1.38	$\pm$ 0.14	69.24	a
Saffron Thistle		3.14	$\pm$ 0.31	29.78	g
Basket Flower		3.92	$\pm$ 0.28	12.37	bcfh
Cypress Vine		3.38	$\pm$ 0.31	24.53	bfgi

Values are mean $\pm$ standard deviation.

[z] Different letters indicate statistically significant differences ($P < 0.05$) according to a multiple range test.

Table 2. Growth parameters of 10 plant species cultivated in a Cs solution for 7 days

Plants	Shoot height (cm)	The maximum length of root (cm)	Shoot dry weight (g)	Root dry weight (g)
Rapeseed	14.65 ± 1.60 c	20.20 ± 7.69 a	0.424 ± 0.030 b	0.04 ± 0.005 b
Sensitive plant	3.15 ± 1.07 d	14.20 ± 4.51 b	0.037 ± 0.002 d	0.009 ± 0.001 cd
Gazania	4.80 ± 1.74 d	11.30 ± 3.64 bc	0.028 ± 0.002 d	0.003 ± 0.001 d
Cosmos	23.23 ± 3.02 b	24.80 ± 6.02 a	0.211 ± 0.022 c	0.037 ± 0.002 b
Zinnia	6.75 ± 1.44 d	19.00 ± 5.16 ab	0.08 ± 0.005 d	0.017 ± 0.001 c
California poppy	4.75 ± 0.51 d	7.70 ± 1.01 c	0.021 ± 0.002 d	0.003 ± 0.001 d
Common sunflower	22.00 ± 4.99 b	23.05 ± 2.44 a	0.759 ± 0.040 a	0.129 ± 0.007 a
Saffron thistle	13.44 ± 3.07 c	15.75 ± 3.31 b	0.202 ± 0.027 c	0.031 ± 0.004 b
Basket flower	4.50 ± 1.58 d	14.35 ± 1.30 b	0.059 ± 0.006 d	0.015 ± 0.002 c
Cypress vine	37.7 ± 4.58 a	16.6 ± 4.56 b	0.192 ± 0.018 c	0.033 ± 0.003 b

Values are mean ± standard deviation.

[z] Different letters indicate statistically significant differences (P < 0.05) according to a multiple range test.

Table 3. Growth parameters of rapeseed, cosmos and sunflower cultivated in 3 Cs solutions for 7 days

Cs concentration of solution (mg/L)	Shoot height (cm)	The maximum length of root (cm)	Shoot dry weight (g)	Root dry weight (g)
		Rapeseed		
0.5	17.16 ± 1.17 a	28.19 ± 2.48 a	0.457 ± 0.022 a	0.042 ± 0.005 a
2	14.41 ± 1.28 a	23.84 ± 0.93 b	0.443 ± 0.018 a	0.042 ± 0.003 a
5	16.45 ± 0.84 a	24.32 ± 1.84 ab	0.468 ± 0.015 a	0.046 ± 0.003 a
		Cosmos		
0.5	28.75 ± 2.71 a	19.27 ± 0.51 a	0.276 ± 0.011 a	0.034 ± 0.004 a
2	28.14 ± 3.19 a	18.59 ± 1.66 a	0.289 ± 0.015 a	0.034 ± 0.006 a
5	29.95 ± 2.86 a	21.66 ± 1.46 a	0.295 ± 0.014 a	0.037 ± 0.004 a
		Common sunflower		
0.5	30.15 ± 2.55 a	26.48 ± 1.70 a	0.756 ± 0.032 a	0.189 ± 0.013 a
2	28.29 ± 2.19 a	22.03 ± 0.88 b	0.698 ±0.030 a	0.176 ± 0.0012 a
5	30.00 ± 2.58 a	24.62 ± 0.44 ab	0.784 ± 0.036 a	0.166 ± 0.010 a

Values are mean ± standard deviation.

[z] Different letters indicate statistically significant differences (P < 0.05) according to a multiple range test.

4. Discussion

In this study we evaluated the potential of 10 ornamental species for the remediation of [133]Cs in hydroponic solutions. Sunflower, rapeseed, and cosmos showed the greatest capacities for [133]Cs uptake, and they showed higher growth rates than the other 7 species. These results suggest that the rate of [133]Cs uptake is correlated with plant biomass, since larger plants absorb greater amounts of water than smaller plants. Previous reports have indicated that sunflower (Soudek et al., 2004) and rapeseed (Chou et al., 2005) have high [133]Cs uptake capacities and our study supports those results.

Borghei et al. (2011) grew *Calendula alata* plants hydroponically in CsCl solutions with concentrations of 0.6, 2, and 5 mg/L, and found that 47%, 41%, and 52%, respectively, of the Cs was remediated after 15 days. Moogouei et al. (2011) reported that when *Chenopodium album* and *Amaranthus chlorostachys* were exposed to 0.5 mg/L of CsCl in hydroponic solutions for 15 days, 68% of the Cs was remediated by the C. album plants. Moreover, when *Chromolaena odorata* were exposed for 15 days to solutions with three different levels of ^{137}Cs (1×10^3, 5×10^3, and 10×10^3 kBq/L), remediation was observed at rates of 89%, 81%, and 51%, respectively (Singh et al. 2009). It is known that plants belonging to the family Asteraceae are efficient accumulators of ^{134}Cs (Tang & Willey, 2003; Singh et al., 2009). Singh et al. (2009) have found that plants reside in the families Amaranthaceae, Asteraceae, and Chenopodiaceae are efficient remediators of ^{137}Cs.

Singh et al. (2009) proved that the rate of remediation was greatest during the first seven days of exposure of *C. odorata* plants to ^{137}Cs. In this study, when sunflower was cultivated in a 5 mg/L solution of ^{133}CsCl for 7 days, nearly 70% of the ^{133}Cs was removed from the solution. Moreover, rapeseed, sunflower, and cosmos were all able to remediate at least 48% of the ^{133}Cs present in the hydroponic solution in 7 days, regardless of the initial ^{133}Cs concentration. Therefore, these species may be useful for the remediation of radioactive Cs isotopes from contaminated sites. In Japan, rapeseed flowers in spring, sunflower flowers in summer, and cosmos flowers in autumn. By combining these three species, the remediation will be also attained simultaneously, enjoying the flower from spring to autumn.

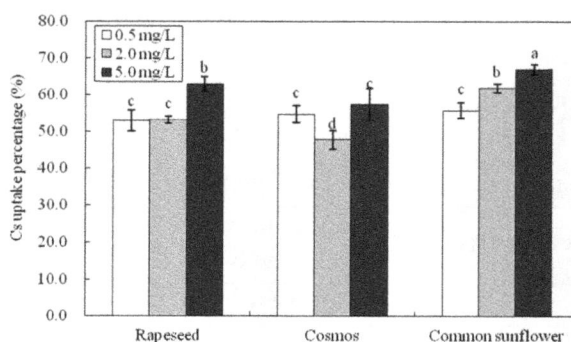

Figure 1. Percentages of ^{133}Cs taken up by rapeseed, cosmos, and sunflower cultivated in three ^{133}Cs solutions for 7 days. The starting CsCl concentrations were 0.5, 2.0, ans 5.0 mg/L. Values are mean ± standard deviation. Different letters indicate statistically significant differences (P < 0.05) according to a multiple range test.

References

Borghei, M., Arjmandi, R., & Moogouei, R. (2011). Potential of Calendula alata for phytoremediation of stable cesium and lead from solutions. *Environmental Monitoring and Assessment, 181*, 63-68. http://dx.doi.org/10.1007/s10661-010-1813-9

Broadley, M. R., & Willey, N. J. (1997). Differences in root uptake of radiocaesium by 30 plant taxa. *Environmental Pollution, 97*, 11-15. http://dx.doi.org/10.1016/S0269-7491(97)00090-0

Broadley, M. R., Willey, N .J., & Mead, A. (1999). A method to assess taxonomic variation in shoot caesium concentration among flowering plants. *Environmental Pollution, 106*, 341-349. http://dx.doi.org/10.1016/S0269-7491(99)00105-0

Bystrzejewska-Piotrowska, G., & Bazala, M. L. (2008). A study of mechanisms responsible for incorporation of cesium and radiocesium into fruitbodies of king oyster mushroom (*Pleurotus eryngii*). *Journal of Environmental Radioactivity, 99*, 1185-1191. http://dx.doi.org/10.1016/j.jenvrad.2008.01.016

Chou, F. I., Chung, H. P, Teng, S. P., & Sheu, S. (2005). Screening plant species native to Taiwan for remediation of ^{137}Cs-contaminated soil and the effects of K addition and soil amendment on the transfer of ^{137}Cs from soil to plants. *Journal of Environmental Radioactivity, 80*, 175-181. http://dx.doi.org/10.1016/j.jenvrad.2004.10.002

Dabbagh, R., Ebrahimi, M., Aflaki, F., Ghafourian, M., & Sahafipour, M. H. (2008). Biosorption of stable cesium by chemically modified biomass of *Sargassum glaucescens* and *Cystoseira indica* in a continuous flow system. *Journal of Hazardous Materials, 159*, 354-357. http://dx.doi.org/10.1016/j.jhazmat.2008.02.026

Kang, D. J., Seo, Y. J., Saito T., Suzuki H., & Ishii Y. (2012). Uptake and translocation of cesium-133 in napiergrass (*Pennisetum purpureum* Schum.) under hydroponic conditions. *Ecotoxicology and Environmental Safety, 82*, 122-126. http://dx.doi.org/10.1016/j.ecoenv.2012.05.017

Kelly, M. S., & Pinder, J. E. III. (1996). Foliar uptake of ^{137}Cs from the water column by aquatic macrophytes. *Journal of Environmental Radioactivity, 30*, 271-280. http://dx.doi.org/10.1016/0265-931X(95)00027-8

Lasat, M. M., Norvell, W. A., & Kochian, L. V. (1997). Potential for phytoextraction of ^{137}Cs from contaminated soil. *Plant and Soil, 195*, 99-106. http://dx.doi.org/10.1023/A:1004210110855

McCutcheon, S. C., & Schnoor, J. L. (2003). Overview of phytotransformation and control of wastes. In S. C. McCutcheon & J. L. Schnoor (Eds.), *Phytoremediation: Transformation and control of contaminants* (pp. 27-58). New Jersey: Wiley. http://dx.doi.org/10.1002/047127304X.ch1

Moogouei, R., Borghei, M., & Arjmandi, R. (2011). Phytoremediation of stable Cs from solutions by *Calendula alata, Amaranthus chlorostachys* and *Chenopodium album. Ecotoxicology and Environmental Safety, 74*, 2036-2039. http://dx.doi.org/10.1016/j.ecoenv.2011.07.019

Pinder, J. E. III, Hinton, T. G., & Whicker, F. W. (2006). Foliar uptake of cesium from the water column by aquatic macrophytes. *Journal of Environmental Radioactivity, 85*, 23-47. http://dx.doi.org/10.1016/j.jenvrad.2005.05.005

Soudek, P., Tykva, R., & Vanek, T. (2004). Laboratory analyses of ^{137}Cs uptake by sunflower, reed and poplar. *Chemosphere, 55*, 1081-1087. http://dx.doi.org/10.1016/j.chemosphere.2003.12.011

Soudek, P., Valenova, S., Vavrıkova, Z., & Vanek, T. (2006). [137]Cs and ^{90}Sr uptake by sunflower cultivated under hydroponic conditions. *Journal of Environmental Radioactivity, 88*, 236-250. http://dx.doi.org/10.1016/j.jenvrad.2006.02.005

Singh, S., Thorat, V., Kaushik, C. P., Raj, K., Eapen, S., & D'Souza, S. F. (2009). Potential of Chromolaena odorata for phytoremediation of ^{137}Cs from solution and low level nuclearwaste. *Journal of Hazardous Materials, 162*, 743-745. http://dx.doi.org/10.1016/j.jhazmat.2008.05.097

Tang, S., & Willey, N. J. (2003). Uptake of [134]Cs by four species from Asteraceae and two variants from chenopodiaceae grown in two types of Chinese soil. *Plant and Soil, 250*, 75–81. http://dx.doi.org/10.1023/A:1022873930771

White, P. J., & Broadley, M. R. (2000). Mechanisms of caesium uptake by plants. Tinsley review no. 113. *New Phytologist, 147*, 241-256. http://dx.doi.org/10.1046/j.1469-8137.2000.00704.x

Zhu, Y. G., & Shaw, G. (2000). Soil contamination with radionuclides and potential remediation. *Chemosphere, 41*, 121-128. http://dx.doi.org/10.1016/S0045-6535(99)00398-7

6

Environmental 'Low Dose' Mesotheliomas and Their Relationship to Domestic Exposures-Preliminary Report

Edward B Ilgren[1], Drew Van Orden[2], Richard Lee[2], Yumi Kamiya[3] & John A Hoskins[4]

[1] Formerly, University of Oxford, Faculty of Biological & Agricultural Sciences, Department of Pathology, Oxford, UK

[2] RJ Lee Group, Monroeville, PA, USA

[3] Formely, Bryn Mawr College for Women, PA, USA

[4] Formerly, Medical Research Council, Leicester, UK

Correspondence: Edward B Ilgren, MA (Oxford), MD, DPhil (Oxford), FRCPath (London), Suite No 808, 840 Montgomery Avenue, Bryn Mawr, PA, USA. Tel: 1-610-624-1510. E-mail: dredilgren@aol.com

Abstract

Crocidolite is a fibrous mineral with unusual crystallography in that the fiber width varies according to geographical origin. The fibers from the mining areas of Western Australia and the Cape Province, in South Africa are 'thin' compared to those from Bolivian and some other mines. Regardless, the mineral is a well known causative agent for mesothelioma. The literature contains many reports of the disease occurring as a consequence of so called environmental exposures due to what some believe to be 'low' dose, ambient exposure. However, closer examination of these exposure conditions generally reveals an unrecognized (or simply ignored) strong domestic (para-occupational) exposure component. Oftentimes, the importance of such domestic exposures is not considered since they too are thought to be too low to contribute to risk. We have, for the first time, applied state of the art measurement methods to evaluate domestic (shake out) conditions created by workers in a historical unregulated crocidolite-cement based operating plant in Bolivia. Our results show that exposures can reach levels more usually associated with historic occupational exposures. The reason for studying the exposures in such a plant is that epidemiological studies have shown that 'thin' crocidolite fibers are associated with a high mesothelioma risk while exposure to the 'thick' Bolivian fibers are associated with a much lessened mesothelioma risk.

Keywords: crocidolite, mesothelioma, Bolivia, domestic exposure, environmental exposure

1. Introduction

Crocidolite is a fibrous mineral with unusual crystallography in that the fiber width varies according to geographical origin and it may be a unique mineral in this respect. This property was first studied by Shedd (1985) from the US Bureau of Mines who showed that the fiber dimensions of crocidolites from crocidolite mining regions in Western Australia and the Cape Province, in South Africa are 'thin' compared to those from Bolivian and some other mines.

Crocidolite is a major cause of the tumor mesothelioma and the relevance of fiber width to its pathogenicity has been reviewed by the authors (Ilgren et al., 2012a).

Environmental mesotheliomas are said to arise under 'low dose' conditions without occupational or para-occupational (domestic) exposure. In truth, environmental mesotheliomas thus labeled are actually domestic mesotheliomas which arise after high dose exposures (Browne & Wagner, 2001). Whilst there is strong evidence to suggest domestic exposures associated with asbestos disease are consistent with high occupational exposure levels (Sawyer, 1979; Browne, 1983; Gardner & Saracci, 1989; Gibbs, Jones, Pooley, Griffiths, & Wagner, 1989; Castleman, 1988 (Note 1); Huncharek, Capotorto & Muscat, 1989; Gaensler, McCloud, & Carrington, 1985; Newhouse, 1991 personal commun), data from direct measurements of historical operating facilities are lacking (Health Effects Institute, 1991). It is commonly presumed levels of exposure in households are lower than those in the workplace, though measurements are not available documenting such levels. We thus document, for the first time, using modern measurement methods in a fully operational historic unregulated manufacturing plant, that the exposures produced during domestic ('shake out') activities, are indeed very high. Taken together, these data provide further evidence that environmental mesotheliomas as currently defined do not exist (Browne & Wagner, 2001).

2. Materials and Methods

2.1 Study Location

Bolivia has several small crocidolite mines spread along both flanks of the headwaters of the Minasmayu river in the Alto Chapare District in the Cochabamba Department, Bolivia (Mindat, 2013).

2.2 Simulation Study and Measurement Details

A shake out simulation study was done in the setting of a historical crocidolite manufacturing plant. The operating conditions, lay out and other details of the plant as well as the methods used for fiber collection and measurement have already been described in detail (Ilgren, Van Orden, Lee, Kamiya, & Hoskins, 2012; Van Orden, Lee, Zock, Sanchez, & Kamiya, & Ilgren, E., 2012; Van Orden, Lee, Zock, & Sanchez, 2012). A worker sieving crocidolite in the fiberizing unit of the plant six days a week, 8 hours a day, was asked to put on new work clothes in the morning before coming to work. He wore a personal sampler throughout the day. The company changing room was cleaned at the start of the day. It was closed until all work at the plant was done 8 hours later. The worker from the sieving department then changed his clothes in the changing room at 5:00 pm and shook them out continuously for two minutes. A sampler was placed less than one foot from his breathing zone and another was placed three meters away at the same height on the other side of the changing room.

3. Findings

The concentrations recorded from the personal sampler and those found during shake out are shown below (Table 1). All of the readings are well into the occupational range. The fiber size dimensional profiles from each sample are also shown (Table 2) are not statistically different. Background concentration data were not available so the lowest reading in the plant has been used in its place. This was from a sample taken upwind of a worker drilling a crocidolite panel (Table 1).

Table 1. Observed PCM and TEM fiber concentrations (f/ml) for samples collected following the shake out of work clothes. (July 2011) (also see [11])

Operation	Distance, (m)	Crocidolite fibers only[3]			
		PCM mean conc. (Conc Range)	All ≥ 5 μm conc. (Conc Range)	PCME conc (Conc Range)	Stanton fibers conc. (Conc Range)
Sieving (SIE 1 and 2)[1]	0	188 (88-278)	514 (148-1391)	303 (73-826)	146 (25-434)
Shake out	0	101.9	116.7	87.9	40.8
Shake out	3	35.6	33.5	21.9	6.1
Background (Bol 042)[2]	2 (upwind of drill panel	0.009	0.028	0.018	0.009

[1] Personal samplers. Workers coded 1 and 2.

[2] No background readings were taken when the plant was not operating. This area sample was the lowest reading taken during operation [10].

[3] All concentrations are f/ml.

Table 2. Observed PCM and TEM fiber sizes for samples collected following the shake out of work clothes [July 2011] (also see [12])

Operation	Distance (m)	Crocidolite fibers only[2]		
		All ≥ 5 μm GM L/W GMAR	PCME GM L/W GM AR	Stanton GM L/W GM AR
Sieving (SIE 1 and 2)[1]	0	9.5 / 0.28 33.7	10.1 / 0.44 23.2	12.1 / 0.17 72.8
Shake out	0	13.4 / 0.42 32.0	14.5 / 0.56 26.0	13.2 / 0.20 66.2
Shake out	3	12.8 / 0.39 32.9	15.6 / 0.62 25.3	12.7 / 0.17 76.4
Background	2 (upwind)	10.6/0.21	6.1/0.26	18.4

[1] Personal samplers. Workers coded 1 and 2.

[2] Fiber parameters expressed as geometric means of length to width ratio and aspect ratio.

4. Discussion

4.1 Shake out Exposures

The domestic 'shake out' exposures described in this report done under simulated conditions produced exposures in the occupational range. It should be noted that the worker was exposed to airborne fiber levels in excess of 1000 f/ml of pure crocidolite and actively handled the crocidolite throughout the day. There is no reason to believe the conditions of fiber generation (that is, clothes shaking) produced in the plant would differ materially from that found in the home. The background fiber levels in the plant are assumed to be representative of those found in the cleaned changing room. As they were very low, these would not have contributed significantly to what was found after shaking out the workers' clothes.

4.2 Domestic Exposures

Significant domestic exposure can also arise in ways aside from the shaking out of clothes. This is well illustrated in New Caledonian homes covered with 'Po', an asbestos containing whitewash (Figures 1 & 2) (Luce et al., 1994) (Figure 3), in which sweeping up can produce extremely high fiber levels (e.g. 78 to 200 f/ml - Luce et al., 1995; Pelletier, 2010, personal commun.) capable of producing mesotheliomas.

Figure 1. A village home in the New Caledonian highlands covered by a tremolite rich Po whitewash

Figure 2. A photomicrograph of the whitewash demonstrating the vast number of long thin tremolite fibers

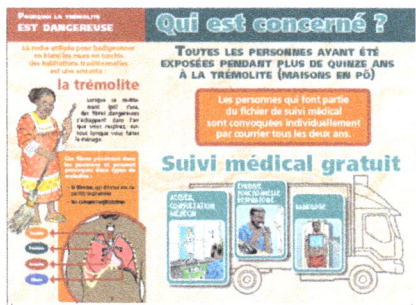

Figure 3. The warning issued to the New Caledonian villagers of the hazards of sweeping in their homes from the Po

4.3 Airborne Fibers

Fibers released after shake out and sweeping may remain airborne in the home for long periods of time (Sawyer, 1979; Newhouse, 1991, personal commun.). This is particularly so for the thinnest fibers (0.06 μm diameter, Note 2) (Dixon et al., 1985) that can also become readily re-entrained (Sawyer, 1977). Some also claim fibers released after shake out may be thinner than those from the source materials (Browne, 1991, personal commun.) but our preliminary data do not appear to support this proposal.

4.4 Environmental Exposure Measurements

4.4.1 Wittenoom: Environmental Mesotheliomas From Domestic Exposure

Roggli (2007) claims the 'environmental mesotheliomas' at Wittenoom were produced by exposures as low as 0.015 f-yrs. However, the risk estimates for these cancers (Hansen, de Klerk, Musk, Eccles & Hobbs, 1993; Hansen, 1995; Hansen, de Klerk, Musk, Hobbs, 1997; Hansen, de Klerk, Musk & Hobbs, 1998; Reid et al., 2007; Reid, Heyworth, & de Klerk, 2008) have never incorporated domestic exposures into their risk models. Indeed, at Wittenoom, virtually all of these so called 'environmental' cases lived with an asbestos worker (Hansen, 1995). Similarly, virtually none were found in residents who only lived there after the mining and milling operations ceased in 1966 (Hansen et al., 1997; de Klerk, 1993). In fact, given the extraordinarily high exposure conditions that prevailed throughout the entire 30 year (1937-1966) operating period at Wittenoom, vast numbers of 'environmental mesotheliomas' would have been noted if Roggli's fiber year estimates were correct.

4.4.2 Wittenoom: Environmental Exposure

The studies of environmental mesothelioma at Wittenoom and in other parts of the world ignored various sources of gross domestic exposure. This confounds their claims that so called 'environmental mesotheliomas' arise from low dose exposure. One can readily see why this is so. At Wittenoom everything was 'blue' (Wagner, 1996 personal commun.; McNulty, 1997, personal commun.). From 1937 to 1950, the residential area (the 'Settlement') was next to the mine. It was constantly shrouded in blue dust from the mill. The dust from the mill was so bad aviators used the blue plume as a navigational guide. When the residents shifted to the town of Wittenoom 10 miles away from the Settlement in 1950, crocidolite-containing ore and tailings were used to build homes, schools, the hospital, airport runway, the golf course, and the race track amongst other things. One of us (EI) had the opportunity to spend several days at Wittenoom in 1997 to interview various long term residents. According to their accounts and those of others (McNulty & Wagner, personal commun., 1996, 1997), clouds of blue dust leapt from the runway as planes landed or as horses ran around the race track. Children's sandboxes were topped up with blue tailings (Doust, personal commun., 1997). Their nappies (diapers) were of consequence frequently 'blue' (Doust, personal commun., 1997). Crocidolite was also used as a garden mulch to 'keep the tenacious red (iron ore) dust down' (Doust, personal commun., 1997). The contamination was so bad that 12 years after the mine had closed, levels as high as 10 f/ml of crocidolite could be measured as a car drove through the town (Cumpston, 1978).

4.4.3 Wittenoom and South African Mines Compared

The Wittenoom story is fully supported by the 'environmental' situation in South Africa notable for the discovery of the association between asbestos and mesothelioma (Wagner, Sleggs, & Marchand, 1960). Indeed, so called environmental mesotheliomas comprised many of Wagner et al.'s original series. This is because the conditions of gross exposure were nearly identical in both places (Wagner, personal commun., 1996). This follows from the fact that the geological conditions under which South African and the Western Australian crocidolite arise are basically the same. As Hodgson (1986) has stated "there may well be some correlation between the amphibole asbestos deposits of South Africa, India and Western Australia in the Archaen Gondwana platform" as "much evidence exists in the form of Paleozoic stratigraphical correlations and in the trends of basement rock structures to indicate the unity of Gondwanaland". The aridity and attendant dusty conditions were instrumental in the production of the asbestos related mesotheliomas initially recognized by Wagner et al. (1960). One of the first cases reported by Wagner et al. (1960) was a woman (Figure 4) cobbing crocidolite sitting in an area literally strewn with blue asbestos. This would not have been her only exposure. The husbands of such women, who worked in the mines, also brought home very dusty clothes that seriously contaminated their living quarters further exposing not only themselves but their children from birth onwards on a continual basis.

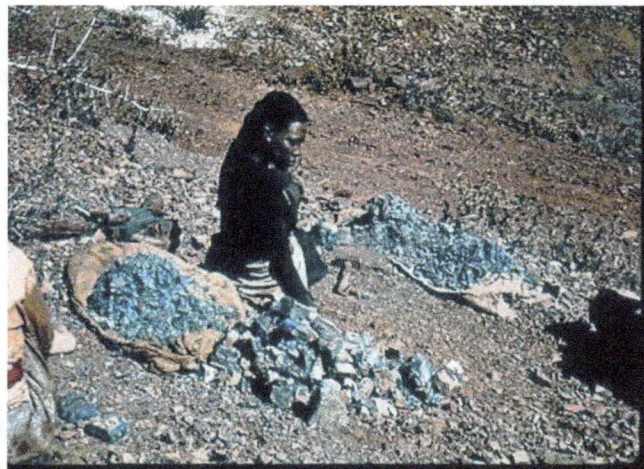

Figure 4. A photo of a lady who was one of the first mesothelioma cases reported by Wagner et al. [27] cobbing crocidolite whilst sitting in a field strewn with blue asbestos. Given to EI by Dr. Wagner in 1996

5. Conclusions

5.1 Rationale of the Study

This is not an epidemiological study. It is a first attempt to study the asbestos fiber concentrations generated in an historical asbestos facility using the most modern measurement methods. The concentrations measured simulate those generated when a heavily exposed worker takes off and shakes out his clothes at the end of a work day. Measurements showed that his occupational exposure in a historic manufacturing plant was to crocidolite at fiber concentrations exceeding 1000 f/ml. Such exposure appears to be towards the assumed (Note 3) historical limit for sieving and shaking pure blue asbestos to place it into different grades. Such exposures have never been directly measured either before or after shake out. Time/cost constraints limited the number of measurements that could be done. Nevertheless, they provide a unique record of conditions in a real rather than simulated environment.

5.2 Exposure Perspectives

Although the study is limited it is a contribution to our understanding of the causation of mesothelioma following shakeout. A mesothelioma risk cannot be due solely to environmental exposures absent significant domestic exposure. The vast majority of environmental mesotheliomas are really 'forme frustes' of domestic mesotheliomas (as discussed by Browne and Wagner, 2001). This is because fibers must reach and remain in the breathing zone for significant periods of time to produce the lung fiber burdens necessary to produce disease. Such exposures are not possible with environmental exposure alone. Therefore, a significant domestic component is required to produce so called environmental mesotheliomas. We hope to be able to further confirm and extend this work in the future in this and other settings in the area including workers' homes.

References

Browne, K. (1983). Asbestos related mesothelioma: epidemiological evidence for asbestos as a promoter. *Arch. Environ. Health, 38*, 261-268. http://dx.doi.org/10.1080/00039896.1983.10544004

Browne, K., & Wagner, J. (2001). Environmental exposure to amphibole asbestos and mesothelioma. The health effects of chrysotile asbestos. *Can. Min. Spec. Pub., 5*, 21-28.

Castleman, B. (1988). *Asbestos: Medical and Legal aspects*. Harcourt, Wash. DC.

Cumpston, A. G. (1978). *The health hazard at Wittenoom*. Perth: Public Hlth. Dept. of W. Australia.

De Klerk, N. (1993). *Testimony to the Select Committee on Wittenoom* (22.10.93).

Dixon, G., Dorla, J., Freed, J., Wood, P., May, I., Chambers, T., & Gesal, P. (1985). *Exposure assessment for asbestos contaminated vermiculite*. EPA OTS Contract No 68-01-6221. EPA 560/5-85-013. Feb, 1985.

Gaensler, E., Mc Cloud, T., & Carrington, C. (1985). Thoracic surgical problems in asbestos related disease. *Annals Thoracic Surgery, 40*, 82-92. http://dx.doi.org/10.1016/S0003-4975(10)61179-4

Gardner, M., & Saracci, R. (1989). Effects on health of non-occupational exposure to airborne mineral fibres. *IACR Sci, 90*, 375-383.

Gibbs, A., Jones, J., Pooley, F., Griffiths, D., & Wagner, J. (1989). Non-occupational malignant mesotheliomas. *IARC Sci pub, 90*, 219-233.

Goldberg, P., Luce, D., Billon-Galland, M. A., Que'nel, P., Salomon-Nekiriai, C., Nicolau, J., Brochard, P., & Goldberg, M. (1995). Potential role of environmental and domestic exposure to tremolite in pleural cancer in New Caledonia. *Rev Epidemiol Santé Publique, 43*, 444-50.

Hansen, J. (1995). *Mortality and morbidity after environmental exposure to crocidolite at Wittenoom, WA*. MPH thesis. Table 5.11.

Hansen, J., de Klerk, N., Musk, A., Eccles, J., & Hobbs, M. (1993). Malignant mesothelioma after environmental exposure to blue asbestos. *Int. J. Cancer., 54*, 578-581. http://dx.doi.org/10.1002/ijc.2910540410

Hansen, J., de Klerk, N., Musk, A., Eccles, J., & Hobbs, M. (1997). Mesothelioma after environmental crocidolite exposure. *Ann. Occup. Hyg., 41*(Inhaled Particles VIII), 189-193.

Hansen, J., de Klerk, N., Musk, A., & Hobbs, M. (1997). Individual exposure levels in people environmentally exposed to crocidolite. *Appl. Occup. Environ., 12*, 485-490. http://dx.doi.org/10.1080/1047322X.1997.10390032

Hansen, J., de Klerk, N., Musk, A., & Hobbs, M. (1998). Environmental exposure to crocidolite and mesothelioma. Exposure response relationships. *Amer. J. Resp. Crit. Care. Med., 157*, 69-75. http://dx.doi.org/10.1164/ajrccm.157.1.96-11086

HEI. (1991). *Health Effects Institute - Asbestos Research - Asbestos in Public and Commercial Buildings: A Literature Review and Synthesis of Current Knowledge*. Cambridge, Mass.

Hodgson, A. (1986). *Scientific Advances in Asbestos: 1967 to 1985*. Reading, Anjalena press, UK.

Huncharek, M., Capotorto, J., & Muscat, J. (1989). Domestic asbestos exposure, lung fibre burden and pleural mesothelioma in a housewife. *Brit. J. Indus. Med., 46*, 354-355.

Ilgren, E., Ramirez, R., Claros, E., Fernandez, P., Guardia, R., Dalenz, J., ... Hoskins, J. (2012a). *Fiber Width as a Determinant of Mesothelioma Induction and Threshold - Bolivian Crocidolite: Epidemiological Evidence from Bolivia - Mesothelioma Demography and Exposure Pathways*. Ann Resp Med. (in press).

Ilgren, E., Van Orden, D., Lee, R., Kamiya, Y., & Hoskins, J. (2012). *Further Studies of Bolivian Crocidolite – Part IV: Fibre Width, Fibre Drift and their relation to Mesothelioma Induction.* Ann Resp. Med (in press).

Luce, D., Brochard, P., Que'nel, P., Salomon-Nekiriai, C., Goldberg, P., Billon-Galland, M. A., & Goldberg, M. (1994). *Malignant pleural mesothelioma associated with exposure to tremolite.* (Letter). Lancet 344:1777. http://dx.doi.org/10.1016/S0140-6736(94)92919-X

Mindat. (2013). Retrieved from http://www.mindat.org/loc-215988.html

Reid, A., Berry, G., de Klerk, N., Hansen, J., Heyworth, J., Ambrosini, G., Fritschi, L., ... Musk, W. (2007). Age and sex differences in malignant mesothelioma after residential exposure to blue asbestos. *Chest 131*, 376-382. http://dx.doi.org/10.1378/chest.06-1690

Reid, A., Heyworth, J., & de Klerk, N. (2008). The mortality of women exposed to environmentally and domestically to blue asbestos at Wittenoom, Western Australia. *Occup Environ Med, 65*, 743-749. http://dx.doi.org/10.1136/oem.2007.035782

Roggli, V. (2007). Environmental asbestos contamination. *What are the risks? Chest, 131*, 337-338. http://dx.doi.org/10.1378/chest.06-2649

Sawyer, R. (1977). Asbestos exposure in a Yale building. *Environ Res, 13*, 146-169. http://dx.doi.org/10.1016/0013-9351(77)90013-5

Sawyer, R. (1979). Indoor asbestos pollution: application of hazard criteria. *Ann NY Acad Sci, 330*, 579-586. http://dx.doi.org/10.1111/j.1749-6632.1979.tb18762.x

Shedd, K. B. (1985). US Department of the Interior: fiber dimensions of crocidolite from Western Australia, Bolivia, and the Cape and Transvaal Provinces of South Africa. US Bureau of Mines Report of Investigations *8998. Washington, DC: United States Department of the Interior. Bureau of Mines; 1985.

Van Orden, D., Lee, R., Zock, A., & Sanchez, M. (2012). Evaluation of Airborne Crocidolite Fibers at an Asbestos-Cement Plant: Part 2 - The Size Distribution of Airborne Bolivian Crocidolite Fibers. Ann Resp Med, Case Study, July 26, 2012.

Van Orden, D., Lee, R., Zock, A., Sanchez, M., Kamiya, Y., & Ilgren, E. (2012). *Evaluation of Airborne Crocidolite Fibers at an Asbestos-Cement Plant.* Ann Resp Med, Case Report, July 26, 2012.

Wagner, J., Sleggs, C., & Marchand, P. (1960). Diffuse pleural mesothelioma and asbestos exposure in the North Western Cape Province. *Br. J. Indust. Med, 17*, 260-271.

Notes

Note 1. Citing 1976 OSHA Testimony of Dr. Paul Kotin.

Note 2. "Fibers 1.6 um in diameter would theoretically fall three meters in about one hour while single fibrils would require over 15 days". (Dixon et al., 1985) see Figure 2, p. 22.

Note 3. Early measurements were in millions of particles per cubic foot which cannot be accurately converted to the modern fibers/ml.

7

Mining Activities and Associated Environmental Impacts in Arid Climates: A Literature Review

Douglas B. Sims[1], Peter S. Hooda[2] & Gavin K. Gillmore[2]

[1] College of Southern Nevada, Department of Physical Sciences, North Las Vegas, Nevada, USA

[2] Centre for Earth and Environmental Sciences Research, Kingston University London, Kingston upon Thames, UK

Correspondence: Douglas B. Sims, Department of Physical Sciences, College of Southern Nevada, North Las Vegas, NV 89030, USA. E-mail: douglas.sims@csn.edu

Abstract

Mining operations have released measurable levels of geogenic trace metals (e.g. Cd, Cr, Pb), metalloids (e.g. As, Se), and anthropogenic chemicals (e.g. CN^-, Hg) into surrounding sediments. Abandoned mining sites in hyperarid climates has not been the focus of much research compared to wet and temperate areas. Research has focused on historical mining sites in semiarid and wetter regions in the United States, south pacific and Europe. Those areas have obvious risks associated with them including aqueous phase mobilization as a result of abundant precipitation. However, many mining areas in the American Southwest and aboard are located in hyperarid regions and viewed as not having a potential for mobilization of contaminants. Seasonal storm events can mobilize sediments containing contaminates beyond a small, localized area and into the wider environment. Literature indicates that arid and hyperarid mining regions have not been studied as extensively as those in wetter climates.

Keywords: mining, trace metals, metalloids, cyanide, migration, arid climates, sediments

1. Introduction

Storage and mobilization of contaminants at abandoned mining sites in arid to hyperarid climates are key areas of inquiry to furthering our understanding of the potential pollution, ecological and human impacts caused by such sites. Mining tends to release geogenic metals (e.g. Ag, Cd, Cr, Pb, Hg), metalloids (e.g. As, Se), and anthropogenic chemicals (e.g. CN^-, Hg) into sediments that can be mobilized beyond a small, localized area, and into the surrounding environments (Purves, 1985; Sims, 2010, 2011; Miranda-Aviles et al., 2012; Sims et al., 2013). These contaminates have been the focus of interest due to their long residence time in the environment and can be toxic to biota (Khalid et al., 1981; Frignana & Bellucci, 2004; Hutchinson & Ellison, 1992).

During the past 20 years mine research has focused on historical mining sites in semiarid and wetter regions in the United States (i.e. California, Colorado, Idaho, Utah), south pacific (i.e. Australia) and Europe. These areas have obvious risks associated with them including more abundant precipitation. However, in arid and hyperarid regions, environmental issues and contamination related to historical mines have not been studied as extensively as those in wetter climates. Mine sites in arid regions are not seen as an immediate or potential risk to the environment due to their location (Ross, 2008).

Anthropogenic and geogenic inputs of trace elements to surficial sediments can impact a greater area than previously considered (Lottermoser & Ashley, 2005; Lottermoser et al., 1999; Coulthard & Macklin, 2003; Mackenzie & Pulford, 2002). Abandoned mines in temperate climates have been a focus for the United States Environmental Protection Agency since the mid 1980s (USEPA, 1995). There has been a great interest in the impacts of abandoned mining operation in semi-temperate climates in areas of Northern California and Nevada, Colorado, Montana, and Idaho (e.g. Engle et al., 2001; Horowitz et al., 1993).

Mining and milling activities are known to concentrate trace elements of geogenic origin into waste materials (Besser & Leib, 2000; Berger et al., 2000; Blowes et al., 1994). Historical mining operations used CN^- to remove precious metals from floated slurries of water and crushed ore, and Hg to form an amalgam with other precious metals so that they could be removed (Bonzongo et al., 1996; Churchill, 1999). Milling processes tend to concentrate naturally occurring elements (e.g. As, Pb, Cr, Cd) during the process by separating the precious metals from the ore and concentrating residual trace elements in the tailings (Prusty et al., 1994). Subsequent disposal of mine waste containing the concentrated geogenic elements (e.g. Ag, As, Cd, Cr, Pb, Se) can greatly add to the anthropogenic contaminants burden to the surrounding area (Nicholson et. al., 2003; Miranda-Aviles et al., 2012).

It has been suggested that mines in hyperarid climates are of little threat to the wider environment because of a lack of precipitation (Ross, 2008). In hyperarid environments however, sediment redistribution from storm events can be the dominant form of transport for mining wastes (Reheis, 2006). Studies have suggested, for example, that contamination from mining activities related to dust (i.e. wind blown and fly ash) has the potential to impact the wider environment many kilometers from the source (Davis & White, 1981; Camm et al., 2003; 2004; Petaloti et al., 2006).

Although it is well documented that mining contamination in wetter climates has the possibility of migrating long distances by water and other transport mechanisms, there has been significantly less research in arid and hyperarid regions. Investigations of mine waste in wet temperate to wet climates are important however, it is clear that similar methods in hyperarid climates would not provided similar data. In mining regions with minute precipitation, contaminants migration occurs with sediment transport during storm events, not overland flow or groundwater movement. Thus it is imperative that research be conducted in these arid and hyperarid regions where significant amounts of mining and milling have occurred to evaluate the environmental distribution of contaminants to better understand the processes.

This paper examines literature concerning transport and distribution of trace elements in arid and hyperarid environments to assess the impact to the wider environment as compared to sites located in wetter regions. In order to fully evaluate relevant processes this review examines trace metals, metalloids, and anthropogenic inputs with wind transport of contaminates; transport of contaminates in wet climates; speciation and chemical behavior of trace metals and contaminates; siltation transport; potential for flora, fauna and human health affects; uptake and effects on flora; uptake and effects on fauna and on human health.

2. Trace Metals, Metalloids, and Anthropogenic Inputs

Historic mining operations in the American West date back to the middle 1800s with some sites having continuous operations into the mid-1950s (Greene, 1975). Many of these mines are small excavations that transported their ore to milling facilities in the area for processing. Some mining operations had their own milling systems adjacent to the mine, many within large ephemeral wash or stream systems (Greene, 1975). Historic milling operations employed the use of CN⁻, Hg, and other chemicals to extract metals from ore however, this process extracted both previous and geogenic metals by concentration techniques. Once tailings were processed, wastes containing high levels of concentrated geogenic trace elements (e.g. As, Ag, Cr, Cd, Hg, Pb, Se) and anthropogenic contaminants (e.g. CN⁻ and Hg) were released into nearby areas for disposal.

Geogenic metals (e.g., Cr, Pb) and metalloids (e.g., As, Se) are those contained within rock, ore and sediments and include major crustal metals such as Al, Fe, and Mn. It is known that mining and milling activities concentrate these metals into waste materials and potentially causing environmental impacts at a later time (Besser & Leib, 2000; Berger et al., 2000; Blowes et al., 1994). On the other hand historical mining operations used CN⁻ to remove precious metals from floated slurries of water and crushed ore, and Hg to form an amalgam with other precious metals so that they could be removed (Bermejo et al., 2003; Bonzongo et al., 1996; Churchill, 1999). Disposal of mine waste containing these geogenic elements (e.g. Ag, As, Cd, Cr, Pb, Se) can greatly add to the anthropogenic inputs (CN⁻, Hg) to the surrounding area.

Historical mining activities are known for their adverse impacts on the local environments, from esthetic to environmental. Typically, acid mine drainage (AMD) is a principal source of environmental impact; however, in an arid climate, a major potential source of contaminants is from mine tailings and waste rock where geogenic and anthropogenic sources are enriched in mine wastes and distributed across the landscape by the operators (Earman, 1996; Lu et al., 2012; Miranda-Aviles et al., 2012). In hyperarid regions wastes are generally left to the elements as it has been suggested that there is little risk due to a lack of precipitation (Ross, 2008).

Tailings of small particle sizes containing contaminants from milling processes are readily transported by wind and in over-land flow during storm events. Crushed ores are processed to particle sizes < 1 mm prior to flotation and extraction. After flotation, spent tailings waste was usually pumped to areas near a mill site such as ephemeral washes and basins. The transport of this material is limited to aeolian transport, aqueous, chemical behavior, flora and fauna and siltation/sediment transport during storm events. In arid regions, transport is limited to aeolian and sediment transport during storm events.

Cyanide and Hg are anthropogenic contaminants in mining areas that have been enriched in wastes and released to surrounding environments at a number of mining sites throughout the world (e.g. Engle et al., 2001; Besser et al., 2008; Yeddou et al., 2010). Most studies involving contaminated milling wastes are primarily in temperate climates where more rainfall occurs, and transport of anthropogenically derived CN⁻ and Hg is facilitated by both sediment and aqueous phase transport (Navarro et al., 2008).

In a study by Ismail et al. (2009), it was shown that the mobility of CN⁻ rich sediments is also controlled by pH of the surrounding sediments and the availability of metals such as Fe, Cu, and Zn to form complexes with CN⁻. For example, between sediment pH 6 and 8, CN⁻ complexes of Zn are stable and less soluble, whereas, in an environment where the pH is > 8, Zn/CN⁻ complexes were found to be more soluble (Rennert & Mansfeldt, 2002).

Jambor et al. (2009) examined the solubility of CN⁻ from metal rich sediments with similar results to Rennert and Mansfeldt (2002) and Ismail et al. (2009). Jambor et al. (2009) illustrated that in metal rich sediments, CN⁻ complexes of Fe and Cu were less soluble from sediments, whereas, CN⁻ complexes of Ni and Zn are more soluble at pH values greater than 8. Thus, the dissolution of CN⁻ salts and metal complexes from sediments, followed by aqueous transport during storm events, was likely to occur in addition to transport with suspended sediment (Jambor et al., 2009; Yngard et al., 2007).

The transport of Hg from mining areas is most the result of trapped elemental Hg (Gemici & Oyman, 2003). Craw (2005) reported that Hg is strongly adsorbed to sediment particles and also to organic matter. Furthermore, Lechler et al. (1997) reported that surface contamination of Hg around mine sites is most likely as elemental Hg, as a result of its release from the Hg-Au amalgam used in the mining processing.

It has been shown that the release of anthropogenic contaminants such as CN⁻ and Hg in arid and semi-arid climates can impact the surrounding areas to a greater extent than previously believed (Navarro et al., 2008; Figueroa et al., 2008). The purpose of examining CN⁻ and Hg in an arid environment can determine the present-day distribution of these mining wastes in arid environments where their potential environmental impacts are not yet fully understood as compared to those in wet climates.

3. Wind Transport of Contaminants

There are few mechanisms of transport that impact the area adjacent to and downgradient/downwind from mining sites in arid climates. One such mechanism is wind (aeolian) dispersion. Aeolian dispersion has been well documented concerning transport of contaminated materials great distances from sources, impacting areas not directly affected by mining activities (e.g. Camm et al., 2003, 2004; Petaloti et al., 2006; Lopez et al., 2005; Yadav & Rajamani, 2006).

Aeolian dispersion can be one of the more dominant forms of contaminant transport of mining wastes in arid climates. Many of the environmental impacts associated with mine wastes can be attributed to the wind-blown transport of fine tailings to the surrounding areas (Davies, 1976; Reheis, 2006; Conko et al., 2013). A number of researchers showed that contamination from mining activities is related to dust, processed tailings, and fly ash, with the potential to impact the environment many kilometers down-wind from the source (Davis & White, 1981; Camm et al., 2003, 2004; Petaloti et al., 2006). For example, Davis and White (1981) have extensively investigated transport of particulate matter (PM) by wind. They showed that wind-blown mineral particles containing Pb, and other mine related wastes, travel great distances from the source. They further noted that particulates < 2 mm and > 2 μm in size are susceptible to aeolian dispersion, and the particles are evenly distributed directionally along wind patterns.

Meteorological data are important when performing such aeolian transport studies to understand the number of windy days, the wind velocity, and ground conditions, including grain size and land cover. Davis and White (1981) used this approach in a study to illustrate the dispersion of contaminated waste at a site in Western Wales. This study looked at 193 days of force 4 winds in a calendar year. Their findings demonstrated that Pb contaminated sediment, found down gradient, originated from the mining activities 1.8 km up gradient from the sampling area. The study of directional dispersion of wind-blown contaminated materials can provide information on the extent of environmental contamination from source areas, even in areas where transport is limited due to a lack of precipitation.

Work by Davis and White (1981) implies that the dispersion of contaminated materials will not only impact sediments and waters around a site but also has the potential of affecting the food-chain. To illustrate further, it has been demonstrated that produce grown (e.g. lettuce) down wind from a source of mine waste in the U.K. has been impacted with Pb above natural background levels (Davis & White, 1981). Wind-blown material from mining areas therefore, can constitute a direct environmental hazard through both inhalation and ingestion of foods by humans and animals in the vicinity of such sites or, beyond (Davis & White, 1981; Camm et al., 2003).

By measuring metals or metalloids such as As, wastes from abandoned mines can be characterized and mapped (Camm et al., 2003, 2004; Conko et al., 2013). They demonstrated that contaminated sediments can be tracked down wind from a known source. Their research showed that As contaminated dust from milling facilities can be transported up to 2 kilometers downwind of source area and ultimately impacting a wider area (Camm et al., 2003, 2004; Davis & White, 1981).

Aeolian dispersion of As at milling facilities also presents a potential environmental hazard. The process of removing metals from ore involves heating in a calciner system which produces contaminated fly ash. Sediments at processing facilities have been found to contain As ranging between 200 and 3,325 mg kg^{-1}, and residue directly around a calciner can be 12% As by weight (Camm et al., 2004). The dominant form of As found at calciner facilities is As-Fe oxides in a range of textural wastes. Particulate matter $\leq$ 20 μm (PM$_{20}$) containing As-Fe oxides, with a good portion of this material in the PM$_{10}$ size fraction, are known to be a potential inhalation exposure hazard (Lopez et al., 2005; Yadav & Rajamani, 2006). The study of PM fractions at mine and mill sites provides a baseline for long-term dispersal of materials containing potentially hazardous materials, and, the potential exposure to humans in relation to locations of milling facilities.

Similar findings regarding transport of trace metals (e.g. Pb, Cd, Cu, Cr, and Zn) in emissions from industrial regions of China have been studied by researcher (e.g. Kim et al., 1998; Hsu et al., 2005; Petaloti et al., 2006). Hsu et al. (2005) found PM$_{10}$ containing Pb, Cd, Cu, Cr, and Zn transported from industrial areas in China to as far away as the United States. This is possible, according to their findings, because, during the winter, particulate matter will adsorb Pb and Cd and be transported by winds blowing from China to Taiwan and beyond (Hsu et al., 2005; Kim et al., 1998). In contrast, metal-containing particulate matter originating from mining sites has a similar but limited range of transport (Petaloti et al., 2006; Tasdemir et al., 2006).

Particle size fractionation has been used to investigate land use and fluvial deposition of contaminated mine waste (Horowitz & Stephens, 2008; Horowitz, 1993, 1988, 2008). Horowitz (2008) and others used air elutriation to evaluate each fraction of contaminated sediment and found that the finer aeolian fraction contained 10 times more metals than the larger fractions (Callender, 1988; Lu et al., 2012). When using this method of evaluating the lighter fractions, it was obvious that contaminants associated with the smaller particles can be transported farther (compared to shorter transport for larger size particulates) and deposited by aeolian processes in areas where there is no anthropogenic activities (Horowitz & Stephens, 2008; Lu et al., 2012). While their study of aeolian transport is extremely valuable in arid regions it was not seen as useful for this study due to the top layer (10-20 mm) of the tailings having a hard crust as a result of evaporation, evaporation in arid regions bring what little dissolved salts to the surface that produces a thin but stable crust trapping sediments beneath (Slowey et al., 2007a). However, studies by Mackenzie and Pulford (2002) of milling and calciner facilities in Scotland showed that wind distribution can transport materials from source areas to areas not directly impacted by industrialization. While literature has shown that wind dispersion has been studied at great lengths around industrial areas, studies do not show that it has been a focus in arid and hyperarid mining regions because of location of sites to population centers.

4. Transport of Contaminates in Wet Climates

Surface water influence on transport of contaminated sediments is a focus of much research. Mine sites located in areas of high precipitation are investigated for overland flow as a major means of transport, and the conveyance of contaminated materials in streams, rivers and subsurface is also a means of transport. Surface water and storm runoff as means of transport are the pathways by which contaminated sediments can move beyond their site of origin. Although it is known that overland flow and aeolian transport are major processes conveying contamination from a mine site, overland flow is typically ignored in arid regions because of the lack of precipitation and resulting infrequency of such an event, not to mention the location of much of these sites.

Abandoned mines in temperate climates have been a focus for the United States Environmental Protection Agency (USEPA) since the mid 1980s (USEPA, 1995) and others (e.g. Slowey et al. 2007a, 2007b; Stover, 1996; Lottermoser & Ashley, 2005). During the 1980's, the USEPA investigated abandoned mines in Colorado to determine if there was potential for environmental issues (Stover, 1996). Many of the mines were located next to permanent water sources, such as Chalk Creek, which received contamination from nearby mining sites. Since that time, the USEPA's research has evaluated precipitation and its potential influence on the transport of contamination into surface waters (USEPA, 1995).

Precipitation promotes the movement of trace elements into surface waters in dissolved and particulate forms and, therefore, can create adverse habitat conditions for fish populations in receiving bodies of water (Stover, 1996). Two sources of mine waste entering the environment are leachates from mine tailing waste and AMD from mine tunnels and shafts (USEPA, 1995). The first source is caused by the leaching of waste materials with high acid-generating potential; second is the surface runoff from contaminated mine sites producing AMD that adversely affect nearby surface waters.

Mines that used gravity separation to remove cinnabar (HgS) from ore for later processing tend to have larger amounts of Hg contamination (Slowey et al., 2007b). Due to environmental issues related to Hg (i.e. methalization), it has been a main focal point for investigation. It has been found that Hg concentration in sediments tends to be higher adjacent to, and down gradient from abandoned mining and milling sites located in either volcanic areas or when Hg was used in processing (Dolenec et al., 2005; Moncur et al., 2005, 2006). High levels of Hg in sediments down gradient from the source area are the direct result of processes of erosion and overland flow as shown by Macklin et al. (2001). Research has shown that in areas where the dominant Hg source is cinnabar, the Au extraction processes alter cinnabar to meta-cinnabar, a more soluble form (Gemici & Oyman, 2003; Moncur et al., 2005). Such a change allows Hg to be more susceptible to physical weathering, permitting an enrichment of Hg in fine-grained sediments and possible methalization later due to environmental effects (Slowey et al., 2007a, 2007b; Kim et al., 2004; Harikumar & Jisha, 2010).

Research shows that higher ratios of Hg between water and sediments are a direct result of a higher solubility of Hg originating from AMD sources adjacent to mines (Nishida et al., 1982; Slowey et al., 2007a, 2007b; Rampe & Runnells, 1989). However, Slowey et al. (2007a) reported that water sources with appreciable amounts of dissolved bicarbonate will buffer the system limiting acid-soluble metals. They reported that AMD receiving waters containing as much as 430 mg L^{-1} $CaCO_3$ can provide sufficient buffering capacity to neutralize acidity that will result in sorption or precipitation of soluble metals. Ultimately, water chemistry controls Hg, and other trace metals, partitioning between the aqueous and particulate phases in an aquatic environment. However, in an arid environment water chemistry is not an issue but, once transported materials can be deposited in an area or location where water can provide an environment where Hg or other trace metals can chemically change.

Release of trace elements into surface sediments and their subsequent transport by storm-water is likely to be influenced by the adsorptive capacity of the sediments, e.g. the texture and the amount of organic matter. For example, Rieuwerta and Farago (1996) found that levels of Pb and Hg decreased in sediments with decreasing organic matter. This work shows that organic matter can control Pb and Hg retention in sediments and subsequent transport. Davis (1976) and others have conducted similar investigations seeking a correlation between base metal mining and pollution in sediment (Rowan et al., 1995; Davis & White, 1981). Davis (1976) found correlations between Pb-Hg and Cu-Zn in sediments with similar organic matter content. It was reported that in areas where base metal mining occurred, high levels of Pb and Hg were more likely to be present (Macklin et al., 1997). In fact, in some historical mining areas, Hg was reported to be as much as 2 mg kg^{-1} in sediments derived from a floodplain that was downgradient from historical mining areas (Macklin et al., 1997).

It was found that water contamination caused by an influx of sediment from mining sites has the capability of impacting a wider area (Zhuo et al., 2007; Luo et al., 2006; Singhal & Islam, 2008; Rowan et al., 1995). Studies suggest that floodplain sediments that contain particulate-bound trace elements (e.g. Pb, Zn, Cd, and As) were most likely the result of repeated fluvial scouring and re-suspension of sediments (Rowan et al., 1995; Luo et al., 2006). Sediments originating from a mine site can spread contaminants primarily by their transport with water (Luo et al., 2006). Rowan et al. (1995) suggested that (water) metals contamination and subsequent transport is however limited to the inevitable precipitation and sorption of metals onto sediments.

Research shows that trace elements (e.g. Hg, Pb, Cu, Zn, Cr, Ni, As) can be used to illustrate movement of sediments in runoff from mining sites (Sanghoon, 2006; Luo et al., 2006; Singhal & Islam, 2008). Additionally, contamination is not restricted to a small geographic area when transport redistributes sediments enriched with trace elements. Materials that are normally stable in sediments will be mobilized in storm flow and transported great distances from the source (Sanghoon, 2006). In areas with high precipitation, the mobilization of metals begins with leaching of tailings, particularly where the pH is low and SO_4^{2-} content is high (Moncur et al., 2005).

At Lake Coeur D'Alene, Idaho, USA Horowitz et al. (1993) found trace elements (e.g. Ag, As, Pb, Cd) were being deposited into lake sediments from adjacent mining activities. Horowitz et al. (1993) illustrated that trace metal-enriched fine sediments from abandoned mining and milling areas were mobilized from the ground surface to Lake Coeur D'Alene during periods of high precipitation. It was establish that association with Fe oxyhydroxide coatings on grains made the trace metals more environmentally available than if they were associated with other constituents such as sulphide minerals. With relations identified between trace metals and

Fe oxides, Horowitz et al. (1993), were able to determine that the trace metals deposited into Lake Coeur D'Alene can be ascribed to the abandoned mining operations rather than geogenic deposits.

More recently, research has focused on water chemistry and the transport of metals and metalloids in mine drainages (Kwong et al., 2007; Slowey et al., 2007a). For example, As in mine drainage from tailings in Ontario (Canada) was an order of magnitude above the Canada drinking water standard of 0.05 mg L^{-1}. Studies by Riemer and Toth (1970) showed that the chemistry of sediment pore-water will directly influence the transport of As and possibly other trace elements. In the short term, as dissolved trace elements enter a system, they tend to adsorb to suspended matter; however, they could re-dissolve farther down the system when the geochemistry changes due to system variability (Riemer & Toth, 1970).

Ashley and Lottermoser (2003, 2004) and others studied mine waste and the distribution of As in surface waters (Lottermoser & Ashley, 1999). They found AMD with a low pH (4.1) had the highest concentration of dissolved As (up to 13.9 mg L^{-1}) that was entering the Mole River, New South Wales, Australia. Once AMD entered the Mole, surface waters entering the river further diluted the As-rich drainage to background concentration (0.0086 mg L^{-1}) beyond 2.5 km downstream. While their study showed that As was eventually diluted to < background, it was evident that environmental conditions have the ability to alter trace element species over time and distance.

In a semi-arid climate environmental impacts by trace metals are influenced by surface runoff during storm events. Taylor and Kesterton (2002) showed that the distribution of trace metals depended on their solubility and more importantly, their mobility. The movement of trace metals within the sediment profile is also influenced by sedimentation, allowing the distinction between pre-mining, mining, and post mining in semi-arid climates (Meza-Figueroa et al., 2009). However, the above processes are lacking in arid locations where contaminant transport is limited primarily to sediment interactions and subsequent storm-water transport of the affected sediment rather than aqueous mechanisms (Taylor & Kesterton, 2002).

Further research of abandoned mine sites determined whether sites were sources of water contamination caused by influx of sediment containing Pb (Zhuo et al., 2007; Luo et al., 2006; Singhal & Islam, 2008; Rowan et al., 1995). Floodplain sediments were found to contain particulate-bound contaminants such as Pb, Zn, Cd, and As, that were most likely the result of repeated fluvial scouring and re-suspension of sediments, further dispersing trace element contaminated particles (Luo et al., 2006; Singhal & Islam, 2008). According to Luo et al. (2006) movement of sediment will impact the degree of traceability while lateral movement accounts for the age of deposition (Giuliano et al., 2007).

Some contaminants (e.g., Hg, Pb, Cr, As), both soluble and particulate forms, in wetter climates can be used to illustrate movement of contamination in surface waters and sediments adjacent to historical mines (Sanghoon, 2006; Luo et al., 2006; Singhal & Islam, 2008; Da Silva et al., 2005). In areas with high precipitation, the mobilization of metals begins with leaching of tailings (Moncur et al., 2005). More recent research has focused on water chemistry and the transport of metals and metalloids in mine drainage (Kwong et al., 2007; Harris et al., 2003; Leinz et al., 2006). Speciation and mobilization of metalloids and the mechanisms of transport in sediments and water are a direct result of the amount of available moisture (Kwong et al., 2007). As dissolved trace elements enter a system, they tend to adsorb to suspended matter; however, they could re-dissolve farther down the system (Lewis, 1977; Riemer & Toth, 1970).

Research clearly shows that the mobilization of contaminates from processed tailings through surface and ground water flow has been a focus of studies (Macklin et al., 2001; Hughes & Diaz, 2008; Bonzongo et al., 1996). Furthermore, studies of historic mining and the resulting contamination has been the focus in regions that receive generous amounts of precipitation rather than arid or hyperarid areas. In an arid environment water is not an issue. However, once transported to an area or location where water is present such as a desert lake or perched aquifer, Hg, or other trace metals, can transform from a minor environmental issue to a major concern such as methyl-mercury.

5. Chemical Behavior of Trace elements

Speciation and geochemical processes can have a major influence on transport of trace elements at abandoned mine sites. Furthermore, studies have shown that the transformation of trace elements by microbes can influence transport (e.g. Macklin et al., 2001; Chen et al., 2006; Conko et al., 2013). An excellent example of how transformation because of microbes can affect their surrounding environment is the methylation of elemental Hg to methyl-mercury (CH_3Hg^+), a more soluble, bioavailable, and highly toxic form. When CH_3Hg^+ enters a water system it can produce fish kills and large amounts of toxic Hg uptake by multiple species resulting in an impairment in higher species and the food web, thus affecting the surrounding environment more so than elemental Hg (Conaway et al., 2004; Hughes & Diaz, 2008; Bonzongo et al., 1996; Mastrine et al., 1999). In the

Western United States, historically Hg was used in the extraction of precious metals by amalgamation. During this process, excess Hg was routinely dumped into the immediate environment (Greene, 1975). Although portions of Hg will volatilize into the atmosphere, it is possible for Hg to bind with organic matter and remain in soils and sediments for longer periods of time, until transported by erosion.

Methyl-mercury has been a focus of investigations to provide the modes of conversion from elemental to CH_3Hg^+, and the ratios of Hg to CH_3Hg^+ (Hughes & Diaz, 2008; Bonzongo et al., 1996; Mastrine et al., 1999). Macklin et al. (2001) found that methylation occurs in sediments when Hg is < 15.3 mg kg^{-1}, than decreasing above 15.3 mg kg^{-1} because of toxicity (Macklin et al., 2001). Decreased methylation is attributed to reduced activity of microbes as a result of higher levels of total Hg; i.e. more toxic to microbes. A reduced activity of microbes during a winter and spring flooding is also directly related to the presence of Hg at > 15.3 mg kg^{-1} because of increased transport (Macklin et al., 2001).

In an aqueous system, CH_3Hg^+ will increase to toxic levels below a point source by microbial mediated methylation. This process is further enhanced as a direct result of additional Hg entering the system during winter and spring runoff events (Bonzongo et al., 1996). Findings have suggests that CH_3Hg^+ decreases in concentration with distance from a source area, in part from volatilization with turbulence of flowing water as it moves downstream, and in part by retention of Hg and CH_3Hg^+ in bed sediments (Bonzongo et al., 1996).

Macklin et al. (2001) and Mastrine et al. (1999) studied several mining areas with similar climates, sediments, vegetation, and rock where Hg was utilized to amalgamate Au during the milling process. They found Hg and iron (Fe) concentrations were directly correlated in sediments. Mastrine et al. (1999) explained that the relation between Hg and Fe was the result of similar mechanisms that control the aqueous forms of both metals. Their findings showed that sorption of Hg to Fe-oxyhydroxides has a direct relation to the amount of Fe and Hg found in waters originating from contaminated sediments at a mining site. They noted that this relation is caused by the mass of total suspended solids (TSS) containing Fe oxyhydroxides that sorb Hg and Fe to TSS in waters.

Waste from middle 1800s mining activities continues to impact the surrounding environment of the Western United States and beyond (Sims, 2013). For example, Hg has been released from historical mill and mine operations to surface waters of the Truckee River and Steamboat Creek systems in Northern Nevada (Bonzongo et al., 1996; Chen et al., 2006). Total Hg and CH_3Hg^+ in sediments along these river systems have been studied to determine the distribution in relation to precipitation and aeolian transport (Stamenkovic et al., 2004; Kim et al., 2000; Jonasson & Boyle, 1972). If levels of Hg in a river system do not decrease with distance in channel sediments then this suggests that a constant source of Hg is being mobilized (Stamenkovic et al., 2004). Such findings are useful in characterizing movement of sediments and the transformation from elemental Hg to CH_3Hg^+ in similar aquatic systems.

Studies have found that anthropogenic Hg at mine sites is one to two orders of magnitude higher than that in pristine streams (Stamenkovic et al., 2004; Macklin et al., 2001). For Hg methylation to occur, conditions favoring an anaerobic environment must be present so that an increase in microbial activity can be available. Stamenkovic et al. (2004) also asserted that minerals and nutrients (i.e. SO_4^{2-}) in sediments might directly correlate high levels of Hg with their reduction. They found that SO_4^{2-} reduction and Hg conversion to CH_3Hg^+ decreases in stream bank sediments containing SO_4^{2-}, which indicated that microbial processes of Hg methylation are inhibited in the presence of significant amounts of dissolved SO_4^{2-}, indicating an aerobic environment.

Speciation is typically used to determine the mobility of Hg in sediments and to identify sources and release potential. Research into Hg speciation in the California Coast Range was undertaken to better understand its mobility in mine waste and how solubility relates to speciation (Kim et al., 2000; Mastrine et al., 1999). However, studies have suggested that Hg and with other anions (i.e. Cl^-) in sediments might have a correlation between solubility and mobility (Kim et al., 2000; Stamenkovic et al., 2004). Determining the correlation between Hg and the anions would provide invaluable information concerning the transport related to Hg in the environment.

Researchers have examined speciation and bioavailability of As, Zn, Cd, Pb, Mn, and Al in mine waste (Wu et al., 2006; Meers et al., 2006; Nair & Robinson, 2000; Chen et al., 2006; Gupta et al., 2007). It was established that the bioavailability of certain metals like Zn, Pb, and Cd are controlled by sediment moisture, pH, and dissolved organic matter. In sediment high in Fe and Mn, it is known that oxidation and adsorption of As (III) and As (V) by pedogenic Fe-Mn rich solid phases at varying sediment pH and organic content controls the speciation of As (Chen et al., 2006; Grafe et al., 2007, 2008; Boyle & Smith, 1994). Slowey et al. (2007b) suggested that As mobilization occurs with the mobility of the particulate phase rather than transport in the aqueous phase. It has been further shown that the dominant As species in the environment is As (V) and it is adsorbed to amorphous iron-hydroxides that co-precipitated with jarosite (Grafe et al., 2008; Chen et al., 2006; Gupta et al., 2007). Grafe et al. (2008) and others showed that As is released from mine waste as As (V) where a low pH and oxidation convert As (III) to As (V), allowing As to migrate as a particulate verses in the aqueous form (Grafe et al., 2008; Slowey et al., 2007a; Breteler et al., 1981). Suspended materials and electrolytes can influence the transport of metals in water bodies. It has been found that suspended particles can remove soluble contaminants from the water column (Al-Busaidi et al., 2005). Stumm and Morgan (1996) and others have shown that high sediment and salinity suppress suspended particles by their flocculation and subsequent settling out from the water column (Al-Busaidi et al., 2005; Sakata, 1987; Zhang & Selim, 2005).

Another factor influencing the mobility of trace metals is the interaction with the sediment surfaces. Cheng et al. (2009) examined arsenic transport and transformation through columns packed with uncoated sand and natural sand coated with inorganic colloids and/or dissolved organic matter (DOM) using arsenate, As (V) in solution. The occurrence of inorganic colloids and/or DOM evidently enhanced As transport, with a larger fraction of As leaching out of the column compared to the total amount added. Cheng et al. (2008) further explained that when a solution of As (V) was combined with dissolved organic matter the mobility was increased. Therefore, As (V) can leach from the sediments to groundwater due to elevated DOM.

It has been shown that in mining areas with high levels of organic matter in sediments, and significant precipitation, the solubility of As (V) increases (Stumm & Morgan, 1996; Chang et al., 2009). For example, when As (V) was spiked in waters similar to mining site seepage, only As (III) was detected in the effluents for uncoated sand columns while both As (III) and As (V) were detected in the coated sand columns (Chang et al., 2009). It was recognized that when monomethylarsonic acid was injected into sediment columns, all As species were present in the effluents (Cheng et al., 2008; Feng et al., 2005; Hutchinson & Meema, 1987). Thus it appears that areas where organic matter is very low, limited to no moisture reduces the possibility for mobilization and transport of metals and other contaminants are poor due to environmental factors other than overland transport (i.e. arid and hyperarid regions).

6. Siltation Transport

Authors have suggested that siltation possesses several issues to the environment; clogging of the water column and trace element leaching among others. James (1991) suggested that high siltation of waterways can impede aquatic life growth because it can effect oxygen uptake by local biota causing a suffocation. Furthermore, high siltation by mine waste will also provide ideal conditions for enriched trace elements leaching into water systems, further impacting biota (Hudson-Edwards et al., 1997). While siltation clogging the water column is an important issue, the potential release of trace elements contained in silt or sediment that originated from mining areas is significant (e.g. Edwards et al., 1997; Hudson-Edwards et al., 1997; Dennis et al., 2002). Siltation of waterways and associated contamination has the potential to impact a larger radius of a stream or river basin by both sorbed and dissolved metals.

Researchers have used Hg and Si to trace siltation from hydraulic mining areas (e.g Edwards et al., 1997; James, 1991). For example, it has been found that elevated silica concentrations in rivers and streams was observed a distance of 60 km downstream from mining sources in just 100 years in the mining areas of northern California (Edwards et al., 1997; James, 1991). It was found that elevated Si was correlated with Hg, suggesting the source Hg was from the same location as the silica. James (1991) stated this material could have moved even farther than 60 km if it was not for dams and other obstructions put in place since the early 1930s.

Siltation in a river or stream can also affect areas adjacent to water sources such as overbanks and low lying areas (Hudson-Edwards et al., 1997). Studies indicate that sediment contamination has produced metal-contaminated overbank river sediments and increased metal concentrations in waters of local rivers and tributaries (Hudson-Edwards et al., 1997; Dennis et al., 2002). By following siltation patterns, scientists have been able to trace sources of mining-related sediments to their original source. Furthermore, it has been found that in some low lying areas, contamination from mining sites have mixed with agricultural runoff. There have been several incidents reported in the United Kingdom in which a major flood occurred on lands once used for base metal mining that are now used primarily for agricultural (Dennis et al., 2002). In these areas, overbank and channel sediments would already have contained high levels of base metals prior to agricultural development. However, in areas where Pb, Cd, and Zn were mined, there were high levels of contamination corresponding to the source areas of the trace elements (Mlayah et al., 2009; Lecce & Pavlowsky, 2001).

Siltation has the potential to clog water ways as the material moves from an abandoned mine site. However, siltation also can have a greater effect on the environment when tailings dams fail and mine waste enters a water system. Although this is a somewhat different release mechanism than storm-water transport, it illustrates the effects of a catastrophic tailings transport event on the environment. Thus, it is possible for sites with contaminated sediments, tailings, or mine waste to impact a much larger area, such as a basin or floodplain (Fuente et al., 2007; Miller, 1997; Hernandez et al., 1999). For example, in early 2000, a dam failed in Maramures County, upper Tisa Basin, Northwest Romania releasing a significant amount of silt into the Tisa River (Macklin et al., 2001). During this dam failure over 40,000 metric tons of contaminated sediment was released into major tributaries of the Danube River, with pollutants of cyanide and metals (Pb, Zn, Cu, and Cd). Although metal concentration in surface waters decreased with distance from the source because of a high buffering capacity and dilution, trace metals in sediments still exceeded acceptable levels downstream (Macklin et al., 2001). The Danube River incident illustrates that flood waters originating from a mining area have the potential to pose a significant environmental threat to its surrounding area.

In Seville, Spain, the Aznalcollar tailing dam at the Boliden Apirsa's mine was breached in April of 1998 (Hernandez et al., 1999). The failure impacted 4600 hectares of land located along the Rios Agrio and Guadiamar with 5.5 million cubic meters of acidic waters and 1.3×10^6 cubic meters of mine waste containing metals (Hernandez et al., 1999; Edwards et al., 2003). It was shown that mine waste contained high concentrations of Ag, As, Cd, Cu, Pb, Sb, Tl, and Zn compared to their background levels (Edwards et al., 2003). Studies showed that Zn and Cd had the highest impact on the environment, whereas Pb and As were not mobile unless reducing conditions that could dissolve greater amounts of the reduced species developed in the mine wastes (Edwards et al., 2003). It was also clear that high mineral sulfide content in these sediments can produce AMD, under aerobic conditions, easily releasing trace metals.

When tailings reach surface waters, however, it can transport large quantities of contaminated sediments to downstream areas such as floodplains, basins, or farm lands. Macklin et al. (1997, 2003, 2006) studied transport of localized contaminated sediments by tracking the material downgradient in what they termed "pulses". They explained that pulses consist of material temporarily suspended in flood waters and ultimately deposited in localized spots beyond the source. Finding showed that contaminated deposits decreased downstream in sediments of rivers and streams (Meijer et al., 2002; Coulthard & Macklin, 2003). It is thus important to account for the secondary deposits when evaluating the overall pollution of a mine area. The materials that make up the isolated spots were the result of contaminant laden sediments being transported (secondary deposits) during storm events rather than the legacy of soluble contaminants in the water column.

6.1 Tracking Movement of Contaminated Material

The issue of geogenic contaminants and their mobilization into the environment has been assessed using different techniques (Zhang & Shan, 2008; Chow, 1970; Frignana & Bellucci, 2004). Sutherland and Tolosa (2000) studied the movement of trace element in humid sediments and found that Pb, Zn and Cu were more mobile in the lower level (7.2-10cm) of the sediment than at the surface level. Once trace elements are in dry sediments, their behavior will be similar to terrestrial environments (Sutherland, 2000; Sutherland et al., 2003; Andrews & Sutherland, 2004).

Sutherland (2000) used mass loading and mass per area enrichment ratios (MAER) to study the impact of trace elements in sediments. It was found that input trace elements such as Pb were 4 to 5 times higher in sediments affected by anthropogenic activities. This statistical technique showed that certain forms of metals tend to be more enriched than background metals, hence, presenting a higher anthropogenic signal in sediments (Sutherland, 2000).

Study of enrichment ratios (ER) has been applied to agricultural fields globally to evaluate anthropogenic enhancement. While agricultural fields are different from mining sites, they both result in significant alteration to the local sediment environment and therefore are a good comparison. In areas where a long history of intensive agriculture is documented, dating back more than 100 years, change in environmental conditions is easily observed (Zhang & Shan, 2008). Researchers used enrichment factors to evaluate agricultural impacts on sediments in China to assess anthropogenic activities such as fertilizer usage (Zhang & Shan, 2008; Agrawal et al., 1981; Qin & Chen, 1996; Yin et al., 2006). It was found that fertilization over long periods of time will impact the sediment environment with the use of phosphorus fertilizers-containing trace metals such as Cd (Inaba et al., 2006; Zhang & Shan, 2008). As shown by Zhang and Shan (2008) with fertilizers, certain concentrated geogenic trace metals (i.e. Cd) tend to bioaccumulate over time, and may be available for plant uptake.

Environmental models have been widely used to predict mass movement of sediments, as "moving waves" at mine sites (Pickup et al., 1983). Input variables to models have included sediment velocity, dispersion coefficients, and the mean distributions of loads to simulate the movement of the wave. Pickup et al. (1983) found that modeling the moving wave of sediment with a dispersion model could adequately simulate sediment transport down slope in an area influenced by surface waters.

Pickup et al. (1983) utilized two models for predicting the movement of sediments down stream with one allowing for the distribution of sediment velocity through dispersion co-efficient, the other used a normal mass conservation approach. Their models included datasets collected over a period 43 months, with data from 7 cross-sections of the Kawerong River, Papua-New Guinea, which is polluted with wastes from a copper mine. They were able to predict the distance and concentrations along the river based on precipitation effects.

Over the years, a number of other studies have used modeling techniques to assist in the interpretation of mine waste (Yager & Stanton, 2000; Ge et al., 2005; Mackenzie & Pulford, 2002; Stanton, 2000). Models using topography, precipitation, geophysics, geochemistry, hydrology, and historical records provide the best quantitative results that can be derived from the investigation of historical mine sites. Many of these models (e.g. Coulthard & Macklin, 2003; Stanton, 2000) incorporate numerical and analytical techniques for evaluating the transport and fate of contaminants in surface sediments. Some of the models used today for evaluating contaminant movement are Tracer, Gaussian Distribution Method, and numerical and analytical models to predict contaminant transport and fate in surface sediments.

Coulthard and Macklin (2003) used numerical and analytical approaches for predicting long term contamination issues in rivers and streams in historical mining environments. The approach, Tracer, uses historical mine records, and topographic and hydrologic maps to predict the movement of contaminated sediments in relation to their sources. This approach provides information concerning the possible extent of contamination, and, what Coulthard and Macklin (2003) refer to as "hot spots". Hot spots, as described by Coulthard and Macklin (2003), are secondary sources of contamination that occur some distance from the original source when deposited during periods of high flow. These hot spots can impact other areas that were not historically contaminated (Ge et al., 2005; Coulthard & Macklin, 2003; Mackenzie & Pulford, 2002).

Identifying natural versus anthropogenic input sources of contaminants aids in understanding environmental problems associated with abandoned mine sites. For example, understanding potential sources of Hg will provide information concerning input point source pollution rather then geogenic sources (Coulthard & Macklin, 2003). Studies have evaluated the differences between anthropogenic Hg and geogenic sources in order to gain a better understanding of mine related point- and non-point source pollution. Hg from anthropogenic sources has been found in higher concentration in sediments compared to geogenic sources (Engle et al., 2001; Stanton, 2000; Yager & Stanton, 2000). This higher concentration is due to the use of Hg in large amounts versus geogenic sources, which tend to be much lower and have more time to disperse before moving with transported sediments (Stanton, 2000).

Modeling sediment transport has been used to measure the distance from geogenic and anthropogenic sources in wet climates that impact downstream environments (Engle et al., 2001). Models have been used to illustrate the effects of atmospheric precipitation on Hg sources and the resulting erosion from sediments and alluvial deposits over time in wet climates (Engle et al., 2001; Stanton, 2000; Yager & Stanton, 2000). The Gaussian Distribution Method has been used to calculate the average daily emission of metals from lithologic units with the average metal flux from a given area. It was shown that 89% of Hg released into the environment will come from naturally enriched sources (geogenic) and 11% will most likely originate from anthropogenic sources (Engle et al., 2001). These methods are useful for predicting sources of contamination. However, they are limited in that the variables used will only reflect the data collected and will not account for unknown variables, i.e., most such models are site specific.

Siltation is arid and hyperarid regions are not a major concern because there is little precipitation however, when precipitation occurs, it will transport large amount of sediments due to mass transport rather than a constant siltation affect on nearby surface waters. When transport occurs in arid and hyperarid regions it is usually associated with a short but intense storm event that transports materials into the wider environment, potentially impacting the food web.

7. Biological Environment

7.1 Potential for Flora, Fauna and Human Health Affects

One of the biggest concerns with the transport of contaminants is when it enters the food web. Impacts of mining waste extend beyond the potential effects on water and sediments and ultimately into the biological environment. Biological impacts include the health and well being of plants, animals, and humans exposed to contaminated waters, sediments and food supplies. The location of mining facilities and their associated contaminated soils, sediments, and wastes, in relation to population and farming, can have an adverse impact on the overall environment through the transport of trace metals (Appanna, 1991; Nicholson et al., 2003).

Mining activity sparked population growth in towns and cities that supplied the labor to mines and milling facilities. Today, many of these towns and cities are larger in population, although the nearby mining and milling sites have been abandoned and left unattended for decades. It should be considered, however, that abandoned mining areas adjacent to towns could still pose potential risks locally to flora, fauna, and ultimately human health (Nicholson et al., 2003; Sierra et al., 2009). The environmental and biological impacts of mining wastes on ecosystems have been studied to determine what, if any, ill effects humans have suffered in areas surrounding abandoned mining sites (Lee et al., 1998; Suner et al., 1999; Appleton et al., 2000).

Thus it appears that siltation and is very low, the mobility and transport of metals and other contaminants are limited due to environmental factors (i.e. arid and hyperarid regions). Although it has been shown that mining can produce adverse effects beyond the mining area itself, the impact of mining activities on local vegetation, fauna, and, eventually, human health in arid environments is not well documented. To understand effects from vegetation, it is important to understand the mechanism in which vegetation becomes a source of contamination. To illustrate, an animal that has ingested contaminated water or food or inhaled contaminated airborne material, can roam long distances before human consumption.

7.2 Uptake and Effects on Flora

Base metal and uranium mining sites have been investigated for contaminants uptake by local vegetation within the continental United States and abroad (Kipp et al., 2009; Kovalchuk et al., 2005). These investigations have studied the uptake of Cu, Ni, Fe, Co, Zn, U, and Pb in vegetation and animal tissues in the vicinity of mining areas. Plants around mining sites have been found to contain higher levels of trace elements (e.g. Pb, As) correlating to concentrations in local sediments. It was found that plants with the highest level of trace elements (e.g. Pb) tended to have broader leaves than small leaf plants. It was found that higher levels of metals is possible in grasses, surrounding mining sites (Nicholson et al., 2003; Vazquez et al., 2008; Wang et al., 2007). Literature has shown that bioaccumulation of As and resulting phytotoxicity to lower plant species were observed in wet mining areas that had several metal-tolerant plant species (e.g., *Angophora floribunda, Cassinia laevis,* and *Chrysocephalum apiculatum*) that colonized the periphery of the site (Clemente et al., 2005; Boularbah, et al., 2006; Han et al., 2006). Bermejo et al. (2003) investigated Pb mining in the Sierra Madrona Mountains Spain. Their study also showed higher concentrations of trace elements (Pb, Zn, Cd, Cu, As, and Se) in plants where trace element concentration was high in sediments (Bermejo et al., 2003). However, higher concentrations of trace elements in vegetation were more prevalent in areas like stream sediments near tailings dumps than in areas not impacted by mining (Reglero et al., 2008).

7.3 Uptake and Effects on Fauna

Animals (fauna) are likely to suffer greater affects through the food chain by the ingestion of flora. Researchers identified trends in metal uptake in vegetation and animals from trace element enriched substrata, leachates, and surface water sources (Lee, 2001; Pankakoski et al., 1994; Sánchez-Chardi & Nadal, 2007). It has been shown that metals such as Cd are of great concern because of its bioaccumulation from the consumption of vegetation (Krishnamurti et al., 2005).

Uptake of metals can impact the environment by adversely affecting fauna as a result of grazing on local flora in affected areas (Gammon, 1970; Pagenkopf et al., 1974). Local vegetation near streams and rivers can be impacted when mine waste containing dissolved elements (e.g Cu, Zn, Pb, and As) enters water-bodies (Lewis, 1977; Gammon, 1970; Pagenkopf et al., 1974; Warnick & Bell, 1969). An influx of dissolved trace elements can affect organisms like insect larvae, bivalve mollusks, and microorganisms, which in turn affect larger organisms up the food chain (Wright & Zamuda, 1987; Qiu et al., 2007). As shown in literature (Krishnamurti et al., 2005; Reglero et al., 2008; Moreno-Jimenez et al., 2009), bioaccumulation in local vegetation will directly affect local fauna by the transfer of metals from one organism to another. The level of tolerance for specific metal accumulation is species-specific, resulting in the further transfer of metals from one species to another with the impact being species specific (Moreno-Jimenez et al., 2009).

Studies have shown that Cu, Ni, and Pb are likely to be found in the stomach linings of local fauna (goats), whereas Ni and Fe are more prevalent in skin and fur of some fauna (Cloutier et al., 1985). Studies have shown, for example, that high concentration of Pb found in animals near mining sites can be caused by animals foraging for plant material that are known to uptake Pb into their tissues (Chopin & Alloway, 2007; Reglero et al., 2008). Studies (Reglero et al., 2008) showed that in lead mining areas of the Sierra Madrona Mountains 13 species of plants contaminated with trace metals resulted in a transfer to local red deer (*Cervus elaphus*) foraging in the area. Reglero et al. (2008) illustrated that in mining impacted soil con tainingh Pb and As at 100 times background concentrations, leafy plants contained Pb and As concentrations 3 times their background levels (Reglero et al., 2008). They further examined the livers of local red deer and found that Pb and As concentrations were 12 times higher than livers from control areas not impacted by mining.

7.4 Uptake and Effects on Human Health

Contamination of the environment and resulting impact to human health is well known and has been noted as far back as 2000 years in the Roman Empire. In ancient Rome, Romans used lead, a documented neurotoxin, for making water pipes, cooking utensils, water tanks, and storage vessels (Bishop, 1989). Lead water pipes were used in most major cities in the empire. In ancient Rome, wine was contaminated with Pb from as many as 14 sources during its preparation. Romans also used Pb as a preservative and a flavor enhancer (Bishop, 1989; Nriagu, 1990). Rome was not the only culture that was contaminating itself with toxic metals, as even the Christian sacramental cups of wine were made of Pb or leaded bronze (Nriagu, 1990). Other trace elements (e.g. Hg, Cd, As, Se, Cr) are well-documented as toxic to ecosystems and humans, and modern health agencies such as the World Health Organization and USEPA have published guidelines for limiting ingestions of these elements in food and drinking water.

Throughout most of the 20th century, the mining in the United States has resulted in impacts on the local and wider environment, such as the USEPA Superfund site from Coeur d'Alene, Idaho that dates back to the 1870s. The Coeur d'Alene area had more than 100 mining and mill operations producing Ag, Pb, Zn, and other metals (USEPA, 2005). According to the USEPA one of the most notorious sites was the Bunker Hill Mine and Smelting Complex located in Kellogg, Idaho. Many of the mine tailings in the region were discharged directly to the Coeur d'Alene River and its tributaries until 1968 when the practice was banned by the US Government. Smelting operations at Bunker Hill discharged large quantities of sulfur dioxide waste containing Pb and other metals that affected local communities and the environment (USEPA, 2005).

The Bunker Hill site was investigated to evaluate the impact to water, ecological and human health downstream from the source of pollution to the river. The USEPA studied the human health risk of Pb intake by local population to determine future population impacts to children from the groundwater and vegetation (USEPA, 2005). It was found that downstream and adjacent farms to the Coeur d'Alene River contained Pb 5 to 10 times higher than regulatory levels (0.05 mg kg^{-1}) as a result of contaminated soils and sediments. The pollution was found a far as 80 km downstream, impacting more than 40 square km (USEPA, 2005).

The effects that mining sites have on surrounding environments also have been studied in great detail in Japan (Vahter et al., 2007). Inaba et al. (2006) and others reported that the disposal of mine waste from a nearby mining operation impacted the local chain (Yamagami et al., 2006). They found that Cd-contaminated waste water from mining operations was used for growing rice (Vahter et al., 2007; Inaba et al., 2006; Yamagami et al., 2006; Takagi et al., 2004). It was found that locals that consumed 3.1 g to 3.8 g of Cd in rice over a lifetime, developed mild to severe mitochondrial dysfunction called Itai–itai disease (Inaba et al., 2005). Studies have shown that over time bioaccumulation of trace metals (e.g. Cd) in tissues of local indigenous peoples affected their tubular epithelial cells after 80-weeks of exposure by ingestion (Takagi et al., 2004; National Academy of Science, 1973). This disease was found to be the direct result of Cd-enriched mine waste water that discharged to local rivers, and ultimately used in the rice fields.

Impacts of mining activities on local flora have shown that vegetation uptake will transfer contaminants to fauna and ultimately up the food chain. The uptake of contaminants into local vegetation will eventually impact humans due to biomagnifications and that the true effect is based on the specific species rather than the actual dose or level of contaminants found in the food chain.

8. Summary and Conclusions

Although it is well documented that mining contamination in wetter climates has the possibility of migrating long distances by water transport, there has been limited research in arid and hyperarid regions. Studies have shown that mining activities can influence the mobilization of geogenic metals into the surrounding environment. Furthermore, trace metals (Cu, Zn, Pb, Ni, Cd, and Cr) have been the focus of interest because of their long residence times and significant toxicities to biota in wet climates. Evaluating sediments contaminated with geogenic and anthropogenic inputs in dry regions will provide valuable data on the extent, origin, and distribution of contaminants from source areas in hyperarid regions. Although there is limited research in hyperarid climates, studies have shown that mining is of great concern because of possible transport. It has become quite clear that trace metals and contaminates are mobile in both the particulate and soluble phase and are transported long distances from source areas contained in sediments. Thus it is imperative that research be conducted in these arid and hyperarid regions where significant amounts of mining and milling have occurred to further understanding in the mechanism of transport in similar environments.

References

Agrawal, Y. K., Patel, M. P., & Merh, S. S. (1981). Lead in soil and plants: Its relationship to traffic volume and proximity to highway (Ialbag, Baroda City). *International J. of Environ. Studies, 16*, 222-224. http://dx.doi.org/10.1080/00207238108709873

Al-Busaidi, A., Cookson, P., & Yamamoto, T. (2005). Methods of pH determination in calcareous soils: use of electrolytes and suspension effect. Aust. *J. Soil Res., 43*, 541-545. http://dx.doi.org/10.1071/SR04102

Andrews, S., & Sutherland, R. A. (2004). Cu, Pb, and Zn contamination in Nuuanu watershed, Oahu, Hawaii. *Science of the Total Environment, 324*, 173-182. http://dx.doi.org/10.1016/j.scitotenv.2003.10.032

Appanna, V. (1991). Impact of heavy metals on an arctic rhizobium. *Bull Environ. Contam. Toxical, 46*, 450-455. http://dx.doi.org/10.1007/BF01688946

Appleton, J., Lee, K. M., Kapusta, K. S., Damek, M., & Cooke, M. (2000). The heavy metal content of the teeth of the bank vole (Clethrionomys glareolus) as an exposure marker of environmental pollution in Poland. *Environmental Pollution, 110*, 441-449. http://dx.doi.org/10.1016/S0269-7491(99)00318-8

Ashley, P. M., Lottermoser, B. G., & Chubb, A. J. (2003). Environmental geochemistry of the Mt. Perry copper mines area, SE Queensland, Australia. *Geochemistry: Exploration, Environment, Analysis, 3*, 345-357. http://dx.doi.org/10.1144/1467-7873/03-014

Ashley, P. M., Lottermoser, B. G., Collins, A. J., & Grant, C. D. (2004). Environmental geochemistry of the derelict Webbs Consols mine, New South Wales, Australia. *Environmental Geology, 46*, 591-604. http://dx.doi.org/10.1007/s00254-004-1063-7

Berger, A. C., Bethke, C. M., & Krumhansl, J. L. (2000). A process model of natural attenuation in drainage from a historic mining district. *Applied Geochemistry, 15*(5), 655-666. http://dx.doi.org/10.1016/S0883-2927(99)00074-8

Bermejo, J. C. S., Beltrán, R., & Ariza, J. L. G. (2003). Spatial variations of heavy metals contamination in sediments from Odiel River (Southwest Spain). *Environmental International, 29*, 69-77. http://dx.doi.org/10.1016/S0160-4120(02)00147-2

Besser, J. M., & Leib, K. J. (2000). *Toxicity of Metals in Water and Sediment to Aquatic Biota.* Chapter E19: Integrated Investigations of Environmental Effects if Historical Mining in the Animas River Watershed, San Juan County, Colorado.

Besser, J. M., Finger, S. E., & Church, S. E. (2008). Impacts of Historical Mining on Aquatic Ecosystems–An Ecological Risk Assessment. *Integrated investigations of Environmental Effects of Historical Ming in the Animas River Watershed, San Juan County, Colorado, 1651*, 87-106.

Blowes, D. W., Robertson, W. D., Ptacek, C. J., & Merkley, L. C. (1994). Removal of agricultural nitrate from tile-drainage effluent using in-line bioreactors. *J. Contam. Hydrol. 15*, 207-221. http://dx.doi.org/10.1016/0169-7722(94)90025-6

Bonzongo, J. C., Heim, K. J., Warwick, J. J., & Lyons, W. B. (1996). Mercury levels in surface waters of the Carson River-Lahontan reservoir system, Nevada: Influence of historic mining activities. *Environmental Pollution, 92*(2), 193-201. http://dx.doi.org/10.1016/0269-7491(95)00102-6

Boularbah, A., Schwartz, C., Bitton, G., Aboudrar, W., Ouhammou, A., & Morel, J. L. (2006). Heavy Metal contamination from mining sites in South Morocco: 2. Assessment of metal accumulation and toxicity in plants. *Chemosphere, 63*, 811-817. http://dx.doi.org/10.1016/j.chemosphere.2005.07.076

Boyle, D. R., & Smith, C. N. (1994). Mobilization of mercury from a Gossan tailings pile, Murry Brook Precious metal vat leaching operation, New Brunswick, Canada. *International Land Reclamation and mine Drainage Conference and the Third International Conference on the Abatement of acidic Drainage*, Pittsburgh, PA, April 24-29, 1994.

Breteler, R. J., Valiela, I., & Teal, J. M. (1981). Bioavailability of mercury in several north-eastern U.S. *Spartina* ecosystems. *Estuarine Coastal and Shelf Science, 12*(2), 155-166. http://dx.doi.org/10.1016/S0302-3524(81)80093-X

Camm, G. S, Glass, H. J., Bryce D. W., & Butcher, A. R. (2004). Characterization of a mining-related arsenic-contaminated site, Cornwall, UK. *Journal of Geochemical Exploration, 82*(1-3), 1-15. http://dx.doi.org/10.1016/j.gexplo.2004.01.004

Camm, G. S., Butcher, A. R., Pirrie, D., Hughes, P. K., & Glass, H. J. (2003). Secondary mineral phase associated with a historic arsenic calciner identified using automated scanning electron microscopy; a pilot study from Cornwall, UK. *Minerals Engineering, 16*(11), 1269-1277. http://dx.doi.org/10.1016/j.mineng.2003.07.005

Chang, J.-S., Yoon, I.-H., & Kim, K.-W. (2009). Heavy metal and arsenic accumulating fern species as potential ecological indicators in As-contaminated abandoned mines. *Ecological Indicators, 9*, 1275-1279. http://dx.doi.org/10.1016/j.ecolind.2009.03.011

Chen, Y., Bonzongo, J. C., & Miller, G. C. (2006). Levels of Methylmercury and controlling factors in surface sediments of the Carson River system, Nevada. *Environmental Pollution, 92*(3), 281-287. http://dx.doi.org/10.1016/0269-7491(95)00112-3

Chow, T. J. (1970). Lead Accumulation in roadside soil and grass. *Nature, 225*, 295-296. http://dx.doi.org/10.1038/225295a0

Churchill, R. (1999). Insights into California mercury production and mercury availability for the gold mining industry from the historical record. *Geol. Soc. Am. Abs., 21*, 45.

Clemente, R., Walker, D. J., & Bernal, M. P. (2005). Uptake of heavy metals and As by Brassica Juncea grown in a contaminated soil in Aznalcóllar (Spain): The effect of soil Amendments. *Environmental Pollution, 138*(1), 46-58. http://dx.doi.org/10.1016/j.envpol.2005.02.019

Cloutier, N, Clulow, F. V., Lim, T. P., & Dave, N. K. (1985). Metal (Cu, Ni, Fe, Co, Zn, Pb) and Ra-226 levels in meadow voles Microtus Pennsylvanicus living on nickel and uranium mine tailings in Ontario, Canada: environmental and tissue levels. *Environmental Pollution Series B, Chemical and Physical, 10*(1), 19-46. http://dx.doi.org/10.1016/0143-148X(85)90014-X

Conaway, H. H., Watson, E. B., Flanders, J. R., & Flegal, A. R. (2004). Mercury deposition in a tidal marsh of south San Fransisco Bay downstream of the historic New ALdaden mining district, California. *Marine Chemistry, 90*(1-4), 175-184. http://dx.doi.org/10.1016/j.marchem.2004.02.023

Conko, K. M., Landa, E. R., Kolker, A., Kozlov, K., Gibb, H. J., Centeno, J. A. . . . Panov, Y. B. (2013). Arsenic and mercury in the soils of an industrial city in the Donets Basin, Ukraine. *Soil and Sediment Contamination: An Int. J., 22*(5), 574-593. http://dx.doi.org/10.1080/15320383.2013.750270

Coulthard, T. J., & Macklin, M. G. (2003). Modeling long-term contamination in river systems from historical metal mining. *Geology, 31*(5), 451-454. http://dx.doi.org/10.1130/0091-7613(2003)031<0451:MLCIRS>2.0.CO;2

Craw, D. (2005). Potential anthropogenic mobilization of mercury and arsenic from soils on mineralized rocks, Northland, New Zealand. *Journal of Environmental Management, 74*, 283-292. http://dx.doi.org/10.1016/j.jenvman.2004.10.005

Da Silva, E. F., Fonseca, E. C., Matos, J. X., Patinha, C., Reis, P., & Oliveira, J. M. S. (2005). The effect of unconfined mine tailings on the geochemistry of soils, sediments and surface waters of the Lousal area (Iberian Pyrite Belt, Southern Portugal). *Land Degrad. Develop., 16*, 213-228. http://dx.doi.org/10.1002/ldr.659

Davis, B. E. (1976). Mercury Content of Soils in Western Britain with Special Reference to Contamination from Base Metal Mining. *Geoderma, 16*(3), 183-192. http://dx.doi.org/10.1016/0016-7061(76)90020-3

Davis, B. E., & White, H. M. (1981). Environmental Pollution by Wind Blown Lead Mine Waste: A Case Study in Wales, U.K. *The Science of the Total Environment, 20*(1), 57-74. http://dx.doi.org/10.1016/0048-9697(81)90036-X

Dennis, I. A., Macklin, M. G., Coulthard T. J., & Brewer, P. A. (2002). The impact of the October-November 2000 floods on contaminated metals dispersal in the River Swale catchment, North Yorkshire, U.K. *Hydrological Processes, 17*(8), 1641-1657. http://dx.doi.org/10.1002/hyp.1206

Dolenec, T., Serefimvski, T., Dobnnikar, M., Tasev, G., & Dolenec, M. (2005). Mineralolgical and heavy metal signature of acid mine drainage impacted paddy soil from the western part of Kocani field (Macedonia). *Materials and Environment, 52*, 397-402.

Earman, S. (1996). *The Impact of Non-point Source Pollution from Mining Waste on Water Quality, Elko County, Nevada* (Unpublished Master thesis in the Geology Department, University of Nevada, Las Vegas).

Edwards, D. P., Lamarque, J., Attié, J., Emmons, L., Richter, A., Cammas, J., . . . Burrows, J. (2003). Tropospheric ozone over the tropical Atlantic: A satellite perspective. *J. Geophys. Res., 108*(D8), 4237. http://dx.doi.org/10.1029/2002JD002927

Edwards, K., Macklin, M. & Taylor, M. (1997). Historic metal mining inputs to Tees River basin. *Science of the Total Environment, 194-195*, 437-445. http://dx.doi.org/10.1016/S0048-9697(96)05381-8

Engle, M. A., Gustin. M. S., & Zhang, H. (2001). Quantifying natural source mercury emissions from the Ivanhoe Mining District, north-central Nevada, U.S.A. *Atmospheric Environment, 35*(23), 3987-3997. http://dx.doi.org/10.1016/S1352-2310(01)00184-4

Feng, M., Schrlau, J. E., Snyder, R., Snyder, G. H., Chen, M., Cisar, J. L., & Cai, Y. (2005). Arsenic transport and transformation associated with MSMA application on a golf course green. *J. Agric Food Chem, 53*, 3556-3562. http://dx.doi.org/10.1021/jf047908j

Figueroa, J. A. L., Wrobel, K., Afton, S., Caruso, J. A., Corona, J. F. G., & Wrobel, K. (2008). Effect of some heavy metals and soil humic substances on the phytochelatin production in wild plants from silver mine areas of Guanajuato, Mexico. *Chemosphere, 70*, 2084-2091. http://dx.doi.org/10.1016/j.chemosphere.2007.08.066

Frignana, M., & Bellucci, L. G. (2004). Heavy metals in marine coastal sediments: Assessing source, flux, history and trends. *Ann. Chim., 94*, 479-486. http://dx.doi.org/10.1002/adic.200490061

Fuente, C. de la, Clemente, R., & Bernal, M. P. (2007). *Changes in metal speciation and pH in olive processing waste and sulphur-treated contaminated soil.* Ecotoxicology and Environmental Safety, Corrected Proof.

Gammon, J. R. (1970). *Effect of Inorganic Sediment on Stream Biota.* USEPA Rep. U.S. Government Printing Office, Washington D.C.

Ge, Y., MacDonald, D., Sauvé, S., & Hendershot, W. (2005). Modeling of Cd and Pb speciation in soil solutions by WinHumicV and NICA-Donnan model. *Environmental Modeling & Software, 20*(3), 353-359. http://dx.doi.org/10.1016/j.envsoft.2003.12.014

Gemici, Ü., & Oyman, T. (2003). The influence of the abandoned Kalecik, Hg mine on water and stream sediments (Karaburun, Izmir, Turkey). *The Science of the Total Environment, 312*, 155-166. http://dx.doi.org/10.1016/S0048-9697(03)00008-1

Giuliano, V., Pagnanelli, F., Bornoroni, L., Toro, L., & Abbruzzese, C. (2007). Toxic elements at a disused mine district: Particle size distribution and total concentration in steam sediments and mine tailings. *Journal of Hazardous Materials, 148*, 409-418. http://dx.doi.org/10.1016/j.jhazmat.2007.02.063

Gräfe, M., Singh, B., & Balasubramanian, M. (2007). Surface speciation of Cd(II) and Pb(II) on kaolinite by XAFS spectroscopy. *Journal of Colloid and Interface Science, 315*(1), 21-32. http://dx.doi.org/10.1016/j.jcis.2007.05.022

Gräfe, M., Tappero, R. V., Marcus, M. A., & Sparks, D. L. (2008). Arsenic speciation in multiple metal environments II. Micro-spectroscopic investigation of a CCA contaminated soil. *Journal of Colloid and Interface Science, 321*(1), 1-20. http://dx.doi.org/10.1016/j.jcis.2008.01.033

Greene, J. M. (1975). *Life in Nelson Township, Eldorado Canyon, and Boulder City* (Unpublished Master Thesis, History Department, University of Nevada, Las Vegas).

Gupta, P. K., Jha, A. K., Koul, S., Sharma, P., Pradhan, V., Gupta, V., . . . Singh, N. (2007). Methane and Nitrous Oxide Emission from Bovine Manure Management ractices in India. *Journal of Environmental Pollution 146*(1), 219-224. http://dx.doi.org/10.1016/j.envpol.2006.04.039

Han, F. X., Su, Y., Monts, D. L., Waggoner, C. A., & Plodinec, M. J. (2006). Binding, distribution, and plant uptake of mercury in a soil from Oak Ridge, Tennessee, USA. *Science of The Total Environment, 368*(2-3), 753-768. http://dx.doi.org/10.1016/j.scitotenv.2006.02.026

Harikumar, P. S., & Jisha, T. S. (2010). Distribution pattern of trace metal pollutants in the sediments of an urban wetland in the southwest coast of India. *International Journal of Engineering Science and Technology, 2*(5), 840-850.

Harris, D. L., Lottermoser, B. G., & Duchesne, J. (2003). Ephemeral acid mine drainage at the Montalbion silver mine, north Queensland. *Australian Journal of Earth Sciences, 50,* 797-809. http://dx.doi.org/10.1111/j.1440-0952.2003.01029.x

Harrison, R. M., & Laxen, D. P. H. (1981). *Lead Pollution–Causes and Control.* UK: Cambridge University Press. http://dx.doi.org/10.1007/978-94-009-5830-2

Hernández, L. M., Gómara, B., Fernández, M., Jiménez, B., González, M. J., Baos, R., . . . Montoro, R. (1999). Accumulation of heavy metals and As in wetland birds in the area around Donñana National Park affected by the Aznalcollar toxic spill. *The Science of The Total Environment, 242*(1-3), 293-308. http://dx.doi.org/10.1016/S0048-9697(99)00397-6

Horowitz, A. J. (2008). Contaminated Sediments: Inorganic Constituents. In M. G. Anderson (Ed.), *Encyclopedia of Hydrological Sciences.* Wiley Inter-science. http://dx.doi.org/10.1002/0470848944.hsa317

Horowitz, A. J., & Stephens, V. C. (2008). The effects of land use on fluvial sediment chemistry for the conterminous U.S.—Results from the first cycle of the NAWQA Program: Trace and major elements, phosphorus, carbon, and sulfur. *Science of The Total Environment, 400,* 290-314. http://dx.doi.org/10.1016/j.scitotenv.2008.04.027

Horowitz, A. J., Elrick, K. A., & Callender, E. (1988). The Effect of Mining on The Sediment-Trace Element Geochemistry of Cores from The Cheyenne River Arm of Lake Oahe, South Dakota (USA). *Chemical Geology, 67,* 17-33. http://dx.doi.org/10.1016/0009-2541(88)90003-4

Horowitz, A. J., Elrick, K. A., & Cook, R. B. (1993). Effect of Mining and Related Activities on The Sediment Trace Element Geochemistry of Lake Coeur D'Alene, Idaho, USA. Part 1: Surface Sediments. *Hydrological Processes, 7,* 403-423. http://dx.doi.org/10.1002/hyp.3360070406

Hsu, S.-C., Liu, S. C., Jeng, W.-L., Lin, F.-J., Huang, Y.-T., Lung, S.-C. C., . . . Tu, J.-Y. (2005). Variations of Cd/Pb and Zn/Pb ratios in Taopei aerosols reflecting long-range transport or local pollution emissions. *Science of the Total Environment, 347*(1-3), 111-121. http://dx.doi.org/10.1016/j.scitotenv.2004.12.021

Hudson-Edwards, K., Macklin, M., & Taylor, M. (1997). Historic metal mining inputs to Tees river sediment. *Science of The Total Environment, 194-195,* 437-445. http://dx.doi.org/10.1016/S0048-9697(96)05381-8

Hughes, M. K., & Diaz, H. F. (2008). Climate variability and change in the drylands of Western North America. *Global and Planetary Change, 64,* 111-118. http://dx.doi.org/10.1016/j.gloplacha.2008.07.005

Hutchinson, I., & Ellison, R. (1992). *Mine Waste Management.* London: Lewis Publishers.

Hutchinson, T. C., & Meema, K. M. (1987). Lead Mercury, Cadmium and Arsenic in the Environment. *Scientific Committee on Problems of the Environment (SCOPE), (Series): 31.*

Inaba K., Murakami, S., Suzuki, M., Nakagawa, A., Yamashita, E., Okada, K., & Ito, K. (2006). Crystal structure of the DsbB=DsbA complex reveals a mechanism of disulfide bond generation. *Cell, 127,* 789-801. http://dx.doi.org/10.1016/j.cell.2006.10.034

Ismail, I., Abdel-Monem, N., Fateen, S.-E., & Abdelazeem, W. (2009). Treatment of a synthetic solution of galvanization effluent via the conversion of sodium cyanide into an insoluble safe complex. *Journal of Hazardous Materials, 166,* 978-983. http://dx.doi.org/10.1016/j.jhazmat.2008.12.005

Jambor, J. L., Martin, A. J., & Gerits, J. (2009). The post-depositional accumulation of metal-rich cyanide phases in submerged tailings deposits. *Applied Geochemistry, 24*(12), 2256-2265 http://dx.doi.org/10.1016/j.apgeochem.2009.09.011

James, L. A. (1991). Quartz concentration as an index of sediment mixing: hydraulic mine-tailings in the Sierra Nevada, California. *Geomorphology, 4*(2). 125-144. http://dx.doi.org/10.1016/0169-555X(91)90024-5

Jonasson, I. R., & Boyle, R. W. (1972). Geochemistry of Mercury and origins of natural contamination of the environment. *CIM Bulletin, 65*(717), 30-49.

Khalid, A. M., El-Tawil, A. H., Ashy, M., & Elbeih, F. K. (1981). Constituents of local plants. Part-8: Distribution of some coumarins in plants of different plant families grown in Saudi Arabia. *Pharmazia, 36*(8), 569-571.

Kim, C. S., Brown Jr., G. E., & Rytuba, J. J. (2000). Characterization and speciation of mercury-bearing mine waste using X-ray absorption spectroscopy. *The Science of the Total Environment, 261*(1-3), 157-168. http://dx.doi.org/10.1016/S0048-9697(00)00640-9

Kim, K. W., Lee, H. K., & Yoo, B. C. (1998). The environmental imact of gold mines in the Yugu-Kwangcheon

Au-Ag metallogenic province, Republic of Korea. *Environmental Technology, 19*, 291-298. http://dx.doi.org/10.1080/09593331908616683

Kim. C. S., Rytuba, J. J. & Brown Jr., G. E. (2004). Geological and Anthropogenic factors influencing mercury speciation in mine waste: an EXAFS spectroscopy study. *Applied Geochemistry, 19*(3), 379-393. http://dx.doi.org/10.1016/S0883-2927(03)00147-1

Kipp, G. G., Stone, J. J., & Stetler, L. D. (2009). Arsenic and uranium transport in sediments near abandoned uranium mines in Harding County, South Dakota. *Applied Geochemistry, 24*, 2246-2255. http://dx.doi.org/10.1016/j.apgeochem.2009.09.017

Kovalchuk, I., Titov, V., Hohn, B., & Kovalchuk, O. (2005). Transcriptome profiling reveals similarities and differences in plant responses to cadmium and lead. *Mutation Research, 570*, 149-161. http://dx.doi.org/10.1016/j.mrfmmm.2004.10.004

Krishnamurti, G. S. R., McArthur, D. F. E., Wang, M. K., Kozak, L. M., & Huang, P. M. (2005). Chapter 7–Biogeochemistry of soil cadmium and the impact of terrestrial food chain contamination. *Biogeochemistry of Trace Elements in the Rhizosphere*, 197-257. http://dx.doi.org/10.1016/B978-044451997-9/50009-X

Kwong, Y. T. J., Beauchemin, S., Hossain, M. F., & Gould, W. D. (2007). Transformation and mobilization of arsenic in the historic Cobalt mining camp, Ontario, Canada. *Journal of Geochemical Exploration, 92*(2-3), 133-150. http://dx.doi.org/10.1016/j.gexplo.2006.08.002

Lecce, S., & Pavlowsky, R. T. (2001). Use of mining-contaminated sediment tracers to investigate the timing and rates of historical flood plain sedimentation. *Geomorphology, 38*, 85-108. http://dx.doi.org/10.1016/S0169-555X(00)00071-4

Lechler, P. J., Miller, J. R., Hsu, L.-C., & Desilets, M. O. (1997). Mercury mobility at the Carson River Superfund Site, west-central Nevada, USA: Interpretation of mercury speciation data in mill tailings, soils, and sediments. *Journal of Geochemical Exploration, 58*(2-3), 259-267. http://dx.doi.org/10.1016/S0375-6742(96)00071-4

Lee, J. S., Chon, H. T., Kim, J. S., Kim, K. W., & Moon, H. S. (1998). Enrichment of potentially toxic elements in areas underlain by black shales and slates in Korea. *Environmental Geochemistry and Health, 20*, 135-147. http://dx.doi.org/10.1023/A:1006571223295

Leinz, R. W., Sutley, S. J., Desborough, G. A., & Briggs, P. H. (2006). *An Investigation of the Partitioning of Metals in Mine Wastes Using Sequential Extractions*. U.S. Geological Survey.

Lewis, M. (1977). *Aquatic Inhabitants of a Mine Waste Stream in Arizona*. United States Forest Service; U.S. Department of Agriculture, Issue 349.

López, J. M., Callén, M. S., Murillo, R.. García, T., Navarro, M. V., de la Cruz. M. T.. & Mastral, A. M. (2005). Levels of selected metals in ambient air PM_{10} in an urban site of Zaragoza (Spain). *Environmental Research, 99*, 58-67. http://dx.doi.org/10.1016/j.envres.2005.01.007

Lottermoser, B. G., & Ashley, P. M. (2005). Tailings dam seepage at the rehabilitated Mary Kathleen Uranium mine, Australia. *Journal of Geochemical Exploration, 85*, 119-137. http://dx.doi.org/10.1016/j.gexplo.2005.01.001

Lottermoser, B. G., Ashley, P. M., & Lawie, D. C. (1999). Environmental geochemistry of the Gulf Creek copper mine area, north-eastern New South Wales, Australia. *Environmental Geology, 39*(1). http://dx.doi.org/10.1007/s002540050437

Lu, W., Ma Y., & Lin, C. (2012). Spatial variation and fractionation of bed sediment-borne copper, zinc, lead and cadmium in a stream system affected by acid mine grainage. *Soil and Sediment Contamination; An Int. J. 21*(5), 831-849. http://dx.doi.org/10.1080/15320383.2012.697933

Luo, X.-S., Zhou, D.-M., Liu, X.-H., & Wang, Y.-J. (2006). Solids/solution partitioning and speciation of heavy metals in the contaminated agricultural soils around a copper mine in eastern Nanjing city, China. *Journal of Hazardous Materials, 131*(1-3), 19-27. http://dx.doi.org/10.1016/j.jhazmat.2005.09.033

Mackenzie, A. B., & Pulford, I. D. (2002). Investigation of contaminant metal dispersal from a disused mine site at Tyndrum, Scotland, using concentration gradients and stable Pb isotope ratios. *Applied Geochemistry, 17*(8), 1093-1103. http://dx.doi.org/10.1016/S0883-2927(02)00007-0

Macklin M. G., Brewer, P. A.. Belteanu, D., Coulthard, T. J., Driga, B., Howard, A. J., & Zaharia, S. (2001). The long term fate and environmental significance of contaminated metals released by the January and Mack 2000 mining tailings dam failures in Maramures County, Upper Tisa Basin, Romania. *Applied Geochemistry, 18*, 241-257. http://dx.doi.org/10.1016/S0883-2927(02)00123-3

Macklin, M. G., Brewer, P. A., Hudson-Edwards, K. A., Bird, G., Coulthard, T. J., Dennis, I. A., . . . Turner, J. N. (2006). A Geomorphological approach to the management of rivers contaminated by metal mining. *Geomorphology, 79*, 423-447. http://dx.doi.org/10.1016/j.geomorph.2006.06.024

Macklin, M. G., Brewer, P. A., Hudson-Edwards, K. A., Bird, G., Coulthard, T. J., Dennis, I. A. . . . Turner, J. N. (2003). A geomorphological approach to the management of rivers contaminated by metal mining. *Geomorphology, 79*(2006), 423-447.

Macklin, M. G., Hudson-Edwards K. A., & Dawson, E. J. (1997). The significance of pollution from historic metal mining in the Pennine Orefields on river sediment contaminant fluxes to the North Sea. *Science of the total Environment, 194-195*, 391-397. http://dx.doi.org/10.1016/S0048-9697(96)05378-8

Mastrine, J. A., Bonzongo, J. J., & Lyons, W. B. (1999). Mercury concentrations in surface waters from fluvial systems draining historical precious metals mining areas in Southeastern U.S.A. *Applied Geochemistry, 14*(2), 147-158. http://dx.doi.org/10.1016/S0883-2927(98)00043-2

Meers, E., Unamuno, V. R., Du Laing G., Vangronsveld, J., Vanbroekhoven, K., Samson, R., . . . Tack, F. M. G. (2006). Zn in the soil solution of unpolluted and polluted soils as affected by soil characteristics. *Geoderma, 136*(1-2), 107-119. http://dx.doi.org/10.1016/j.geoderma.2006.03.031

Meijer, R. J., Bosboom, J., Cloin, B., Katopodi, I., Kitou, N., Koomans, R. L., & Manso, F. (2002). Gradation effects in sediment transport. *Coastal Engineering, 47*, 179-210. http://dx.doi.org/10.1016/S0378-3839(02)00125-4

Meza-Figueroa, D., Maier, R. M., de la O-Villanueva, M., Gónez-Alvarez, A., Moreno-Zazueta, A., Rivera, J., . . . Palafox-Reyes, J. (2009). The Impact of unconfined mine tailings in residential areas from a mining town in a semi-arid environment: Nacozari, Sonora, Mexico. *Chemosphere, 77*, 140-147. http://dx.doi.org/10.1016/j.chemosphere.2009.04.068

Miller, J. R. (1997). The role of fluvial geomorphic processes in the dispersal of heavy metals from mine sites. *J. Geochem Explor, 58*, 101-118. http://dx.doi.org/10.1016/S0375-6742(96)00073-8

Miranda-Aviles, R., Puy-Alquiza, M. J., & Arvizu, O. P. (2012). Anthropogenic metal content and natural background of overbank sediments from the Mining District of Guanajuato, Mexico. *Soil and Sediment Contamination; An Int. J., 21*(5), 604-624. http://dx.doi.org/10.1080/15320383.2012.672488

Mlayah, A., da Silva, E. F., Rocha, F., Hamza, Ch. B., Charef, A., & Noronha, F. (2009). The Oued Mellègue: Mining activity, stream sediments and dispersion of base metals in natural environments, North-western Tunisia. *Journal of Geochemical Exploration, 102*, 27-36. http://dx.doi.org/10.1016/j.gexplo.2008.11.016

Moncur, M. C., Ptacek, C. J., Blowes, D. W., & Birks, S. J. (2006). Groundwater and surface water discharge from an abandoned tailings impoundment: Implications for watershed water quality. *American Geophysical Union Fall meeting, San Francisco, CA*, 11-15 December.

Moncur, M. C., Ptacek, C. J., Blowes, D. W., & Jambor, J. L. (2005). Release, transport and attenuation of metals from an old tailings impoundment. *Applied Geochemistry, 20*(3), 639-659. http://dx.doi.org/10.1016/j.apgeochem.2004.09.019

Moreno-Jiménez, E., Peñalosa, J. M., Manzano, E., & Carpena-Ruiz, R. O. (2009). Heavy metals distribution in soils surrounding an abandoned mine in NW Madrid (Spain) and their transference to wild flora. *Journal of Hazardous Materials, 162*, 854-859. http://dx.doi.org/10.1016/j.jhazmat.2008.05.109

Nair, P. S., & Robinson, W. E. (2000). Cadmium speciation and transport in the blood of the bivalve Mytilus edulis. *Marine Environmental Research, 50*(1-5), 99-102. http://dx.doi.org/10.1016/S0141-1136(00)00097-0

Navarro, M. C., Pérez-Sirvent, C., Martínez-Sánchez, M. J., Vidal, J., Tovar, P. J., & Bech, J. (2008). Abandoned mine sites as a source of contamination by heavy metals: A case study in a semi-arid zone. *Journal of Geochemical Exploration, 96*, 183-193. http://dx.doi.org/10.1016/j.gexplo.2007.04.011

Nicholson, F. A., Smith, S. R., Alloway, B .J., Carlton-Smith, C., & Chambers, B. J. (2003). An inventory of heavy metals inputs to agricultural soils in England and Wales. *The Science of the Total Environment, 311*, 205-219. http://dx.doi.org/10.1016/S0048-9697(03)00139-6

Nishida, H., Miyai, M., Tada, F., & Suzuk, I. S. (1982). Computation of the index of pollution caused by heavy metals in river sediment. *Environmental Pollution Series B, Chemical and Physical, 4*(4), 241-248. http://dx.doi.org/10.1016/0143-148X(82)90010-6

Pagenkopf, G. K., Russo, R. C., & Thurston, R. V. (1974). Effects of complexation on toxicity of copper to fishes. *Journal of Fish. Res. Bd. Canada, 31*, 462-465. http://dx.doi.org/10.1139/f74-077

Pankakoski, E., Koivisto, I., Hyvarinen, H., & Terhivuo, J. (1994). Shrews as indicators of heavy metal pollution. (pp. 137-149). In J. F. Merritt et al. (Eds.), *Advances in the biology of shrews.* Pittsburgh, PA.: Spec. Publ. 18. Carnegie Museum of Nat. Hist.

Patterson, J. W., & Passino, R. (1987). *Metals Speciation, Separation, and Recovery.* MI.: Lewis Publishers, Inc.

Petaloti, C., Triantafyllou, A., Kouimtzis, T., Samara, C. (2006). Trace elements in atmospheric particulate matter over a coal burning power production area of western Macedonia, Greece. *Chemosphere, 65*(11), 2233-2243. http://dx.doi.org/10.1016/j.chemosphere.2006.05.053

Pickup, G, Higgins, R. J., & Grant, I. (1983). Modeling Sediment Transport as a Moving Wave–The Transfer and Deposition of Mining Waste. *Journal of Hydrology, 60*(1-4), 281-301. http://dx.doi.org/10.1016/0022-1694(83)90027-6

Prusty, B. G., Sahu, K. C., & Godgul, G. (1994). Metal contamination due to mining and milling activities at the Zawar zinc mine, Rajasthan, India 1. Contamination of Stream sediments. *Chemical Geology, 112*, 275-291. http://dx.doi.org/10.1016/0009-2541(94)90029-9

Purves, D. (1985). *Trace Element Contamination of the Environment.* NY: Elsevier.

Qin, D. M., & Chen, X. H. (1996). The agriculture development of Yantze-Haihe region in six dynasties period. *Agric. Hist. China, 15*, 9-12.

Qiu, J.-W., Tang, X., Zheng, C., Li, Y., & Huang, Y. (2007). Copper complexation by fulvic acid affects copper toxicity to the larvae of the polychaete Hydroides elegans. *Marin Environmental Research, 64*(5), 563-573. http://dx.doi.org/10.1016/j.marenvres.2007.06.001

Rampe, J. J., & Runnells, D. D. (1989). Contamination of water and sediment in a desert stream by metals from an abandoned gold mine and mill, Eureka District, Arizona, U.S.A. *Applied Geochemistry, 4*(5), 445-454. http://dx.doi.org/10.1016/0883-2927(89)90002-4

Reglero, M. M., Monsalve-González, L., Taggart, M. A., & Mateo, R. (2008). Transfer of metals to plants and red deer in an old lead mining area in Spain. *Science of The Total Environment, 406*, 287-297. http://dx.doi.org/10.1016/j.scitotenv.2008.06.001

Reheis, M. C. (2006). A 16-year record of eolian dust in Southern Nevada and California, USA: Controls on dust generation and accumulation. *J. of Arid Environments, 67*(3), 487-520. http://dx.doi.org/10.1016/j.jaridenv.2006.03.006

Rennert, T., & Mansfeldt, T. (2002). Sorption of Iron-Cyanide Complexes in Soils. *Soil Science Society of America Journal, 66*, 437-444. http://dx.doi.org/10.2136/sssaj2002.0437

Riemer, D. N., & Toth, S. J. (1970). Adsorption of copper by clay minerals, humic acids and bottom muds. *J. of American Water Works Association, 62*, 195-197.

Rieuwerta J. S., & Farago, M. (1996). Mercury concentrations in historic lead mining and smelting town in the Czech Republic: A pilot study. *Science of the Total Environment, 188*(2-3), 167-171. http://dx.doi.org/10.1016/0048-9697(96)05167-4

Ross, C. (2008). Preserving the Culture While Closing the Holes-Abandoned Mine Reclamation in Nevada. *Presented at the 30ᵗʰ annual National Association of Abandoned Mine Land Program Conference*, October 26-29, Durango, CO.

Rowan, J. S., Barnes, S. J. A., Hetherington, S. L., & Lambers, B. (1995). Geomorphology and pollution: the environmental impacts of lead mining, Leadhills, Scotland. *Journal of geochemical Exploration, 52*(1-2), 57-65. http://dx.doi.org/10.1016/0375-6742(94)00053-E

Sakata, M. (1987). Relationship between adsorption of arsenic(III) and boron by soil properties. *Environ. Sci. Technol., 21*, 1126-1130. http://dx.doi.org/10.1021/es00164a016

Sanchez-Chardi, A., & Nadal, J. (2007). Bioaccumulation of metals and effects of landfill pollution in small mammals, Part I. The greate white-toothed shrew, Crocidura russula. *Chemosphere, 68*, 703-711.

http://dx.doi.org/10.1016/j.chemosphere.2007.01.042

Sanghoon, L. (2006). Geochemistry and partitioning of trace metals in paddy soils affected by metal mine tailings in Korea. *Geoderma, 135*, 26-37. http://dx.doi.org/10.1016/j.geoderma.2005.11.004

Sierra, M. J., Millán, R., & Esteban, E. (2009). Mercury uptake and distribution in *Lavandula Stoechas* plants grown in soil from Almadén mining district (Spain). *Food and Chemical Toxicology, 47*(11), 2761-2767. http://dx.doi.org/10.1016/j.fct.2009.08.008

Sims, D. B. (1997). The Migration of Arsenic and Lead in Surface Sediments at Three Kids Mine: Henderson, Nevada (Unpublished Master thesis in the Geology Department, University of Nevada, Las Vegas).

Sims, D. B. (2010). Contamination mobilization from anthropogenic influences in the Techatticup wash, Nelson, NV (USA). *Soil and Sediment Contamination; An Int. J., 19*(5), 515-530.

Sims, D. B. (2011). *Fate of Contaminants at an Abandoned Mining site in an Arid Environment* (Unpublished dissertation in environmental science. Kingston University, London).

Sims, D. B., & Bottenberg, B. C. (2007). Environmental Hazards Associated with Historic Mining on Water Quality in Nelson, Nevada: A Test Case for Environmental Issues Associated with Historic Mining in Nevada. *Journal of the Nevada Water Resources Association, 4*(2), 21-27.

Sims, D. B., Hooda, P. S., & Gillmore, G. K. (2013). Sediment Contamination along Desert Wash Systems from Historic Mining Sites in a Hyperarid Region of Southern Nevada, USA. *Soil and Sediment Contamination: An Int. J., 22*(7), 737-752. http://dx.doi.org/10.1080/15320383.2013.768200

Sims, D., B., & Bottenberg, B. C. (2008). Arsenic and Lead Contaminator in Wash Sediments at Historic Three Kids Mine-Henderson, Nevada: The environmental hazards Associated with Historic Mining Sites and Their Possible Impact on Water Quality. *Journal of the Arizona-Nevada Academy of Science*, pending final publication. http://dx.doi.org/10.2181/1533-6085(2008)40[16:AALCIW]2.0.CO;2

Singhal, N., & Islam, J. (2008). One-dimensional model for biogeochemical interactions and permeability reduction in soils during leachate permeation. *Journal of Contaminant Hydrology, 96*(1-4), 32-47. http://dx.doi.org/10.1016/j.jconhyd.2007.09.007

Slowey A. J., Rytube, J. J., Hothem, R. L., & May, J. T. (2007b). Mercury at the Oat Hill Extension Mine and James Creek, Napa County, California; tailings, sediment, water, and biota, 2003-2004. *USGS report: OF 2007-1132, Vol. 60.*

Slowey, A. J., Johnson, S. B., Newville, M., & Brown, G. E. (2007a). Speciation and colloid transport of arsenic from mine tailings. *Applied Geochemistry, 22*(9), 1884-1898. http://dx.doi.org/10.1016/j.apgeochem.2007.03.053

Stamenkovic, J., Gustin, M. S., Marvin-DiPasquale, M. C., Thomas, B. A., & Agee, J. L. (2004). Distribution of total methyl mercury in sediments along Steamboat Creek (Nevada, USA). S*cience of The Total Environment, 322*(1-3), 167-177. http://dx.doi.org/10.1016/j.scitotenv.2003.10.029

Stanton, M. R. (2000). *The Role of Weathering in Trace Metal Distributions in subsurface samples from the Mayday Mine Dump near Silverton, Colorado.*

Stover, Bruce; Colorado Division of Reclamation, Mining & Safety. (1996). *Mine-Drainage 319 Project, Bulkhead Feasibility Investigation, 2008 Reconnaissance Report*, Underground Source Control Sites (1, 2, 3, 4, CD).

Stumm, W., & Morgan, J. J. (1996). *Aquatic chemistry, chemical equilibrium and rates in natural waters.* New York: John Wiley & Sons.

Suner, M. A., Devesa, V., Muñoz, O., López, F., Montoro, R., Arias, A. M., & Blasco, J. (1999). Total and inorganic arsenic in the fauna of the Guadalquivir estuary: environmental and human health implications. *The Science of The Total Environment, 242*(1-3), 261-270. http://dx.doi.org/10.1016/S0048-9697(99)00399-X

Sutherland, R. A. (2000). Bed sediment-associated trace metals in an urban stream, Oahu, Hawaii. *Environmental Geology, 39*(6), 611-627http://dx.doi.org/10.1007/s002540050473

Sutherland, R. A., & Tolosa, C. A. (2000). Multi-element analysis of road-deposited sediment in an urban drainage basin, Honolulu, Hawaii. *Environmental Pollution, 110*, 483-495. http://dx.doi.org/10.1016/S0269-7491(99)00311-5

Sutherland, R. A., Day, J. P., & Bussen, J. O. (2003). Lead Concentration, Isotope Ratios, and Source Apportionment in Road Deposited Sediments, Honolulu, Oahu, Hawaii. *Water, Air, and Soil Pollution, 142*, 165-186. http://dx.doi.org/10.1023/A:1022026612922

Takagi, Y., Nomizu, M., Ui-Tei, K., Tokushige, N., & Hirohashi, S. (2004). Active sites in the carboxyl-terminal region of the laminin a chain in Drosophila neuronal cell spreading. *Arch. Insect Biochem. Physiol., 56*(4), 162-169. http://dx.doi.org/10.1002/arch.20006

Tasdemir, Y., Kural, C., Cindoruk, S. S., & Vardar, N. (2006). Assessment of trace element concentrations and their estimated dry deposition fluxes in an urban atmosphere. *Atmospheric Research, 81*(1), 17-35. http://dx.doi.org/10.1016/j.atmosres.2005.10.003

Taylor, M. P., & Kesterton, R. G. H. (2002). Heavy metal contamination of an arid river environment: Gruben River, Namibia. *Geomorphology, 42*, 311-327. http://dx.doi.org/10.1016/S0169-555X(01)00093-9

USEPA. (1995). *Historic hard rock mining: The west's toxic legacy.* USEPA Region 8; EPA 908-F-95-002

USEPA. (2005). *Five-Year Review Report: Second Five-Year Review for the Bunker Hill Mining and Metallurgical Complex Superfund Site, Operable Units 1, 2, and 3, Idaho and Washington.*

Vázquez, S., Carpena, R. O., & Bernal, M. P. (2008). Contribution of heavy metals and As-loaded lupin root mineralization to the availability of the pollutants in multi-contaminated soils. *Environmental Pollution, 152*(2), 373-379. http://dx.doi.org/10.1016/j.envpol.2007.06.018

Wang, Y.-P., Shi, J.-Y., Lin, Q., Chen, X.-C., & Chen, Y.-X. (2007). Heavy metal availability and impact on activity of soil microorganisms along a Cu/Zn contamination gradient. *Journal of Environmental Sciences, 19*(7), 848-853. http://dx.doi.org/10.1016/S1001-0742(07)60141-7

Warnick, S. L., & Bell, H. L. (1969). The acute toxicity of some heavy metals to different species of aquatic insects. *Journal of Water Pollution Cont. Fed., 4*, 280-285.

Wright, D. A., & Zamuda, C. D. (1987). Copper accumulation by two bivalve molluscs: Salinity effect is independent of cupric ion activity. *Marine Environmental Research, 23*(1), 1-14. http://dx.doi.org/10.1016/0141-1136(87)90013-4

Wu, S. C., Luo, Y. M., Cheung, K. C., & Wong, M. H. (2006). Influence of bacteria on Pb and Zn speciation mobility and bioavailability in soil: A laboratory study. *Environmental Pollution, 144*(3), 765-773. http://dx.doi.org/10.1016/j.envpol.2006.02.022

Yadav, S. K., & Rajamani, V. (2006). Air Quality and trace metal chemistry of different size fractions of aerosols in N-NW India—implications for source diversity. *Atmospheric Environment, 40*(4), 698-712. http://dx.doi.org/10.1016/j.atmosenv.2005.10.005

Yager, B. D., & Stanton, M. R. (2000). Multidimensional Spatial Modeling of the May Day Mine Waste Pile, Silverton, Colorado. *Society for Mining, Metalling & Exploration, 1*, 297-303.

Yeddou, A. R., Nadjemi, B., Halet, F., Ould-Dris, A. A., & Capart, R. (2010). Removal of cyanide in aqueous solution by oxidation with hydrogen peroxide in presence of activated carbon prepared from olive stones. *Minerals Engineering, 23*, 32-39. http://dx.doi.org/10.1016/j.mineng.2009.09.009

Yin, C. Q., Shan, B. Q., & Mao, Z. P. (2006). Sustainable water management by using wetlands in catchments with intensive land use. *Wetlands and natural resource management, 190*, 53-65. http://dx.doi.org/10.1007/978-3-540-33187-2_4

Yngard, R., Damrongsiri, S., Osathaphan, K., & Sharma, V. K. (2007). Ferrate(VI) oxidation of zinc-cyanide complex. *Chemosphere, 69*(5), 729-735. http://dx.doi.org/10.1016/j.chemosphere.2007.05.017

Zhang, H., & Selim, H. M. (2005). Kinetics of arsenate adsorption-desorption in soils. *Environ. Sci. Technol., 39*, 6201-6208. http://dx.doi.org/10.1021/es050334u

Zhang, H., & Shan, B. (2008). Historical records of heavy metal accumulation in sediments and the relationship with agricultural intensification in the Yangtze-Huaihe region, China. *Science of the Total Environment, 399*, 113-120. http://dx.doi.org/10.1016/j.scitotenv.2008.03.036

Zhou, J. M., Dang, Z., Cai, M. F., & Liu, C. Q. (2007). Soil heavy metal pollution around the Dabaoshan mine, Guangdong Province, China. *Pedosfhere, 17*(5), 588-594. http://dx.doi.org/10.1016/S1002-0160(07)60069-1

8

Defining Full-Scale Anaerobic Digestion Stability: The Case of Central Weber Sewer Improvement District

Morris E. Demitry[1] & Michael J. McFarland[1]

[1] Utah Water Research Laboratory, USA

Correspondence: Morris E. Demitry, Utah Water Research Laboratory, USA. E-mail: morris.d@aggiemail.usu.edu

Abstract

A full-scale anaerobic digester receiving a mixture of primary and secondary sludge was monitored for one hundred days. A chemical oxygen demand (COD), volatile solids (VS), and mass balance were conducted to evaluate the stability of the digester and its capability of producing methane gas. The COD mass balance could account for nearly 90% of the methane gas produced while the VS mass balance showed that 91% of the organic matter removed resulted in biogas formation. Other parameters monitored included: pH, alkalinity, VFA, and propionic acid. The values of these parameters showed that steady state had occurred. At mesophilic temperature and at steady state performance, the anaerobic digester stability was defined as a constant rate of methane produced per substrate of ΔVS (average rate=0.40 L/g). This constant rate can be used as stability index to determine the anaerobic digestion stability in an easy and inexpensive way.

Keywords: anaerobic digestion, mass balance, renewable energy, steady state, stability, stability index

1. Introduction

Producing renewable energy is a challenge for the world today because it is often more costly than the harvesting of fossil fuels. Finding new and economically sustainable sources of energy to fulfill the world energy demand is a technological and economic challenge. Use of the anaerobic digestion of sludge may represent a cost-effective approach to generate a sustainable and renewable energy source.

Anaerobic digestion produces biogas, which consists primarily of methane (50 to 75% on a volumetric basis) as well as carbon dioxide (25 to 50%). The methane produced from the anaerobic digestion of municipal sludge, animal and crop wastes can cover up to 20% of the natural gas consumption in the US (McCarthy, 1973). The average energy content of biogas is approximately 600 to 800 British Thermal Units (BTUs) per cubic foot (ft^3), which compares favorably to the energy content of natural gas (approximately 1,000 BTUs per ft^3).

A primary benefit of using anaerobic digestion for the generation of renewable energy is that it is a standard sludge treatment process utilized in many municipal wastewater treatment plants. In the anaerobic digestion process, specific groups of facultative and obligate anaerobic microorganisms act in concert to metabolize organic matter associated with sludge, resulting in the production of methane gas. The important groups of microorganisms found in anaerobic digesters include the hydrolytic, acidogenic and methanogenic bacteria (McCarthy, 1964).

Hydrolytic bacteria convert the complex organic matter, like carbohydrates, fats, and proteins to simple compounds like sugar, fatty and amino acids; the acidogenic bacteria are responsible for converting these intermediate compounds to fermentation products including volatile fatty acids (VFA), hydrogen, and carbon dioxide. The methanogenic bacteria utilized the fermentation products to produce methane. One group of methanogenic bacteria, the aceticlastic methanogens, split acetate into methane and carbon dioxide, while the other group, called hydrogen-utilizing methanogens, uses hydrogen and carbon dioxide to produce methane (Turovskiy & Mathai, 2006).

Defining stability is a challenge; many researchers reported different ways in order to indicate stability, but there is no simple and direct definition of the term "stability." The best way to control the anaerobic digestion process is through studying the anaerobic digester steady state besides defining the term 'stability.' Steady state was assumed to be occurring when digesters were operating at or near their controlled and fixed-variable design levels and when gas production and gas rates were relatively constant (Kroeker et al., 1979). Process stability is dependent upon maintenance of the biochemical balance between acid formers and methane formers while instability is usually indicated by a rapid increase in the concentration of volatile acids with a concurrent decrease in methane gas production (Kroeker et al., 1979). Cohen et al. (1981) have discussed the influence of phase separation on the anaerobic digestion stability. Methane reactors with one-phase system and two-phase systems were subjected to gradually increasing feed rate of glucose until the maximum load was reached. The results pointed to the fact that the stability of the two-phase reactor was more than one phase since all the VFA broke down immediately unlike the one-phase reactor (Cohen et al., 1981).

At any rate, the previous studies for stability are confusing since there are several situations that can play a significant role in the anaerobic process's stability. For example, does the stability of the digestion process depend on the digester temperature, mesophilic or thermophilic? Does stability depend on VFA concentrations or un-ionized VFA concentrations or alternatively, does it depend on ammonia toxicity, and what are the toxic concentrations to the microorganisms such as nitrogen? Clearly, defining stability is a challenge, because there is no simple and direct definition of the term "stability."

Failure to establish a reproducible digester stability metric(s) could result in catastrophic failure of the anaerobic digestion process as well as impairment in the discharged water quality.

In this study, the performance of a full-scale anaerobic digester operating at mesophilic temperatures (i.e., 36 °C or 98 °F) has been monitored for over one hundred (100) days. Collection and analysis of operational data from the anaerobic digesters at Central Weber Sewer Improvement District, Ogden, Utah, served as the scientific basis for defining stability. The goal of the study was to establish and quantify the range of specific operational parameters that could define digester operational stability. Enhancing the production of biogas from the digestion of sludge and other organic matter requires the development of a simple and cost-effective performance tool that can gauge the stability of the digester environment.

2. Objectives

(1) Collecting the digester's operational data including biogas production, percent methane in biogas, total solids, volatile solids destruction, influent and effluent chemical oxygen demand, digester pH, alkalinity, and volatile fatty acid concentrations in order to study steady digester operation.

(2) Using statistical analysis for the operational parameter behavior to determine a universal performance metric (stability index) that reflects steady state for the digester operation.

3. Background about Central Weber Sewer Improvement District

Central Weber Sewer Improvement District (CWSID) is located at 2618 West Pioneer Road, Ogden, Utah, 84404. It provides service for approximately 200,000 people in Weber and Davis counties. The plant was constructed in 1957. The existing treatment facility had a rated capacity of 45 million gallons per day (MGD), using a single-stage trickling filter process. Project upgrades completed in 2011, included construction of a new parallel 30-MGD activated sludge treatment plant, a new headwork's facility and a new raw sludge pump station. Focus was placed on value engineering directed at emerging areas of design where improvements could be made to reduce construction costs without affecting the process design or overall finished product.

The upgrades increased the treatment capacity to 70 MGD, supporting the District's goal of accommodating projected population growth in Davis and Weber Counties until 2025. The facility was also brought into compliance with current Environmental Protection Agency (EPA) and State of Utah regulatory requirements (CWSID, 2011).

4. Literature Review

One of the important parameters is the pH, which is defined as the negative logarithm of the hydrogen-ion concentration (Tchobanoglous et al., 2003). An important environmental parameter, pH indicates if the environment is healthy for the microorganisms in the anaerobic digester. The pH should be around neutral (or pH=7) according to McCarthy (1964), while Turovskiy & Mathai (2006) mentioned that the anaerobic microorganisms are sensitive to changes in pH lower than 6.8 and higher than 7.2. The pH inside the digester should be in the range of 6.8- 7.2 in order to keep the microorganisms in a healthy environment.

Due to the chemical reactions inside the anaerobic digester, the volatile fatty acids like acetic, propionic, valeric and butyric acids may accumulate as a result of a drop in the pH. The drop in the pH may occur because the carbon dioxide ranges between 30-50% of the produced biogas; the carbon dioxide may react with the water and form H_2CO_3, which leads to a drop in pH.

In case an insufficient buffer is present, the pH is subjected to a sudden drop, and that will affect the anaerobic digester's microorganism groups especially methanogensis. Methanogenesis archaea will not be able to convert the hydrogen and acetic acid to biogas and that will cause the accumulation of VFA.

The buffering capacity (alkalinity) of the system is important to avoid a sudden drop in pH. Alkalinity in water and wastewater results from the presence of hydroxide [OH$^-$], carbonate [CO$_3^{-2}$], and bicarbonate [HCO$_3^-$]. Alkalinity concentration is an important factor for the anaerobic digester; alkalinity in the range between 2000 to 4000 mg/L as CaCO$_3$ is typically required to maintain the pH at or near the optimum value for the anaerobic digester (Turovskiy & Mathai, 2006).

Another important parameter for the anaerobic digester is temperature. Usually anaerobic microorganisms are sensitive to the temperature in the anaerobic digester. Anaerobic digesters can be operated at different ranges of temperature like mesophilic (30-40°C), for best results. The important factor is to avoid sharp and frequent fluctuations in temperature in order to keep the methanogen microorganisms working in a healthy environment (Arsova, 2010).

Wang et al. (2009) discussed the effects of VFA concentration on methanogen microorganisms and methane yield within anaerobic digesters. The result from this study confirmed that, when the highest concentrations of ethanol, acetic and butyric acid were 2400, 2400 and 1800 mg/L respectively, there was no significant inhibition in the activity of the methane bacteria. However, when the propionic acid concentrations had been increased from 300 to 900 mg/L, a significant inhibition appeared, and due to that, the methanogens bacteria concentration decreased from $6*10^7$ to $1*10^7$ mg/L. Wang et al. (2009) also discussed the effects of VFA concentration on methane yield and methanogen microorganism; these effects demonstrated the accumulation of ethanol and VFA while methane yield becomes very low. Galert & Winter (2006) additionally examined the propionic acid accumulation and degradation during restart of full-scale anaerobic digesters. Their results confirmed that an increase in VFA due to the increase of the organic wastes between the periods of 7 to 28 days, leads to a decrease of methane gas and a drop in pH values from 7.5 to 7.1. During this period of the restart of the full-scale anaerobic reactor, the propionic acid reached its maximum concentration of 6.2 g/L; after 5 days the propionic acid was degradable completely and the pH increased from 7.1 to 7.4; and methane content increased from 60% to 65%.

Other researchers concluded that VFA themselves are toxic to methane archaea at concentrations above 2000 mg/L (Buswell and McKinney, 1962), while McCarthy confirmed that VFA were not toxic to methanogensis bacteria at concentrations that occur in malfunctioning digesters (1964).

Accumulation of propionic acid above 300 mg/L may affect the methanogens. On the other hand, if propionic acid concentration is less than 300 mg/L, that indicates the methanogens are not stressed and are functioning under good conditions (Gallert & Winter, 2006).

In the anaerobic digestion process, Chemical Oxygen Demand (COD) usually is the best way to track the energy flow during biological oxidation of sludge; the test uses oxidize agent to oxidize organic compounds to carbon dioxide.

COD mass balance can be used to account for the changes in COD during digestion. The COD removed in the anaerobic digester is accounted for by the biogas production as shown in the mass balance equation below:

$$\text{COD }_{in} - \text{COD }_{out} = \text{COD (Biogas)} \tag{1}$$

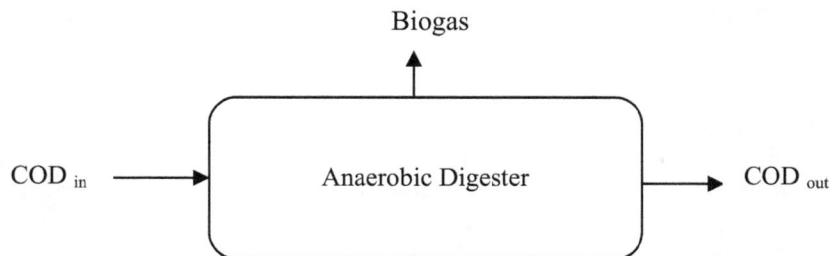

Figure 1. Schematic diagram for the flow through anaerobic digester

The COD mass balance equation is able to estimate methane production if other terms were measured.

Equation 1 is used to determine the methane gas production from the anaerobic digester at CWSID after COD removed was measured.

Sötemann et al. (2010) studied the steady-state model for the anaerobic digestion of sewage sludge, applying mass balance equation and measuring total COD from the sludge flow. The samples were taken from the influent and effluent side of the four laboratory reactors, using wastewater from a plant in Cape Town, South Africa. The results confirmed that 96, 100, 95, and 99% of the total COD had been recovered for the four lab reactors.

In this paper, the total COD concentrations and volatile solids in municipal (primary and secondary) sludge were monitored, in order to study the anaerobic digester performance at CWSID. Other important parameters were also measured for the same purpose (pH, alkalinity, VFA, and propionic acid). All the analysis and measurements are discussed in full detail.

5. Materials and Methods

In order to monitor the performance of the anaerobic digester operation, influent and effluent sludge samples were taken from a mesophilic digester operating at a 20-day hydraulic retention time. Duplicate influent and effluent sludge samples (ca. 500 milliliters) were analyzed for total solids; VS and COD twice per week using Environmental Protection Agency method (EPA, Method 1684). All sludge samples were collected in plastic bottles (500 milliliters) and mixed gently by inverting the bottles several times.

The percent total solids (TS%) consist of the solid residue remaining after the sludge sample had been evaporated and dried at 105°C. To measure percent total solids, approximately fifteen (15) milliliters of sample was placed on a pre-weighted fiberglass pad and then heated to 105°C (for 30 minutes) in a CEM microwave instrument (Model CEM001; Matthews, North Carolina). Percent volatile solids (VS%), which is the percentage of the total solids that can be volatilized at 550°C, was measured by taking the total solids sample and placing it in a muffle furnace set at 550°C for two hours (EPA, 2012). The remaining ash was measured and recorded to determine the percent volatile solids.

COD of influent and effluent sludge samples was measured using a spectrophotometer (HACH 8000), with accuracy ±5%.

In addition to total solids, volatile solids and COD, effluent sludge samples (ca. 500 milliliters) were taken twice a week to monitor digester pH, alkalinity, and volatile fatty acid concentrations. The pH was measured using a pH meter (Orion 001, Model 230 A-Cole Parmer, Inc. Vernon Hills, Illinois) that was calibrated using pH buffer solutions of 4 and 10 (sodium bicarbonate, RICCA Chemical Company). The accuracy of the pH meter was ±0.02 pH units. Alkalinity measurements were conducted according to Standard Methods 2320B using an automated titration system (METER TOLEDO, Columbus, OH) having an accuracy of ± 0.02 milligrams per liter as $CaCO_3$. Prior to the analysis, the pH meter was calibrated using 4, 7, and 10 buffer standards (potassium acid, potassium phosphate and sodium bicarbonate respectively).

Biogas generation (cubic feet per minute), percent carbon dioxide, and hydrogen sulfide concentrations in biogas were measured twice per week. Biogas was measured using a gas flow meter (Sierra instrument company Model 640S-NAA-L09-M1-E2-P3-V4-DD-5 L Monterey, CA 93940). To measure the concentration of carbon dioxide and hydrogen sulfide in biogas, a one-liter sample of biogas was collected from the digester using a sealed polyvinyl fluoride (PVF) TedlarTM sampling bag. Dragger tubes (model, D-23560, Lubeck, Germany) were used to measure the concentration of carbon dioxide and hydrogen sulfide in the biogas. The accuracy of the dragger tube was ±5% for both kinds of tubes.

To measure volatile fatty acids (VFA) (acetic, propionic, butyric and valeric), 500 milliliter sludge samples were taken from the effluent side of the digester. Total volatile fatty acids were measured using a distillation method technique number 5550 C (APHA, 2012). From the sample, 200 ml was centrifuged for five (5) minutes. After that, 100 milliliter supernatant liquid was placed in a 500-milliliter distillation flask. Next, 100 milliliters of distilled water was added to the solution along with 0.3 grams of Polytetrafluoroethylene (PTFE) boiling stones and 5 milliliters of 95.9% sulfuric acid. The solution was mixed by inverting the bottle upside down several times, and then 150 milliliters of solution was placed in a 250 milliliter graduated cylinder. The solution was titrated with 0.1N NaOH and expressed as acetic acid content.

Propionic acid was measured in effluent sludge samples two times every week. The sludge samples were collected in a plastic bottle (500 milliliters) and preserved at 5°C. Within 24 hours, the samples were measured for propionic acid using a ThermoFisherTM ICS-5000 chromatograph equipped with an AS18-4um, 4X150mm capillary column and a thermal conductivity detector. The standards used to determine the detection limits for the various acids ranged from 0.5ppm to 2ppm.

6. Results and Discussion

Table 1 shows the results for pH, alkalinity, propionic and VFA respectively during one hundred days of study.

The results show stable performance during the period of study since the average pH was 7.31±0.12, and alkalinity at 4113±229 mg/L as $CaCO_3$, which indicates stable and optimum performance for the digester. Moreover, average propionic acid was 29.38±7.89 mg/L, while the VFA average was 65.72±14 mg/L, which confirms the supreme performance of the digester at CWSID.

Table 1. pH, Alkalinity, Propionic and VFA results

Process	pH	Alkalinity as $CaCO_3$(mg/L)	Propionic (mg/L)	VFA(mg/L)
Day 1	7.40	4275.00	12.90	26.40
Day 3	7.46	3900.00	27.54	69.40
Day 8	7.32	3550.00	25.80	69.11
Day 10	7.25	3892.00	33.00	55.50
Day 15	7.43	4125.00	24.96	55.00
Day 17	7.39	4200.00	15.84	41.60
Day 22	7.21	4450.00	26.40	52.00
Day 24	7.26	4325.00	32.64	55.50
Day 29	7.34	4350.00	41.47	66.20
Day 31	7.30	3825.00	31.20	69.11
Day 36	7.39	4125.00	20.40	41.40
Day 38	7.39	3562.50	24.18	54.40
Day 44	7.30	3992.00	41.64	86.11
Day 46	7.42	4200.00	31.20	69.40
Day 52	7.39	4430.00	45.60	78.23
Day54	7.26	4120.00	30.60	72.25
Day 59	7.36	4245.00	24.84	69.40
Day 63	7.37	4075.00	33.30	80.30
Day 68	7.29	3994.00	33.00	70.50
Day 70	7.05	4275.00	22.59	76.98
Day 75	7.34	4170.00	29.16	78.19
Day 77	7.07	4215.00	23.64	65.34
Day 82	7.00	4214.00	30.11	81.18
Day 84	7.35	4200.00	42.26	80.35
Day 90	7.41	4140.00	30.18	79.21
Average	*7.31*	*4113.98*	*29.38*	*65.72*
SD±	*0.12*	*229.14*	*7.89*	*14.71*

Note: SD = Standard Deviation

Table 2. COD, Equivalent CH_4, Actual CH_4 and Percentage recovery

Process	[1]COD inf (mg/L)	[2]COD eff (mg/L)	[3] net COD (lb/L)	[4]Equivalent CH_4 (Ft^3/d)	[5]Actual CH_4 (Ft^3/d)	[6]Percentage %
Day 1	64930	30850	16371.49	102338.00	90923	88.85
Day 3	65610	33450	15449.15	96572.00	87043	90.13
Day 8	85160	24490	29144.90	182184.00	166011	91.12
Day 10	71294	24895	22289.34	139330.00	122821	88.15
Day 15	73000	23147	23948.58	149702.00	128999	86.17
Day 17	81245	27450	25842.26	161539.00	143999	89.14
Day 22	79745	25575	26022.40	162666.00	145003	89.14
Day 24	83230	27860	26598.86	166269.00	148215	89.14
Day 29	87450	32125	26577.24	166134.00	153031	92.11
Day 31	92090	33970	27919.92	174527.00	160762	92.11
Day 36	90950	33815	27446.74	171569.00	158037	92.11
Day 38	87845	29375	28088.05	175578.00	153035	87.16
Day 44	98400	31375	32197.74	201268.00	179413	89.14
Day 46	89175	35125	25964.76	162305.00	139859	86.17
Day 52	88067	34075	25936.89	162131.00	141315	87.16
Day54	85800	28295	27624.48	172680.00	157351	91.12
Day 59	77125	23200	25904.71	161930.00	145951	90.13
Day 63	81500	22890	28155.31	175998.00	158631	90.13
Day 68	68500	28875	19035.22	118980.00	107247	90.13
Day 70	84437.5	26925	27628.08	172703.00	152239	88.15
Day 75	74593	22754.5	24902.39	155664.00	138762	89.14
Day 77	96580	32393	30834.41	192745.00	173725	90.13
Day 82	91885	32916.5	28327.52	177075.00	159601	90.13
Day 84	91880	31883	28371.26	177505.85	160128	90.21
Day 90	90905	30831	28407.67	177733.66	161276	90.74
Average	*83255.86*	*29141.6*	*25959.57*	*162285.06*	*145335.08*	*89.51*
SD±	*9365.5*	*4029.20*	*3984.32*	*24914.64*	*22950.07*	*1.68*

Note: SD = Standard Deviation

The mass balance for COD has been calculated in order to determine the methane gas from COD (equivalent COD) and to compare it with the actual methane gas produced from the digester. As mentioned before, the actual gas has been measured using the flow meter. Percentage recovery between equivalent and actual methane was calculated as shown in Table 2.

The average percentage recovery was 89.51% ±1.68; the anaerobic digester was successful in producing a renewable energy (methane gas beside carbon dioxide) and converting the organic wastes (COD) to methane with 89.51% recovery. This ratio, (89.51%) of recovery, demonstrated the superior performance of the digester and also demonstrated the stability and steady state for the anaerobic digester.

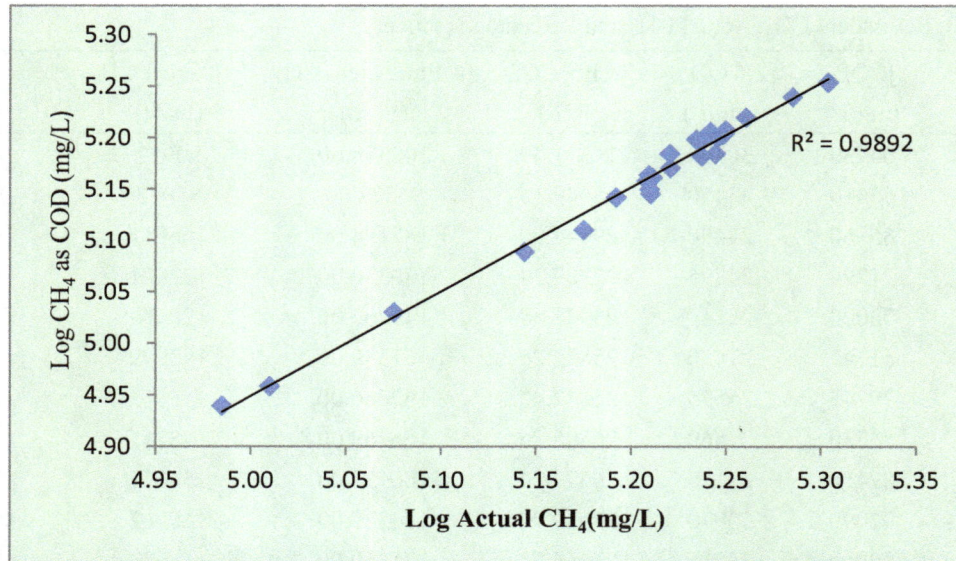

Figure 2. The relationship between theoretical (CH₄ as COD) and actual methane gas

Figure 2 shows the relationship between theoretical (CH$_4$ as COD) and actual CH$_4$; linear relationship and high correlation between the two variables were noticed. The data was transformed to log transformation to improve data interpretability. The difference between the two averages has been calculated by applying t-test function using R programming.

At 99.96% confidence, there was no difference between theoretical and actual mean of the methane determined and produced, demonstrating the steady state and stability of the anaerobic digester at CWSID.

The percentage of VS destroyed (ΔVS) was determined and converted to equivalent CH$_4$ during the period of study; the results and percentage recovery of methane gas were determined and displayed in Table 3.

Table 3. Equivalent CH_4, Actual CH_4 and percentage recovery results

Process	CH_4 as VS (L/d)	Equivalent CH_4 (Ft^3/d)	Act CH_4(Ft^3/d)	Recovery %
Day 1	2,982,435.00	105,386	90923	86.28
Day 3	2,853,506.22	100,831	87043	86.33
Day 8	4,850,354.40	171,391	166011	96.86
Day 10	3,994,021.14	141,131	122821	87.03
Day 15	3,935,884.01	139,077	128999	92.75
Day 17	4,233,591.29	149,597	143999	96.26
Day 22	4,194,233.60	148,206	145003	97.84
Day 24	4,579,952.00	161,836	148215	91.58
Day 29	4,745,208.00	167,675	153031	91.27
Day 31	5,076,446.40	179,380	160762	89.62
Day 36	4,703,089.51	166,187	158037	95.10
Day 38	4,987,000.00	176,219	153035	86.84
Day 44	5,876,765.00	207,660	179413	86.40
Day 46	4,621,902.20	163,318	139859	85.64
Day 52	4,825,382.17	170,508	141315	82.88
Day54	4,533,550.27	160,196	157351	98.22
Day 59	4,911,462.28	173,550	145951	84.10
Day 63	4,621,902.20	163,318	158631	97.13
Day 68	3,726,590.32	131,682	107247	81.44
Day 70	4,396,687.40	155,360	152239	97.99
Day 75	4,848,107.33	171,311	138762	81.00
Day 77	4,963,958.00	175,405	173725	99.04
Day 82	5,392,226.00	190,538	159601	83.76
Day 84	5,187,218.00	183,294	160128	87.36
Day 90	4,678,750.00	165,327	161276	97.55
Average	*4548808.91*	*160735.30*	*145335.08*	*90.41*
SD±	*675158.30*	*23857.18*	*22950.07*	*6.01*

Note: SD = Standard Deviation

The mass balance for the ΔVS was calculated; the equivalent amount of methane gas from ΔVS has been calculated, and the percentage recovery was determined. High percentage recovery was noticed (90±6) %. This result confirmed again the stability and the steady situation of the digester. The relationship between actual and equivalent methane is plotted in Figure 3.

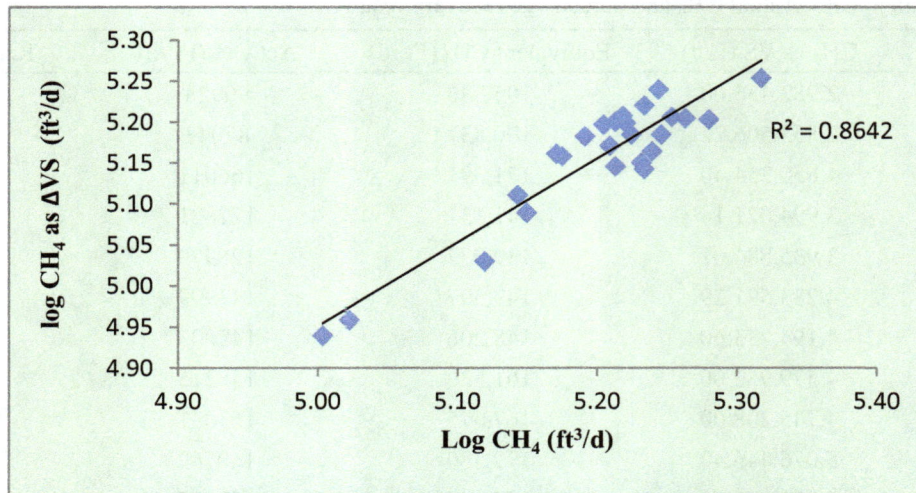

Figure 3. The relationship between theoretical (CH_4 as ΔVS) and actual CH_4

Linear relationship with strong correlation was demonstrated, t-test using R programming was applied in order to determine the difference between the two averages. At 96% confidence, there was no difference between the two means, which leads the research to confirm the stability and steady state of the digester.

Stability means stable performance during period of time. For more clarification about the stability of the digester at CWSID, Figure 4 below shows the variations of the pH, alkalinity, propionic, VFA and COD, respectively, and was plotted with time (days) to study the anaerobic digester stability.

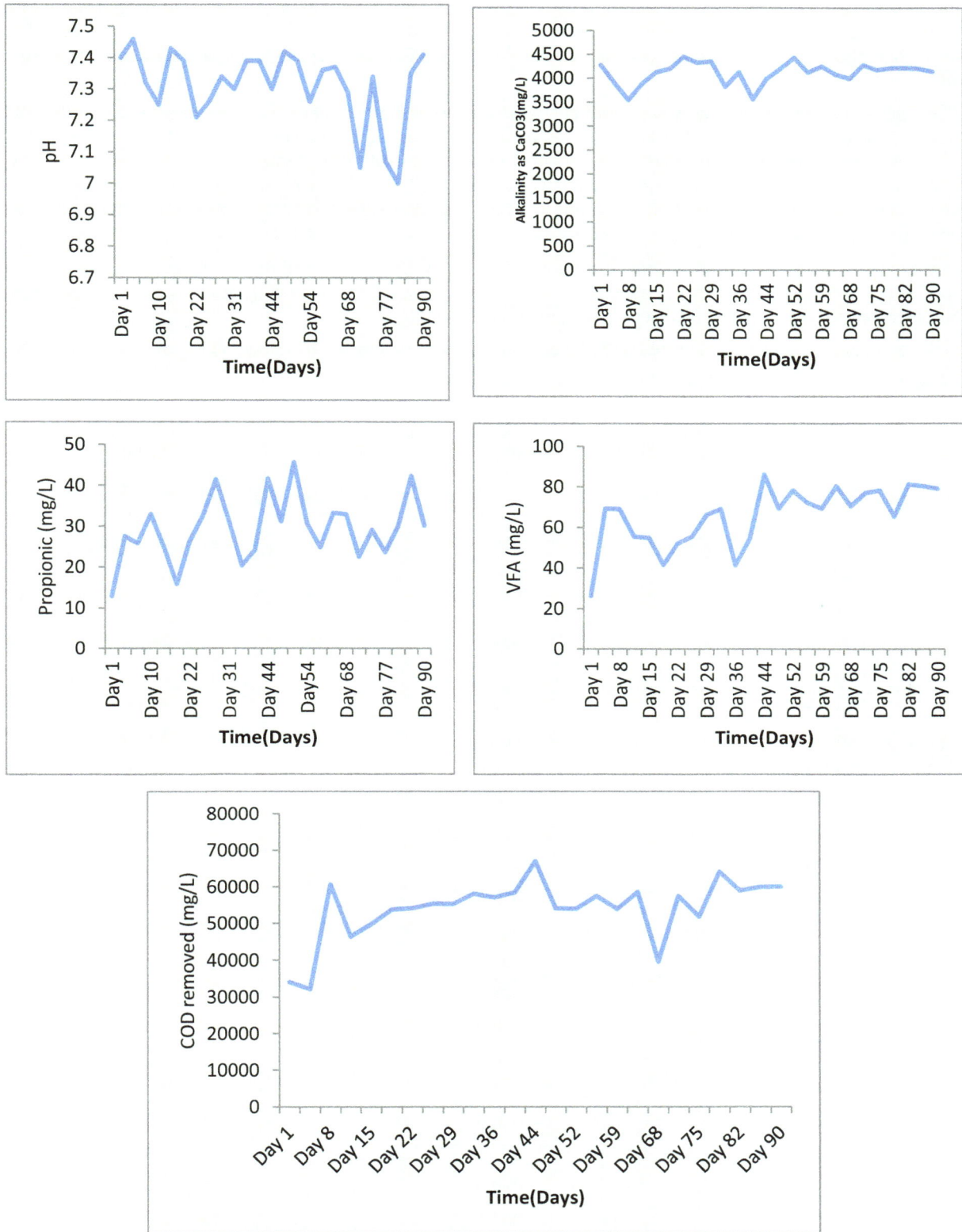

Figure 4. pH, alkalinity, Propionic acid, VFA and COD variation with time

In Figure 4, there is no significant variation noticed for the monitored parameters (pH, alkalinity, propionic acid and COD) over time; all parameters vary within the allowable range for each parameter. For example, maximum pH was 7.46, and minimum pH was 7.0. Alkalinity readings vary between 3500mg/L to 4450mg/L. Accordingly, pH is considered neutral, and the alkalinity results reflected strong buffering capacity to the change in pH inside the digester. Moreover, stable variation in both VFA and propionic acid within the period of time was noticed, which demonstrates the stable rate of converting these intermediate products to acetic acid and hydrogen. The stable rate of conversion keeps the dynamic relationship between the acidogensis bacteria and the methaongensis archaea in good status.

The digester is considered to be at a steady-state condition because it was operating at or near the controlled and fixed-variable designed levels. Furthermore, gas production rates were relatively constant during the period of study. According to that, a universal metric function was determined to define the anaerobic digestion stability. The rate between methane gas produced from the digester and ΔVS in liter per gram has been determined during one hundred days of study as shown in Table 4. Daily rate of (0.40 ± 0.017)L/g has been remarkable, which demonstrates that stability is achievable as long as the constant rate of (0.4 ± 0.017)L/g is reached or maintained.

The rate of CH$_4$/ΔVS (L/g) can be used as a universal metric to indicate the stability of the anaerobic digester as applied at CWSID. Because ΔVS and methane gas are required to be measured daily at the wastewater treatment facilities, only two parameters can be used to examine the stability.

In Table 4 the rate has been calculated and plotted with propionic acids that are shown in Figure 5.

Table 4. Stability index (CH$_4$/ΔVS (L/g)) and propionic acid results

Process	CH$_4$/ΔVS(L/g)	Propionic (mg/L)
Day 1	0.439	12.90
Day 3	0.400	27.54
Day 8	0.404	25.80
Day 10	0.391	33.00
Day 15	0.407	24.84
Day 17	0.436	15.84
Day 22	0.400	26.40
Day 24	0.394	31.200
Day 29	0.388	41.64
Day 31	0.397	30.60
Day 36	0.430	20.40
Day 38	0.420	24.18
Day 44	0.388	42.26
Day 46	0.395	31.20
Day 52	0.386	45.60
Day54	0.398	30.18
Day 59	0.415	24.96
Day 63	0.391	33.30
Day 68	0.394	32.64
Day 70	0.429	22.59
Day 75	0.400	29.16
Day 77	0.424	23.64
Day 82	0.391	33.6
Day 84	0.389	41.47
Day 90	0.391	33.00
Average	*0.404*	*29.51*
SD±	*0.017*	*7.93*

Note: SD=Standard Deviation

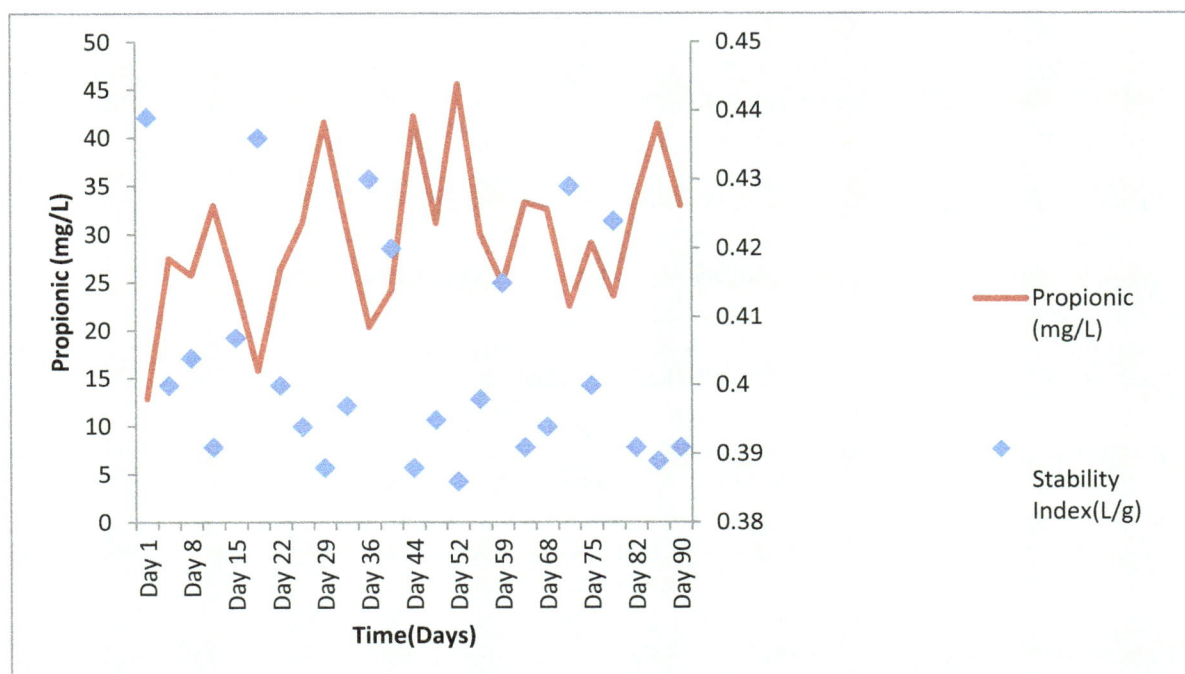

Figure 5. The relationship between propionic acid and stability index during the study

Direct relationship between the stability index ($CH_4/\Delta VS$) and propionic acid was observed, as shown in Figure 5. Inverse proportion between the two variables was noticed. An increase in propionic acid will affect the rate of methane gas produced per ΔVS ($CH_4/\Delta VS$). However, the increase in the stability index indicates low accumulation in the propionic acid inside the digester. Methanogensis archaea may get stressed partially when propionic acid accumulates and reaches 45 mg/L, which causes the low stability index readings as shown in Figure 5.

7. Conclusion

In this paper, full-scale anaerobic digester stability at CWSID was tested and monitored during one hundred days of study. The municipal primary mixed with secondary sludge was characterized as COD.

Snap shots of the anaerobic digester parameters during the period of study were monitored. The COD mass balance was applied to the anaerobic digester in order to study its stability and its capability of producing methane gas. The anaerobic digester mass balance showed promising results in terms of wastewater treatment and energy production. There was a 10% loss of the methane gas (the best gas recorded was 90% of the organic wastes loaded). Mass balance of ΔVS was calculated, and 91% recovery was possible. Essentially, this research indicates that anaerobic digesters are a good source of renewable energy.

The monitored parameters for the anaerobic digester were pH, alkalinity, VFA, and propionic acid. All the results confirmed a superior performance for the anaerobic digester.

Finally, at mesophilic temperature and steady state performance, anaerobic digester stability has been defined as a constant rate of methane produced per substrate of ΔVS (average rate = 0.40 L/g). This definition (the stability index) can be used as a new and inexpensive way to define and examine the anaerobic digestion stability. Since defining "stability" was considered an initial problem, this research also furthered the ability to define or redefine it more simply by using the consistent results of this study.

Acknowledgements

- Utah Water Research Laboratory for their financial support and supervision.
- Central Weber Sewer Improvement District for day-to-day effort to complete the work and for the permission to present the data.

References

APHA. (1989). *Standard Methods for Examination of Water and Wastewater* (17th ed.). American Public Health

Association, Washington, DC.

Arsova, L. (2010). Anaerobic digestion of food waste: Current status, problems and alternative products. (Master's thesis, Colombia University, NY, USA).

Buswell, A. M. (1947). Important Consideration in Sludge Digestion and theory of anaerobic digestion. *Sewer Works Journal, 19,* 28-36. Retrieved from http://www.jstor.org/discover/10.2307/25030399?sid=21105058201931&uid=3739256&uid=2&uid=3739928&uid=4

Central Weber Sewer District, expansion projects. (2011). http://www.centralweber.com

Cohen, A., Breure, A. M., Van Andel, J. G., & van Deursen, A. (1981). Influence of phase separation on the anaerobic digestion of glucose-II. *Journal of water resources, 16,* 449-455. Retrieved from file:///C:/Users/Morris%20Dimitry/Downloads/cohen%20et%20al.%201982%20WR%2016,%20449-455.pdf

Environmental Protection Agency (EPA) method 300. (2014). http://water.epa.gov/scitech/methods/cwa/bioindicators/upload/2007_07_10_methods_method_300_0.pdf

Galert, G., & Winter, J. (2006). Propionic Acid accumulation and degradation during restart of a full-scale anaerobic biowaste digester. *Journal of Bioresource Technology, 99,* 170-178. Retrieved from http://www.sciencedirect.com/science/article/pii/S0960852406006055

HACH company method 8000. (2014). http://www.hach.com/quick.search-product.search.jsa?keywords=Method%208000&pr.parameterId=7639975163&pr.isNew=false&pr.pimContext=USen

Kroeker, E. J., Schulte, D. D., Sparling, A. B., & Lapp, H. M. (1979). *Journal of Water Pollution Control Federation, 51,* 718-727. Retrieved from http://www.jstor.org/discover/10.2307/25039893?sid=21105058375611&uid=3739928&uid=2&uid=3739256&uid=4

McCarthy, P. L. (1973). Methane Fermentation-Future Promise or Relic of the Past. Proceedings of the Bioconversion Energy Research Conference, for the National Science Foundation. Amherst, Mass.

McCarty, P. L. (1964). Anaerobic Waste Treatment fundamentals. *Chemistry and biology, 5,* 107-112. Retrieved from http://seas.ucla.edu/stenstro/Anaerobic%20assignment.pdf

McCarty, P. L. (1964). Anaerobic waste treatment fundamentals. *Public works, 95*(9), 107-112.

Stoweman, S. W., Ristow, N. E., Wentzel, M. C., & Ekama, G. A. (2005). A steady state model for the anaerobic digestion of sewage sludge. *Ethiopian journal of science, 31,* 511-528. Retrieved from http://www.ajol.info/index.php/wsa/article/view/5143

Tchobanoglous, G., Burton, F. L., & Stensel, H. D. (2003). *Wastewater Engineering, Treatment and Reuse.* New York, NY.

Tourovskiy, I. S., & Mathai, P. K. (2006). *Wastewater Sludge Processing.* Hoboken, New Jersey. http://dx.doi.org/10.1002/047179161X

Wang, Y. Y., Zhong, Y. L., Wang, J. B., & Meng, L. (2009). Effects of Volatile Fatty Acids concentration on methane yield and methanogenic bacteria. *Journal of Biomass and Bioenergy, 33,* 848-853. Retrieved from http://www.sciencedirect.com/science/article/pii/S096195340900018X

9

Temporal Changes in Environmental Health Risks and Socio-Psychological Status in Areas Affected by the 2011 Tsunami in Ishinomaki, Japan

Kohei Makita[1], Kazuto Inoshita[1], Taishi Kayano[1], Kei Uenoyama[1], Katsuro Hagiwara[1], Mitsuhiko Asakawa[1], Kenta Ogawa[2], Shin'ya Kawamura[3], Jun Noda[1], Koichiro Sera[4], Hitoshi Sasaki[2], Nobutake Nakatani[2], Hidetoshi Higuchi[1], Naohito Ishikawa[5], Hidetomo Iwano[1] & Yutaka Tamura[1]

[1] School of Veterinary Medicine, Rakuno Gakuen University, 582 Bunkyodai Midorimachi, Ebetsu, Japan

[2] College of Agriculture, Food and Environmental Sciences, Rakuno Gakuen University, Bunkyodai Midorimachi, Ebetsu, Japan

[3] Department of Human Sciences, Graduate School of Letters, Hokkaido University, Kita-ku, Sapporo, Japan

[4] Cyclotron Research Center, Iwate Medical University, Takizawa, Iwate, Japan

[5] Action for Peace, Capability and Sustainability (APCAS), Colombo, Sri Lanka

Correspondence: Kohei Makita, School of Veterinary Medicine, Rakuno Gakuen University, Ebetsu 069-8501, Japan. E-mail: kmakita@rakuno.ac.jp

Abstract

On March 11 2011, a tsunami caused by a magnitude 9.0 earthquake devastated the northeastern coast of Honshu, Japan. The present study was conducted to assess environmental health risks of the areas affected and socio-psychological status of the dislocated people in Ishinomaki.

Samples of sludge, water, flies and rodents were collected in 20 urban neighborhood associations affected by the tsunami in July and August 2011, and in August 2012. A socio-psychological survey was conducted in two urban and one rural temporary housing complexes in 2012. Animal feed concentrates and fish from damaged factories were scattered along the coast which caused a strong odor and great number of flies. Removal of fish and feed along with spraying insecticides reduced the odor and the number of flies by August 2011. The sludge and water samples contained potentially hazardous bacteria, but none were highly pathogenic. Heavy metals in sludge were not in alarming quantities. A rodent was captured in one unit in August 2011, and monitoring in two units found that the log number of rodents captured increased significantly over time (slope=0.08, $p = 0.005$). In temporary housing complexes, those who originally lived in rural fishing villages wished to return to their homes more (64.2%, 9/14) than in urban areas (30.6%, 11/36, $p = 0.06$). Risk factors for depression included absence of friends ($p = 0.011$) or trusted person to counsel in the housing complexes ($p = 0.003$) and illness of the respondent or a family member ($p = 0.003$). In conclusion, overall environmental health risk was acceptable for living, and monitoring of rodents population was recommended. In addition, psychological and economical support was needed for evacuees in temporary housing complexes.

Keywords: The 2011 Tohoku-Oki Tsunami, earthquake, environmental health risk assessment, insect infestation, heavy metals, mental stress, rodent infestation, disaster

1. Introduction

A magnitude 9.0 earthquake occurred offshore of the northeast coast of Honshu, Japan at 14: 46 pm on March 11, 2011, generating a devastating tsunami that destroyed many towns and villages near the seashore in Iwate, Miyagi and Fukushima prefectures (Shibahara, 2011). The earthquake was named the 2011 Off the Pacific Coast of Tohoku Earthquake by the Japan Meteorological Agency (JMA, 2011) and the Japanese cabinet named the collective disasters caused by this earthquake the Great East Japan Earthquake (PMJHC, 2011). The total number of deaths caused by the disaster totaled 15883, and 2681 people were still missing by April 1, 2013 (NPA, 2013). The coastal city of Ishinomaki in Miyagi Prefecture (Figure 1) was severely affected by the tsunami, which flooded about 13% of the city where 70% of the population lived (Ueda, Hanzawa, Shibata, &

Suzuki, 2012). Table 1 summarizes the damage and health problems experienced by inhabitants of Ishinomaki. The large scale tsunami warning was announced at 14: 49 pm and at 15: 26 pm, more than 8.6 m of tsunami reached to Ishinomaki (JMA, 2011). In the affected areas by the tsunami overall in Japan, 62.6% of people started evacuation before the large scale tsunami arrived, 10.6% after arrived, and 26.8% did not evacuate, based on a study with 4,421 respondents (MLITT, 2011). Ishinomaki lost the lives of 3506 people, which accounted for 22% of total deaths in Japan (Table 1). The tsunami necessitated the evacuation of 31.6% of the city population (peak on March 17, 2011) to disaster shelters. Immediately after the tsunami, the Self Defense Force of Japan, together with foreign rescue teams including US military forces, began rescuing people trapped in debris, searched for those who were missing or presumed dead and restored life lines to the affected areas. There were domestic and international financial and material donations through United Nations, Red Cross and other NGOs. Volunteers and NPOs also played critical roles in restoring the city, providing meals and caring for evacuees. Evacuees were encouraged to move to rental flats and rapidly constructed temporary housing complexes, and all disaster shelters were closed by October 11, 2011, 7 months after the earthquake. The supplying period of the temporary houses had been for two years but was extended another year until 2014 (Miyagi Prefecture, 2012). The government is reclaiming uplands for relocation from coastal residential areas and creating employment opportunities (Reconstruction Agency, 2013). The other political supports include partial financial supports for lost and damaged houses, exemption of tax and interests for a double loan (a loan for new house construction in addition to the remaining loan), and supports for families with children and those who lost employment, and mental and physical health (Miyagi Prefecture, 2012).

The clinical phases of natural disasters can be classified as follows (Aghababian & Teuscher, 1992; Kouadio et al., 2012): phase 1 (impact phase, 0-4 days), initial treatment of disaster-related injuries is provided; phase 2 (post-impact phase, 4 days to 4 weeks), the first waves of air-, food-, and/or water-borne infectious diseases emerge as a substantial portion of the population is displaced into unplanned and overcrowded shelters; and phase 3 (recovery phase, after 4 weeks), infections with long incubation periods or latent types can become clinically apparent.

In Ishinomaki, in the phase 1, dehydration, vomiting and diarrhea were frequently observed due to lack of clean water and hygiene. In the phase 2, seven cases of tetanus, two cases of legionellosis, small outbreaks of diarrhea, respiratory diseases and deep vein thrombosis resulted from dehydration and restricted movement in crowded disaster shelters were reported (Table 1). There were 259 disaster shelters in Ishinomaki. In the tsunami-flooded areas, shelters were very crowded with approximately 1000 evacuees per shelter. Most of the shelters provided 10 square feet per person for many weeks and at most 20 square feet per person for several months (Picture 1) (Ueda et al., 2012).

While public health authorities concentrated on the health status and treatment of evacuees in the disaster shelters, environmental hygiene received little focus. Poor environmental hygiene can be a source of infectious and non-infectious diseases not only for evacuees in shelters, but also for those who return to damaged homes or move into rented rooms or temporary housing complexes (Picture 2). The most frequently reported infectious diseases associated with a tsunami and floods in the world are diarrhea, hepatitis A and E, acute respiratory infections (ARIs), measles, meningitis, tetanus, cholera, and leptospirosis (Kouadio et al., 2012). Although international data are scarce, levels of heavy metals were increased by the tsunamis in 2011 in Japan (Baba & Sera, 2012) and in 2004 in India (Ranjan, Ramanathan, Singh, & Chidambaram, 2008) and such contaminants in soil might pose public health risks unless long-term exposure and internal accumulation are prevented. The present study assesses environmental health risks due to microbiological agents, wildlife and chemical hazards in the areas of Ishinomaki affected by the tsunami to provide information that can be used to plan reconstruction policies. As a risk is assessed by the combination of probability of occurrence of a scenario that will affect humans and the size of the impact (Vose, 2008), we also conducted a sociological survey to assess the desires of evacuees in temporary housing complexes to return the areas affected by the tsunami. The mental health status of the residents was also surveyed to provide adequate and timely policy support.

The locations and needs of evacuees, the progress of restoration, and environmental hygiene obviously changed over the course of the study period. The present paper describes the temporal dynamics of general observations, entomology, wildlife, microbiology, chemistry and socio-psychology in the areas affected by the tsunami in Ishinomaki.

Table 1. Damages in Ishinomaki, Miyagi caused by the Great East Japan Earthquake

Damage	Description of damage	Source
Deaths: 3506	Greatest amongst townships in Japan, 2.2% of city population, 22% of total deaths	(MIAC, 2011) (Ishinomaki City, 2013)
Missing: 453	16.9% of total missing	(Ishinomaki City, 2013)
Population lived in the flooded areas: 112000	Greatest amongst townships in Japan	(Statistics Bureau, 2011)
People dislocated to disaster shelters: 50758	Peak on March 17, 2011 31.6% of the city population	(Statistics Bureau, 2011)
Number of disaster shelters: 259	Peak on March 18, 2011. Public shelters 99 and the rest voluntarily set up	Ishinomaki City Council
Health problems among evacuees in the shelters in Ishinomaki	Impact phase: Dehydration, vomiting and diarrhea	(Ueda et al., 2012)
	Post-impact phase: Tetanus 7 cases, legionellosis 2 cases	(IASR, 2011a)
	Gastro-intestinal diseases 92, influenza 15, respiratory diseases 788, rash 5, scabies 6 and injury 2 cases	(IDSC, 2012)
	Deep vein thrombosis, 200 times higher incidence (2.2%)	(Ueda et al., 2012)

Picture 1. A disaster shelter set up in a gym of a primary school (taken on May 27, 2011)

Picture 2. A house in Kaisei temporary housing complex, Ishinomaki

Note: Each entrance on the right side of the house belong to each family (taken in June, 2011).

2. Study Sites

The present study assessed the environmental health risk of areas affected by the tsunami and the socio-psychological status of individuals who were displaced to temporary housing complexes. Environmental health risk was assessed in an 11-km strip of coastline with 2 km width towards inland of Ishinomaki, Miyagi. The socio-psychological study was conducted in one rural (Aikawa) and two urban (Kaisei and Watanoha) temporary housing complexes (Figure 1). We define urban temporary housing complex as the complex located within or peripheral of densely populated areas where receives evacuees from urban areas of Ishinomaki City, whereas rural temporary housing complex as the complex located in a rural area where urbanisation has not yet started. Kaisei temporary housing complex (t2, Figure 1) falls in a category 'peri-urban' in a development context while Watanoha (t3) is located in urban area (Makita et al., 2010). Kaisei temporary housing complex was constructed in a large park which was not inundated. Watanoha complex was located in the flooded area within proximity of severely damaged areas. Aikawa (t1) was located on a hill of a fisherman village, where the areas along the coast were severely affected. The dominant industry around the urban temporary housing complexes was services, while that of Aikawa was fishery.

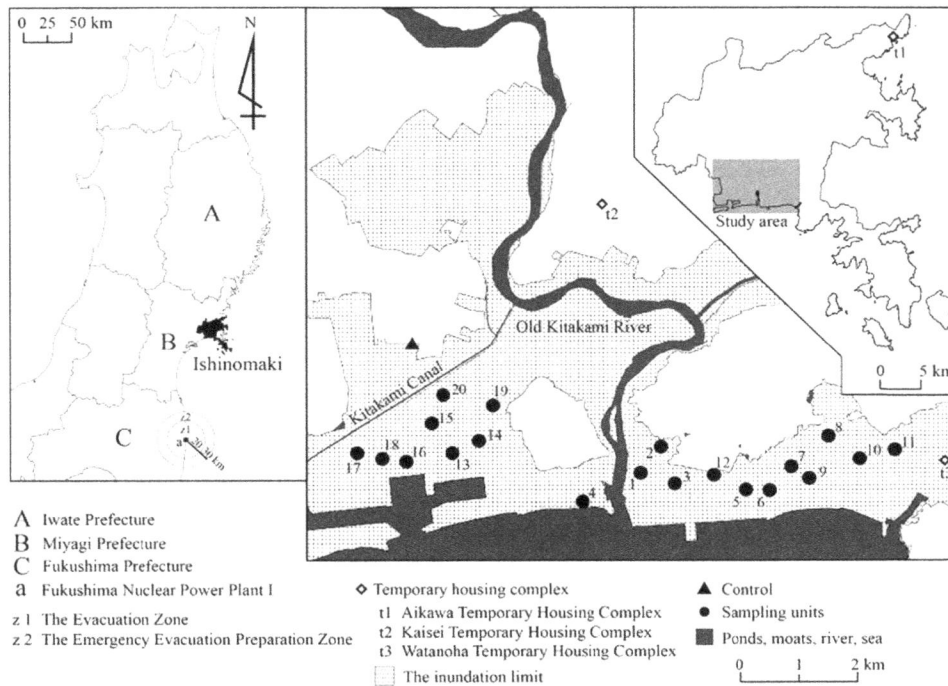

Figure 1. Maps showing the study sites.

Note: The left panel shows Tohoku Region, the northern part of the main land of Japan, with the locations of Iwate, Miyagi and Fukushima Prefectures, Ishinomaki City and Fukushima Nuclear Power Plant I. Right panel shows the locations of environmental study area and Aikawa Temporary Housing Complex (t1). Central panel shows closer view of the study area. Black dots represent 20 sampling units selected (neighborhood associations), and Kaisei (t2) and Watanoha (t3) Temporary Housing Complexes.

Ishinomaki has a cool climate with a mean annual temperature 11.6 °C. The annual rainfall in 2012 was 954.5 mm (JMA, 2013). The annual mean temperature, maximum temperature in summer (July), and minimum temperature in winter (January) in 2012 were 11.6 °C, 32.0 °C, and -8.3 °C, respectively.

3. Materials and Methods

3.1 Sampling Framework for Environmental Health Risk Assessment

A stratified random sampling of *chonai-kai* (neighborhood associations), the smallest administrative units in Japan, was performed in the study areas described above. The study areas were divided into two groups: east and west of the Old Kitakami River. Minamihama-cho, Kadowaki-cho and Hibarino-cho, the flat areas located at the corner of the west bank of the Old Kitakami River and the ocean, were excluded, as all houses in these areas were completely lost or destroyed and no one lived there. Twenty of the 87 *chonai-kai* units were selected (Figure 1), among which 12 and eight were located in the western and eastern strata, respectively. The sample size was determined based on a formula to estimate prevalence.

$$N = \frac{1.96^2 \times P_{exp} \times (1 - P_{exp})}{d^2}$$

Where N is a sample size representative of an infinite population, P_{exp} is an expected prevalence, and d is required precision (Thrusfield, 2005). P_{exp} was set to be 3.6% targeting pathogenic *Vibrio cholerae*, referring prevalence in river water at estuary in Kanagawa Prefecture (toxin producing *V. cholerae* O1 and non-O1 *V. cholerae* were isolated from 30 and 513 (61.1%) samples, respectively, out of 840 samples) between 1989 and 1995 (Yamai, Okitsu & Katsube, 1998). The required precision was set to be 90%. N was calculated as 13.3, but in order to have a robust figure, the sample size was increased by 50% and determined to be 20.

3.2 Entomological, Biological, and Chemical Sampling

Sludge and water were collected in a 500ml light-resistant glass bottle in the selected units in July and August 2011 and August 2012. Flies were sweep-sampled using 42cm diameter sweeping net with 1m shaft for bacteriological tests in five purposively selected units where the numbers of flies were subjectively judged to be great in July 2011. Flies were also captured using water traps with a few drops of surfactant in a yellow plastic container for 1 day and 1 night in one unit each strata where the number of flies was great. All the samples were chilled and packaged in Ishinomaki and sent to Rakuno Gakuen University (RGU) for morphological, microbiological, and chemical analyses.

Rodents were sampled in the 20 units in August 2011, using Sherman Traps (H. B. Sherman Traps Inc., FL, USA). Follow-up sampling was conducted at the unit 17 (Figure 1) in the western stratum, where a rodent was trapped in August 2011 (see Results), in November 2011 and in March and May 2012. In November 2012, sampling was conducted at the unit 17 again, and a grassland with a pine woods near the unit 9 (Figure 1), both of which were ecologically suitable habitats for rodents – proximity to water source such as canal and sea, secondary woods, and compiled debris. Collected rodents were euthanized in the field, chilled, and sent to RGU for microbiological tests.

3.3 Microbiological Tests

In 2011, bacteriological tests were performed on the sludge and water samples collected for total bacteria, enterobacteriaceae, *Salmonella*, spore-forming bacteria, *Vibrio,* and *Aeromonas*. Bacteriological follow-up in August 2012 comprised assessments of spore-forming bacteria.

Total bacteria, enterobacteriaceae, *Salmonella,* and spore-forming bacteria were counted in serially diluted samples of sludge and water. Total bacteria was counted after incubation on brain heart infusion agar in 5% CO_2 at 37 °C for 24 h. Enterobacteriaceae, which can indicate contamination with sewage or feces, were cultured on DHL agar at 37 °C for 24 h. *Salmonella*, which is a common cause of severe food poisoning, was enriched using Hajna tetrathionate broth at 37 °C for 18 h, left at ambient temperature at around 25 °C for 7 days, and inoculated onto Mannitol Lysine Cristal Violet Brilliant Green (MLCB) agar (Nissui, Tokyo, Japan) at 37 °C for 24 h. Aerobic spore-forming bacteria were cultured on Trypticase soy agar (Becton, Dickinson and Company, MD, USA) supplemented with 5% sheep blood (5% sheep blood agar) at 37 °C for 24 h. Anaerobic spore-forming bacteria were cultured on 5% sheep blood agar anaerobically for 72 h. Aerobic spore-forming bacteria were identified as *Bacillus* spp. and the anaerobic spore-forming bacteria were *Clostridium* spp. *Bacillus* spp. includes *B. anthrax*, which causes cutaneous and intestinal diseases, and *B. cereus*, which can cause food poisoning due to enterotoxin. *Clostridium* spp. includes *C. tetanus*, which causes tetanus after invading skin wounds, and *C. perfringens*, which causes food poisoning. *C. perfringens* isolated by anaerobic incubation on 10% egg yolk CW agar (Nissui) at 37 °C for 72 h was identified based on Nagler's reaction and inhibition by antiserum. *Vibrio* and *Aeromonas* can cause food poisoning and *V. cholerae* serogroups O1 biotype El Tor and O139 cause cholera, and they are frequently associated with sea foods in developed countries and contaminated water in developing countries (Sack, Sack, Naire & Siddique, 2004). Sludge samples suspended in phosphate-buffered 2% NaCl and water samples enriched with alkaline peptone broth at 37 °C overnight were inoculated onto ES *Vibrio* agar plate (Eiken Chemical Co., Ltd., Tokyo, Japan) and incubated at 37 °C overnight to determine the presence of *Vibrio* and *Aeromonas*. Isolated colonies were pure cultured on ES *Vibrio* agar plate once again and passaged onto 2% NaCl Trypticase soy agar. The biochemical characteristics of pure colonies were determined using TSI, LIM, and oxidase tests, and bacteria were identified using API20E (bioMérieux Inc., Durham, NC, USA). Bacteria that were identified as *V. cholerae* were tested for sero-groups O1 and O139, as well as the *ctxA* virulence gene using PCR.

A pool of 10 flies sampled by sweeping in July 2011 was prepared for each of the five sites. These five pools were served for bacteria counting of total bacteria, enterobacteriaceae, and *Salmonella.*

Sludge samples were virologically tested for hepatitis E virus (HEV) and Norwalk-like virus (NLV), which are hazardous to human health and can be found in the environment. The HEVs associated with human hepatitis are classified into four genotypes (Lu, Li & Hagedorn, 2006). Genotypes 1 and 2 cause waterborne outbreaks in developing countries and genotypes 3 and 4 are considered to be zoonotic and transmitted through the consumption of uncooked or undercooked contaminated meat (Mitsui et al., 2004; Wong, Purcell, Sreenivasan, Prasad, & Pavri, 1980). Total RNA was extracted from sludge suspensions using QIAamp Viral RNA Kits (QIAGEN, Hilden, Germany). The HEV RNA of the 5′terminal region of ORF1 was detected by semi-nested RT-PCR (Kanai et al., 2009) using the One Step RT-PCR Kit (Qiagen). NLVs which belong to the Caliciviridae family are major causes of acute nonbacterial gastroenteritis and a major public health concern. The NLV GI ORF1-ORF2 junction region was amplified by PCR using three forward primers for G1FF corresponding to nucleotides (nt) 5075 to 5097 in Norwalk/68, and the reverse primer, G1SKR (Kojima et al., 2002). The NLV GII ORF1-ORF2 junction region was also amplified by PCR using three forward primers for G2FB, corresponding to nt 4922 to 4941 in the Camberwell virus, and the reverse primer, G2SKR (Kojima et al., 2002). The RNA samples were reverse-transcribed using Transcriptor reverse transcriptase (Roche, Basel, Switzerland) and a random primer (2.5 μM). All representative cDNA samples were then amplified by PCR.

Apodemus speciosus trapped in August 2011 were tested using RT-nested PCR for Borna disease virus (BDV), which infects a wide range of mammals and causes immune-mediated, neurological Borna disease (BD), a disease that was originally discovered in horses (Staeheli, Sauder, Hausmann, Ehrensperger & Schwemmle, 2000). Total RNA (1 μg) isolated from the hippocampus of brain tissues using TRIzol (Invitrogen, Carlsbad, CA, USA) was reverse-transcribed using 200 units of SuperScript II reverse transcriptase (RT) (GIBCO BRL, Carlsbad, CA, USA) and random hexamers (100 ng). Borna disease virus-specific cDNAs corresponding to the BDV phosphoprotein (BDV-P) ORF were amplified by nested PCR as described elsewhere (Kishi et al., 1995).

3.4 Chemical Analysis

Water samples collected in June and August 2011 were passed through a syringe filter (0.45 μm) and analyzed for anions and cations using DionexIC-20 (Thermo-Fisher Scientific, Waltham MA, USA) and PIA-1000 (Shimadzu, Kyoto, Japan) ion chromatographs.

Sludge samples were dried, homogenized, and analyzed for cadmium (Cd), mercury (Hg), lead (Pb), and arsenic (As) using particle-induced X-ray emission (PIXE) at the Nishina Memorial Cyclotron Center (Iwate, Japan). The estimated detection limits for the toxic elements Cd, Hg, Pb, and As were 20, 3, 3, and 1.4 ppm, respectively. A detailed description of the measurement setup is provided elsewhere (Sera & Yanagisawa, 1992).

A control unit was selected from an unaffected residential area called Hebita (shown in Figure 1) as a reference for chemical analysis. The levels of all radioactive materials measured by the government surveillance remained below the standard levels established by the government and this topic was not further studied.

3.5 Socio-Psychological Survey

Findings of environmental damage and mental health that had been recorded in Excel data spread sheets by students, staff, and faculty members of RGU who had volunteered in Ishinomaki since May 2011 were qualitatively reviewed by the authors, and a summary was prepared in a participatory manner (Mariner & Paskin, 2000). This database was created in Ishinomaki in order to accumulate information so that a new volunteer group arrived in the base camp can efficiently follow up the activities done by the previous group. Participatory appraisals (Mariner & Paskin, 2000) were conducted in one rural and two urban temporary housing complexes and plans for future accommodations, current health status, stress, and the living environment of the residents were assessed in July 2012.

A questionnaire was designed based on the results of the participatory appraisals to understand the factors associated with reduced mental health in detail. K6 (Kessler et al., 2003) values were collected to screen for a serious mental illness (SMI) among residents in the temporary housing complexes. Respondents with K6 values > 13 were considered to have SMI, referring a cut-off point suggested by Kessler et al. (2003). The Japanese translation of questions for K6, which was validated through the backtranslation procedure and with high value of the areas under receiver operating characteristic curves (AUCs): 0.94 (95% confidence interval 0.88-0.99) (Furukawa, 2007), was used in the present study. The questionnaires were distributed to 50 households each at the two urban complexes and to all 44 households in the rural complex. To ensure anonymity, the respondents returned sealed questionnaires to RGU by mail.

3.6 Statistical Analysis

The number of rodents captured in a unit was $\log_e$-transformed, as a count follows Poisson distribution whose link for regression is logarithm, and a regression was performed with the explanatory valuable the month elapsed from occurrence of the tsunami when captured. The units where rodents were not captured were excluded from the analysis.

$\log_{10}$-transformed concentrations of bacteria were compared among samples of water and sludge that had been collected at different times using paired student *t* tests. Means and confidence intervals were calculated in $\log_{10}$ scale. The prevalence of *C. perfringens*, *Vibrio* spp., and *Aeromonas* was compared using Chi-square test and Fisher's exact test was applied when at least one cell in 2×2 tables included an expected frequency below 5.

Ion concentrations were log-normally distributed and $\log_e$-transformed data collected at different times were compared using paired Student's *t* tests. The ratio of Na^+ and Cl^- are useful for determining proximity to seawater (Yoshii et al., 2012) and the value of 0.56 for seawater was calculated based on data provided by Sverdrup, Johnson and Fleming (1961). The degree of proximity to seawater according to the ion composition of the samples is therefore expressed as a ratio of sample and seawater Na^+/Cl^- (a ratio of 1 suggests that the Na^+/Cl^- values of the seawater and the sample are identical). Ratios were also log-normally distributed and compared using paired student *t* tests of $\log_e$-transformed data. A value of 0.1 mg/kg was assigned to samples with values below detection limits to $\log_e$-transform log-normally distributed heavy metal concentrations for statistical analysis. These $\log_e$ means and confidence intervals were transformed back to the original scale for presentation.

For socio-psychological data, proportions were compared using Chi-squared or Fisher's exact tests. The K6 value was analyzed using Wilcoxon rank sum tests. Generalized linear models (GLMs) with Quasipoisson errors were performed for multivariable analysis to determine risk factors associated with mental stress (high K6 value) using statistics software R version 2.14.1.

4. Results

4.1 Transect of the Study Areas

The tsunami destroyed or severely damaged all infrastructures within a few blocks of the coast in the eastern and western strata. The first floors of some housing structures located 800-1000 m inland from the coast were completely filled with sludge and debris. Many people had returned to live on the second floors of such homes by May 2011.

The ground subsided in various coastal areas of the Tohoku region after the earthquake; in Ishinomaki, the ground subsided by 1.2 m, which was the greatest extent in Japan (Suito et al., 2011). Therefore, the lands near the coast had been covered with water pools containing sludge (Picture 3).

Picture 3. The residential areas near the coast covered with water containing sludge

Note: There were houses in this area before the occurrence of the tsunami (taken in June, 2011).

All the storage facilities and freezers at fish-processing factories were damaged and electricity was lost along the coast of the eastern stratum. Rotten fish from these factories were scattered on the roofs of damaged houses and in gardens throughout the areas, creating a foul odor (Picture 4).

Picture 4. Rotten fish scattered in the eastern areas of the study sites (taken in June, 2011)

Disposing of marine products in the ocean is normally prohibited by law, but Miyagi Prefecture changed this policy due to the circumstances and these products were discarded into the ocean and dump sites in Yamagata Prefecture between April 11 and July 6, 2011 (Kahoku Shimpo, 2011). All storage facilities for animal feed along the coast of the western stratum were also damaged, and animal feed concentrates were scattered on the ground throughout the western stratum. These situations attracted innumerable flies to the ground and walls of housing structures (Picture 5).

Picture 5. Meigen flies trapped by a resident using a hand-made trap containing attractant made of sugar and Japanese Sake wine (taken in June, 2011)

4.2 Entomology

Flies captured in a water trap in the eastern stratum in July 2011 comprised *Phormia regina* (Meigen) (n = 17), *Musca domestica,* and *Muscina stabulans* (n = 1 each). Sphaeroceridae spp. are tiny flies whose size is smaller than 1 mm and their population in the eastern stratum was semi-quantified as +, 1-10 individuals captured; ++,11-100 captured; +++, more than 101 captured. Flies in the western stratum comprised *Phormia regina* (Meigen) (n = 3); *Musca domestica* (n = 28), and *Muscina stabulans* (n = 10), and the density of Sphaeroceridae spp. was +++. *Phormia regina* (Meigen) were dominant in the eastern stratum, where fish carcasses were scattered, as this species favors protein, whereas Muscidae, which favor carbohydrates, dominated in the western stratum, where animal feed concentrates were scattered. Sphaeroceridae spp. were particularly evident in pools of water containing sludge in the western stratum. The complete disposal of marine products, removal of sludge, and spraying the area with insecticides remarkably reduced the number of flies by August 2011.

4.3 Rodents

Rodent traps were set in all the 20 sampling units during August 2011; only one *A. speciosus* was trapped in the sampling unit 17, which was close to the Kitakami canal (Figure 1). One *A. speciosus* was captured again at the same site in November 2011. In March 2012, two *Mus musculus* were captured in unit 17. In May 2012, one *A. speciosus* and four *M. musculus* were captured at the same site. In November 2012, traps set at the same site, unit 17, as well as in the pine woods along the coast in the eastern stratum near sampling unit 9 caught eight and 23 *M. musculus*, respectively. The log number of rodents captured in these two units (units 9 and 17) significantly increased over time (slope=0.08, $p = 0.005$).

4.4 Microbiological Tests

Table 2 shows the Log_{10} of the number of colony-forming units (CFU) per mL of total bacteria and the number of enterobacteriaceae, *Bacillus* spp., and *Clostridium* spp. in surface water and sludge samples collected in July 2011, August 2011, and August 2012. The numbers of samples were not 20, because some units lacked surface water or sludge. Samplings for microbiology were conducted by two independent teams. The samples collected by Team A were served for total bacteria, enterobacteriaceae, *Vibrio* and *Aeromonas*, although the numbers of samples slightly differ due to the insufficient quantities. The samples collected by Team B were tested for *Bacillus* spp. and *Clostridium* spp. In July and August 2011, water samples were not collected by Team A at six units (2, 3, 5, 6, 9 and 20). Team B could not collect water samples at eight units (2, 3, 5, 6, 9, 12, 19 and 20) in

July 2011, and at three units (3, 9 and 12) in August 2011. Sludge samples were not collected by Team A at five units (3, 5, 6, 9 and 20) in July 2011 and at five units (3, 5, 6, 9 and 13) in August 2011. Sludge samples were not collected by Team B at nine units (2, 3, 5, 6, 9, 12, 14, 19 and 20) in July 2011 and at five units (3, 7, 9, 12 and 19) in August 2011. These areas had been cleaned up and improved further by August 2012 and water samples could not be collected from 11 units (1-4, 6-9, and 12-14) and sludge samples at slightly different 11 units (1, 3-6, 9, 12-14, and 19-20).

The density of enterobacteriaceae and *Clostridium* in surface water samples significantly decreased (3.4 to 2.7, p = 0.006 and 3.3 to 2.7 mean $Log_{10}CFU/mL$, p = 0.023, respectively) between July and August 2011. In contrast, the total bacterial count (5.4 to 6.3 mean $Log_{10}CFU/mL$, $p < 0.001$) and the density of enterobacteriaceae (4.4 to 4.9, p = 0.046) in sludge significantly increased between July and August 2011. The concentrations of *Bacillus* (mean $Log_{10}CFU/ml$: 4.9 to 3.8, $p < 0.001$) and *Clostridium* (5.0 to 4.5, p = 0.004) significantly decreased between August 2011 and August 2012. *Salmonella* was not identified in any of the water or sludge samples. The prevalence of *Clostridium perfringens* in water samples was not significantly different between July (3/12, 25%) and August 2011 (2/17, 11.8%, p = 0.62), or between August 2011 and August 2012 (2/9, 22.2%, p = 0.59). The prevalence in sludge samples was also not significantly different between July (4/11, 36.4%) and August 2011 (5/15, 33.3%, p = 1.0), or between August 2011 and August 2012 (5/9, 55.6%, p = 0.4).

Flies carried high numbers of bacteria; the mean Log_{10} CFU/mL of total bacteria was 8.7 (95%CI: 7.3-10.1) and that of enterobacteriaceae was 8.0 (95%CI: 6.6-9.5, data not shown).

Table 2. $Log_{10}CFU/ml$ and 95% confidence interval of total bacteria, enterobacteriaceae, *Bacillus* spp. and *Clostridium* spp. in surface water and sludge

	Jul 2011	Aug 2011	Aug 2012	Statistics
Surface water				
Total bacteria	3.7 (3.2-4.2) (n=14)	3.4 (3.0-3.8) (n=14)		t=0.7, df=13, p=0.5
Enterobacteriaceae[a]	3.4 (2.9-3.9) (n=14)	2.7 (2.1-3.4) (n=14)		t=3.3, df=13, p=0.006
Bacillus spp.	3.2 (2.6-3.8) (n=12)	2.7 (1.9-3.5) (n=17)	3.1 (1.4-4.1) (n=9)	Jul - Aug 2011: t=0.9, df=11, p=0.38 Aug 2011 – Aug 2012: t=-0.87, df=8, p=0.41
Clostridium spp.[a]	3.3 (2.6-4.1) (n=12)	2.7 (2.0-3.4) (n=17)	3.6 (2.4-4.7) (n=9)	Jul- Aug 2011: t=2.6, df=11, p=0.023 Aug 2011- Aug 2012 : t=-1.28, df=8, p=0.24
Sludge				
Total bacteria[a]	5.4 (4.9-5.9) (n=15)	6.3 (5.9-6.6) (n=15)		t=-4.4, df=14, p<0.001
Enterobacteriaceae[a]	4.4 (3.8-4.9) (n=15)	4.9 (4.4-5.4) (n=15)		t=-2.2, df=14, p=0.046
Bacillus spp.[b]	5.1 (3.6-6.6) (n=11)	4.9 (3.8-6.1) (n=15)	3.8 (3.4-4.3) (n=9)	Jul-Aug 2011: t=0.18, df=9, p=0.86 Aug 2011- Aug 2012: t=5.3, df=10, p<0.001
Clostridium spp.[b]	6.2 (4.9-7.5) (n=11)	5.0 (3.9-6.1) (n=15)	4.5 (4.3-4.8) (n=9)	Jul-Aug 2011: t=1.8, df=9, p=0.11 Aug 2011- Aug 2012: t=3.7, df=10, p=0.004

a: significantly different between July and August 2011

b: significantly different between August 2011 and August 2012

Table 3 shows the prevalence of *Vibrio* spp. and *Aeromonas hydrophila* in surface water and sludge samples. The numbers of samples are different from those in Table 2 because a different team collected the samples. One extra surface water sample was collected in August 2011. Although *V. cholerae* were isolated, they were not highly pathogenic serotypes O1 or O139 and none had the *ctxA* gene which is associated with virulence. The prevalences of *V. cholerae*, *V. fluvialis,* and *A. hydrophila* were not significantly different.

Table 3. Prevalence of *Vibrio* spp. and *Aeromonas hydrophila*

	July 2011	August 2011	*p*-value
Surface water	n=13	n=14	
*Vibrio cholerae**	3 (23.1%)	3 (21.4%)	1
Vibrio fluvialis	1 (7.7%)	3 (21.4%)	0.60
Aeromonas hydrophila	1 (7.7%)	4 (28.6%)	0.33
Sludge	n=12	n=12	
*Vibrio cholerae**	3 (25.0%)	3 (25.0%)	1
Vibrio fluvialis	4 (33.3%)	4 (33.3%)	1
Aeromonas hydrophila	2 (16.7%)	2 (16.7%)	1

*Highly pathogenic *V. cholerae* serotype O1 or O139 was not detected. None of *V. cholerae* had *ctxA* gene which is associated with the virulence.

Virological tests in July 2011 detected an HEV monoclonal band in one of 12 samples of sludge, but in none of 13 samples of surface water. Norovirus was not detected in any water or sludge samples in July 2011. Virological tests for water and sludge were not conducted in August 2011. Bornaviruses were undetectable in rodents sampled in July 2011.

4.5 Chemical Analysis

Table 4 shows changes over time in mean ion concentrations between July and August 2011. In July, the mean ratio of sample Na^+/Cl^- to seawater Na^+/Cl^- was 1.01 (95%CI: 0.78-1.30) and the composition of the sampled surface water was similar to that of seawater. The surface water with similar composition to seawater was geographically evenly distributed in the studied areas (black dots in Figure 2A). In August 2011, the concentration of Na^+ remained ($p = 0.65$) but that of Cl^- significantly decreased ($p = 0.007$), which resulted in a significant increase in the ratio of sample Na^+/Cl^- to seawater Na^+/Cl^- (2.61, $p = 0.007$). The composition of surface water had considerably changed in all tested areas by August 2011 (Figure 2B).

Table 4. Changes of the mean ion concentrations (mg/L) and 95% confidence intervals in surface water between July and August 2011

	July (n=11)	August (n=12)	Statistics
Na^+	212 (50-894)	227.4 (67.2-769.3)	t=0.49, df=4, p=0.65
NH_4^+	1.0 (0.2-4.7)	4.1 (0.8-21.1)	t=-0.06, df=4, p=0.96
Cl^-	376 (89-1593)	156 (39-629)	t=3.0, df=4, p=0.04
Na^+/Cl^-	0.56 (0.44-0.73)	1.46 (1.07-2.00)	t=-5.2, df=4, p=0.007
Ratio of sample and seawater Na^+/Cl^{-*}	1.01 (0.78-1.30)	2.61 (0.32-3.58)	t=-5.2, df=4, p=0.007

*Seawater Na^+/Cl^- (0.56) was calculated using the values in Sverdrup et al. (1961)

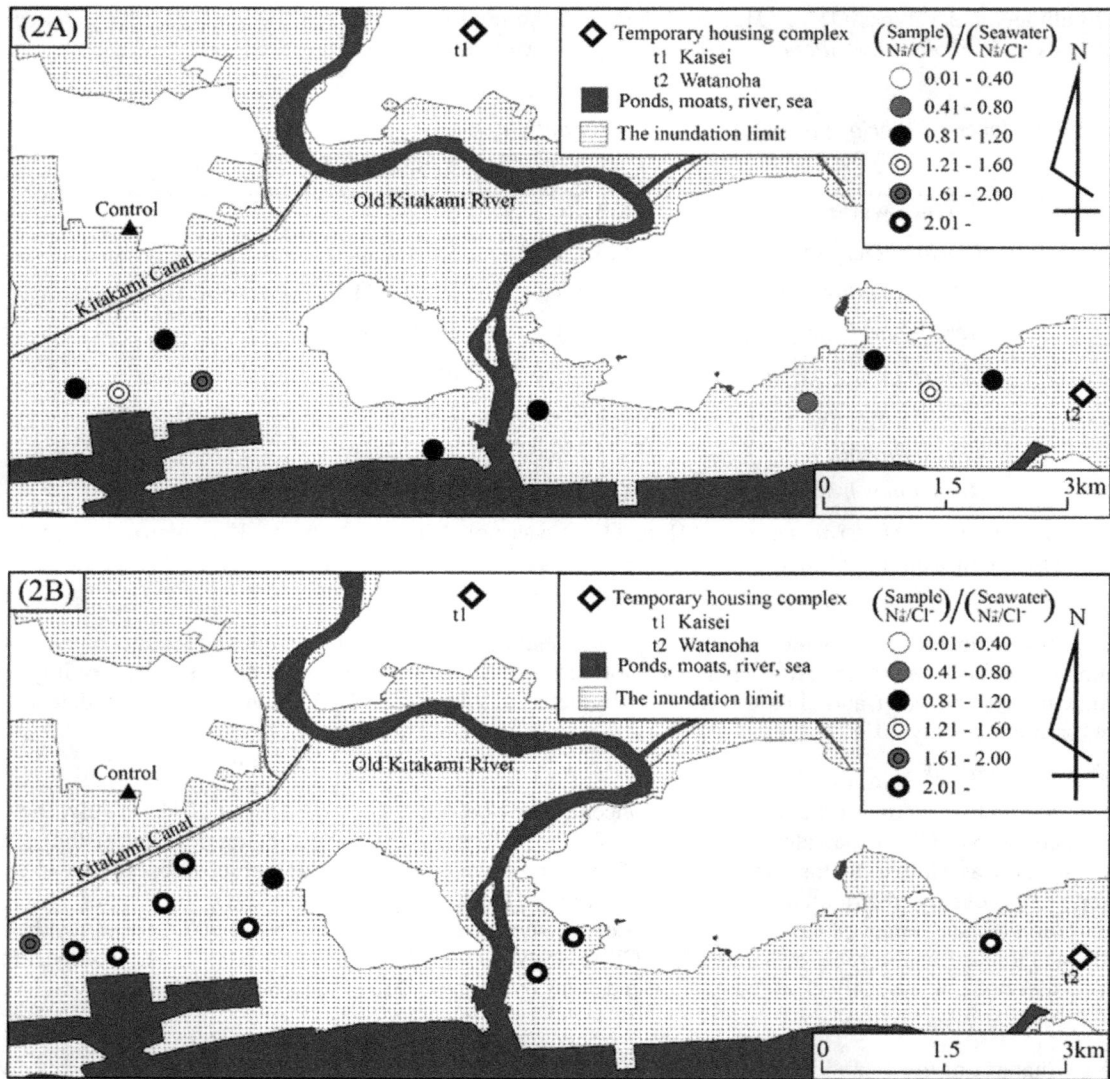

Figure 2. Geographical representation of the ratio of sample and seawater Na^+/Cl^- in July and August 2011

Note: 2A shows the map of July 2011 and 2B August 2011.

The concentration of NH_4^+ did not significantly increase between July and August 2011. However, the maximum NH_4^+ concentration increased from 21 in July to 173 in August 2011 (data not shown). The NH_4^+ concentrations were higher in the damaged residential areas of the western stratum (Figures 3A and B). Both months included surface water with an NH_4^+ concentration below detection limit, 0.10 mg/L, and the proportions (4/11 (36.4%) in July and 2/12 (16.7%) in August) were not significantly different ($p = 0.37$).

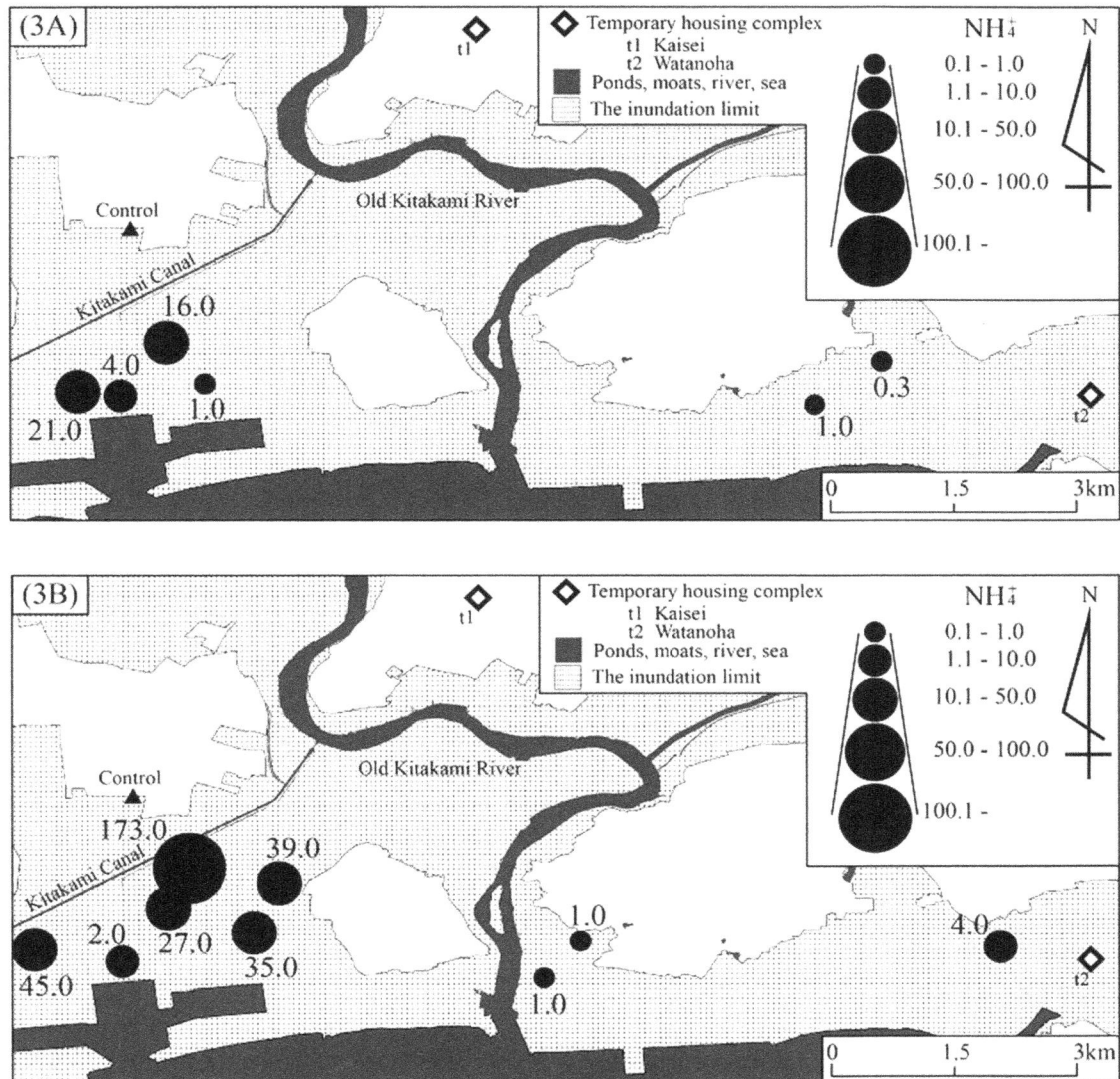

Figure 3. Geographical representation of the NH_4^+ concentrations (mg/L) in July and August 2011

Note: 3A shows the map of July 2011 and 3B August 2011.

Table 5 shows mean heavy metal concentrations in July and August 2011. The detection limit of Cd is 20 mg/kg by PIXE and Cd was undetectable in all samples. The concentrations of Cu, Hg, Pb, Ni and As did not significantly differ between July and August 2011. The standard levels defined in the Soil Contamination Countermeasures Act in Japan for Cd, Hg, Pb and As are 150, 15, 150 and 150 mg/kg, respectively, and those for Cu and Ni are not stipulated (Ministry of Environment, 1991). The means and confidence intervals of Cd, Hg, Pb, and As did not exceed these levels. However, in July 2011, one sample in unit 1 of the eastern stratum (Figure 1) contained 24.9 mg/kg of Hg, which did exceed the standard level. Levels of Pb and As did not exceed standard values in all samples. In a control soil sample (Figure 1), Hg was undetectable and the concentrations of Cu, Pb, Ni, and As were 119, 87.5, 85.3, and 2.3 mg/kg, respectively.

Table 5. Changes of the mean heavy metal concentrations (mg/kg) and 95% confidence intervals in sludge between July and August 2011

	July (n=11)	August (n=17)	Statistics
Cadmium (Cd)*	<20	<20	Not applicable
Cupper (Cu)	23.5 (7.1-78.1)	30.0 (13.1-68.8)	t=-1.1, df=8, p=0.30
Mercury (Hg)	0.6 (0.1-2.4)	0.4 (0.1-0.9)	t=-0.2, df=8, p=0.83
Lead (Pb)	47.0 (32.9-61.1)	66.1 (46.4-85.8)	t=-2.1, df=8, p=0.07
Nickel (Ni)	14.8 (3.1-71.1)	27.7 (9.6-80.3)	t=-1.9, df=8, p=0.10
Arsenic (As)	3.9 (0.8-19.2)	7.5 (3.3-17.2)	t=-1.1, df=7, p=0.33

*Detection limit of Cd is 20 mg/kg and Cd was not detected from all the samples.

4.6 Socio-Psychological Situation in Temporary Housing Complexes

4.6.1 Qualitative Findings

According to the qualitative records collected from the RGU students during volunteer activities, the contents of stress dynamically changed over time. Shocks due to having lost homes, jobs, family, and friends were prevalent among those living in disaster shelters and damaged houses in May 2011. Many people felt guilty for not having died themselves, considering those who had died. The amounts of debris and sludge were overwhelming, but joint cleaning activities with volunteers offered a distraction. Long waiting list to enter in a room in the temporary housing complexes and inconvenience of many of the complexes due to the remote locations induced stress in May and June. Lost community function due to the absence or relocation of community leaders and board members was a huge obstacle for those who returned to damaged homes, where they lived without basic supplies including food, clothing, and hygiene products.

According to the participatory appraisals, in September 2012 the focus of mental stress shifted to the imminent future. Several political leaders announced closing dates of the temporary housing complex, but in reality, planning, upland reclamation, housing construction, and policy support for the relocation of evacuees were delayed. Obstacles for elderly respondents included ineligibility for loans to construct houses. Even middle-aged individuals worried about how to pay for a second loan while still paying original loans for homes that were destroyed by the tsunami.

4.6.2 Questionnaire Results

In September 2012, 44 evacuees in the two temporary urban housing complexes and 16 in the rural complex responded to the questionnaire. The urban and rural response rates were 44% (44/100) and 36.4% (16/44), respectively, and the mean ages of respondents were 63.6 and 58.7 years, respectively. The urban relocation area planned by the authorities was inland Ishinomaki and only 30.6% (11/36) of the respondents wished to return to their own homes in urban areas, whereas 64.2% (9/14) of those in rural areas wished to return (x^2 = 3.48, df = 1; p = 0.06). Regardless the planned destination either original home or inland, 46.2% (12/26) of urban respondents felt that their wish would be unable to be realized and 58.3% (7/12) of these stated that financial problems were the main obstacle. Those in rural areas who chose relocation (35.8%, 5/14) wished to move to nearby residential areas with higher elevation after development of these areas by the authorities, and wished to continue working in the fishing industry. All of the respondents in the rural housing complex felt that this would be realized.

High K6 value which suggests SMI (K6 value > 13) was found in 15.9% (7/44) and 6.3% (1/16) of the urban and rural respondents, respectively. A total of 13.3% (8/60) had severe mental health conditions. This was higher than the prevalence of K6 above 13 at pre-tsunami status, 5% in 2010 (NCNP, 2013). The proportions were not significantly different between urban and rural respondents (p = 0.7). Table 6 shows univariate analysis of K6 values which indicate mental stress. Statistical significance was found in health problems (p<0.001), lack of friends (p = 0.003) or trusted person to counsel (p = 0.005), not participating in events (p = 0.02), and pest infestation (p = 0.03). The overall mean K6 value was 6.5 (data not shown). Multivariable analysis revealed that risk factors for a high K6 value were a lack of friends (p = 0.011) or trusted person to counsel (p = 0.003) and the illness of the respondent or a family member (p = 0.003, Table 7).

Table 6. Univariate analysis for mental health

Factors	Attributes	Sample	Percentage (%)	K6	p-value
Sex	Male	27	48.2	6.1	0.19
	Female	29		7.8	
Age	>=65	28	50.0	6.0	0.99
	<65	28		7.3	
Health problem of the	Exist	26	52.0	3.0	<0.001
respondent or family	Not exist	24		9.9	
Friends in the housing	Exist	42	72.4	5.0	0.003
Complex	Not exist	16		10.6	
Trusted person to counsel	Exist	29	56.9	3.3	0.005
in the housing complex	Not exist	22		10.9	
Participation in events	Participate	43	75.4	5.4	0.02
	Not participate	14		10.2	
Activities to relieve	Have	23	38.3	5.6	0.7
Stress	Do not have	37		7.1	
Room mate	Exist	42	75.0	5.4	0.5
	Not exist	14		7.0	
Children	Exist	6	10.5	9.0	0.1
	Not exist	51		6.8	
Income	Have	7	12.7	6.5	0.7
	Do not have	48		6.7	
Level of urbanization	Urban	44	73.3	5.9	0.7
	Rural	16		10.0	
Pest infestation	Exist	20	34.5	9.2	0.03
	Not exist	38		5.2	
Satisfaction from the	Satisfied	22	39.3	5.2	0.3
Environment	Not satisfied	34		8.0	
Land of origin	Ishinomaki	46	76.7	7.0	0.6
	Other place	14		4.5	

*Note: total numbers of answer are different between the questions because of the answers not provided.

Table 7. Maltivariable analysis for mental health

Factors	Estimate	Standard error	p-value
Health problem of the respondent or family member	0.92	0.30	0.011
Existence of trusted person to counsel in the housing complex	-0.58	0.22	0.003
Existence of friends in the housing complex	-0.79	0.28	0.003

Table 8 compares associations between human relationships and mental stress in urban and rural areas. Respondents with friends in the temporary urban housing complexes had significantly lower K6 values, indicating less mental stress (4.9 vs. 11.9; $p < 0.001$) than those without friends, whereas these values did not significantly differ in the rural complex (5.4 vs. 1.0, $p = 0.4$). Smilarly, K6 values were significantly lower among respondents in urban areas with trusted person to counsel than without (4.1 vs. 12.7, $p = 0.002$), but did not significantly differ in the rural respondents (3.3 vs. 6.2, $p = 0.8$). The proportions of respondents with friends ($p = 0.2$) and trusted counselors ($p = 1$) did not significantly differ between the urban and rural complexes.

Table 8. Comparison of the associations between human relationships and mental stress (K6) in urban and rural areas

Factors	Attributes	Sample	Percentage (%)	K6	p-value
Friends in the complex					
Urban	Exist	29	67.4	4.9	<0.001
	Not exist	14		11.9	
Rural	Exist	13	86.7	5.4	0.4
	Not exist	2		1.0	
Trusted person to counsel					
Urban	Exist	21	56.8	4.1	0.002
	Not exist	16		12.7	
Rural	Exist	8	57.1	3.3	0.8
	Not exist	6		6.2	

*Note: total numbers of answer are different between the questions because of the answers not provided.

5. Discussion

The present study revealed the temporal dynamics of environmental health risks and socio-psychological status in areas of Ishinomaki affected by the tsunami between 2011 and 2012. While public health interests might focus on monitoring the incidence of infectious diseases, public health risks must be understood from the environmental viewpoint of tsunami-affected areas for mid- and long-term reconstruction planning. Japan has been promoting the relocation of residential areas to uplands to avoid future tsunami-related disasters (Reconstruction Agency, 2013). On the other hand, the present study showed that some populations would prefer to live in their original locations. The present environmental and socio-psychological risk assessment is important in providing real-time information to policy makers in Ishinomaki so that they can plan adequate policy support for those who lived in the tsunami-affected areas, as well as those in temporary housing complexes in Ishinomaki.

Common health problems that arose in the disaster shelters comprised respiratory and gastro-intestinal (GI) syndromes (IASR, 2011b). Norovirus infections were reported in Fukushima (IASR, 2011c) and Iwate Prefectures (IASR, 2011b). However, none of these outbreaks were large-scale considering the reported epidemic patterns (IASR, 2011b; ISAR, 2011c) and there might not be many spill-over infections from the environment, except for infections that developed during phases 1 and 2, when clean water was scarce under crowded conditions (Ueda et al., 2012). The risk of food poisoning was high in the affected areas due to the contamination of water, sludge, and flies with potentially hazardous bacteria. The disposal of fish carcasses, cleaning, and insecticide spraying dramatically decreased the number of flies, which adequately decreased health risks by August 2011.

Cholera frequently occurs during natural disasters. The global attack rate of *V. cholerae* O1 Cholera epidemic in Haiti after the 2010 earthquake and hurricanes was 488.9/10,000 inhabitants and the mortality rate was 6.2/10,000 inhabitants (Gaudart et al., 2013). Although *V. cholerae* and *V. vulnificus* was isolated in Ishinomaki, toxigenic *V. cholerae* was not. However in the USA, environmentally acquired *V. vulnificus*, *V. parahaemolyticus,* and non-toxigenic *V. cholerae* caused infections and deaths soon after Hurricane Katrina (CDC, 2005). This suggests that the bacteria isolated in Ishinomaki could have caused infections and deaths. Early establishment of hygiene in the disaster shelters and relocation to temporary housing complexes might have prevented such outbreaks. *Vibrio* outbreak associated with the tsunami has not been reported in any of the affected areas of Japan. After the 2004 tsunami in Banda Aceh, Indonesia, acute jaundice potentially due to infection with water-borne hepatitis A (HA) and E was identified among displaced population (WHO, 2005). The present study detected HEV in only one sample and no outbreak occurred, probably because HA and HE are not endemic in Japan. Although the land in areas affected by the tsunami remained covered with seawater until July 2011, and the NH_4^+ concentration suggested high urine contamination, hygiene at the disaster shelters in the flooded areas might have been maintained at a high enough level to avoid such waterborne outbreaks. On the other hand, the high NH_4^+ concentration suggested a harsh living condition of the people remained in the damaged houses, who were later relocated to temporary housing complexes or restored toilet facilities. The increase of the Na^+/Cl^- ratio was probably due to Na^+ sorption onto soil's clay particles, while Cl^- was leached by rainfall. Evacuees in the disaster shelters and damaged houses were relocated to temporary housing complexes by October 11, 2011, when they were finally released from an overcrowded, uncomfortable environment.

The most dynamic changes observed in the present study were in rodent populations. After the affected areas were cleaned and debris was removed, the original residential areas became covered with grass and the lands became favorable to rodents. These animals are recognized as important mammalian reservoirs of *Leptospira* spp. (Meerburg, Singleton, & Kijlstra, 2009) and shed infectious organisms in urine throughout their lifespan (Li et al., 2013; Vinetz, 2001). Leptospirosis can be fatal and an outbreak in the Philippines after the flood in 2009 was characterized by jaundice, anuria, and hemoptysis (Amilasan et al., 2012). *Leptospira* was not investigated in the present study, but future environmental surveys should include diagnosing leptospirosis in rodents to understand the health risks.

Notably high level of Hg (24.9 mg/kg) was observed in only one sludge sample. This might have been carried from soils beneath the sea. A study of the coast in Aomori, Iwate and Miyagi prefectures found a greater variety of heavy elements in sludge than in inland samples (Baba & Sera, 2012). Sediments in Japanese bays contain considerable amounts of toxic heavy metals (Kabir et al., 2006). A study of hair from individuals in areas affected by the tsunami in Iwate Prefecture did not find clear differences in heavy metal concentrations between before and after the tsunami (Sera et al., 2012). However, no data are available for individuals in Ishinomaki who lived in damaged houses contaminated with sludge, and the effect of exposure to heavy metal particles in dried sludge is unknown. Concentrations of Cd, Cr, Cu, Ni, and Pb increased from 7 to 35, 141 to 617, 32 to 132, 62 to 252, and 11 to 144 mg/kg, respectively, after the 2004 tsunami in India (Ranjan et al., 2008). Although we did not find such alarming concentrations, metal elements might be released into the ecosystem and create a threat in the event of changed geochemical status (Ranjan et al., 2008). Therefore, future monitoring, especially of agricultural lands, is recommended, as elements absorbed from agricultural products can enter the food chain.

The contents of mental stress considerably changed over time, but residents in temporary housing complexes remained highly stressed even at one and a half years after the tsunami. A published review has argued that 30-50% of individuals after a tsunami would experience moderate to severe psychological distress that might resolve with time, or mild distress that could become chronic (Carballo et al., 2005), which supports the findings in the present study. The risk factors for serious mental illness suggested that individuals without close human relationships might be vulnerable. Our comparison of associations between human relationships and mental stress in urban and rural areas also supported this assertion. Human relationships remained close in rural areas, where even those who felt isolated did not have increased mental stress. In contrast, evacuees were from many different places in urban complexes, and weaker human relationships affected mental health. A review paper which summarizes the impacts from tsunamis in the world has indicated that women, children, and elderly individuals are the most vulnerable to mental stress during a tsunami (Carballo, Heal, & Hernandez, 2005). The present study, however, did not find such a tendency. The prevalence of mental illness among such vulnerable populations might have been higher soon after the tsunami.

6. Conclusion

Considering the effect of removing sludge from the soil of affected areas and the limited number of people returning to their original residences, the assessed environmental health risk has been reduced to a level that is acceptable for living. However, the population of rodents is increasing, and these pests may harbor hazards for humans. Monitoring rodent population dynamics and the prevalence of zoonotic agents such as *Leptospira* spp. might be needed. Financial support for evacuees to construct homes is also recommended. Furthermore, mental support is needed for evacuees in temporary housing complexes and efforts should be directed towards enhancing closer relationships in such complexes.

Acknowledgements

We would like to thank Ishinomaki City Council for the advice, help and facilitation in our collaborative activities since the very difficult time of the disaster. We also would like to thank Mitsui & Co., Ltd Environmental Fund and RGU for research funding. Many thanks go to NPOs APCAS (Action for Peace, Capability and Sustainability), Ishinomaki Environmental Net, P-CAT (Primary Care for All) and Agarain, NGO PARCIC (Pacific Asia Resource Center Inter-Peoples Cooperation) and RGU student volunteer service, Raku-Net for the enthusiastic and heartfelt joint activities, advices, logistical assistance and accommodation. We thank Dr. Hajime Takahashi at RGU in March-April 2011 for the coordination with NPOs. Professional advices on psychiatry were provided by Dr. Michiko Watari at the National Center for Psychiatry and Neurological Research, Japan. We thank Miyagi Prefecture Furukawa Agriculture Research Institute for the discussions on comparative data on heavy metal concentration in unaffected soil by tsunami. The inundation limit spatial data was produced by the Earth Environmental Engineering Group, Institute of Industrial Science, the University of Tokyo, under permission by the Geospatial Information Authority of Japan. The biggest thanks and our love go to the people in Ishinomaki participated in this study as well as all those who were affected by the Great East Japan Earthquake. We shall never forget about the victims of the disaster.

References

Aghababian, R. V., &Teuscher, J. (1992). Infectious diseases following disease emergencies in disasters. *Annals of Emergency Medicine, 21*, 4. http://dx.doi.org/10.1016/S0196-0644(05)82651-4

Amilasan, A. T., Ujiie, M., Suzuki, M., Salva, E., Belo, M. C. P., Koizumi, N., … Ariyoshi, K. (2012). Outbreak of leptospirosis after flood, the Philippines, 2009. *Emerging Infectious Diseases, 18*(1), 91-94. http://dx.doi.org/10.3201/eid1801.101892

Baba, F., & Sera, K. (2012). Analysis of contaminated sludge deposited on the land attacked by great tsunami following Tohoku Great Earthquake Disaster. *International Journal of PIXE, 22*(1-2), 231-39. http://dx.doi.org/10.1142/S012908351240027X

Carballo, M., Heal, B., & Hernandez, M. (2005). Psychological aspects of the Tsunami. *Journal of the Royal Society of Medicine, 98*, 396-399. http://dx.doi.org/10.1258/jrsm.98.9.396

CDC. (2005). *Vibrio* illness after Hurricane Katrina – multiple States, August-September 2005. *Morbidity and Mortality Weekly Report, September 14, 2005, 54*, 1-4.

Furukawa, T., Kawakami, N., Saitoh, M., Ono, Y., Nakane, Y., Nakamura, Y., … Kikkawa, T. (2008). The performance of the Japanese version of the K6 and K10 in the World Mental Health Survey Japan. *International Journal of Methods in Psychiatric Research, 17*(3), 152-158. http://dx.doi.org/10.1002/mpr.257

Gaudart, J., Rebaudet, S., Barrais, R., Boncy, J., Faucher, B., Piarroux, M., ... Piarroux, R. (2013). Spatio-temporal dynamics of cholera during the first year of the epidemic in Haiti. *PLoS Neglected Tropical Diseases, 7*(4), e2145. http://dx.doi.org/10.1371/journal.pntd.0002145

IASR. (2011a). Countermeasure against infectious disease outbreaks in Miyagi Prefecture after the 2011 off the Pacific coast of Tohoku Earthquake (in Japanese). *Infectious Agents Surveillance Report, 32*, S3-S4.

IASR. (2011b). Disaster shelter surveillance and countermeasure against infectious diseases in Iwate Prefecture (in Japanese). *Infectious Agents Surveillance Report, 32*, S1-S3.

IASR. (2011c). The emesis and diarrhea outbreak in a disaster shelter in Koriyama, Fukushima (in Japanese). *Infectious Agents Surveillance Report, 32*, S8-S9.

IDSC. (2011). Infectious disease incidence report associated with the Great East Japan Earthquake as of April 13th 2011 (in Japanese). *Japan Infectious Disease Surveillance Center.*

IDSC. (2012). Infectious disease surveillance of safe shelter in Miyagi Prefecture after Tohoku earthquake (March 11, 2011) occurrence (in Japanese). *Annals of Miyagi Prefecture Health and Environment Center, 30*, 52-57.

Ishinomaki City. (2013). Damage situations report as of 31st March, 2013. Information associated with Great East Japan Earthquake (in Japanese). *Ishinomaki City.*

JMA. (2011). *The 2011 off the Pacific Coast of Tohoku Earthquake – Portal -. Japan Meteorological Agency (JMA).* Retrieved from http://www.jma.go.jp/jma/en/2011_Earthquake/2011_Earthquake.html

JMA. (2013, May 14). *Statistics in Ishinomaki. Japan Meteorological Agency (JMA).* Retrieved from http://www.data.jma.go.jp/obd/stats/

Kabir, H. M., Narusawa, T., Nishiyama, F., & Sumi, K. (2006). Elemental analysis of Uranouchi Bay seabed sludge using PIXE. *International Journal of PIXE, 16*(3-4), 221-230. http://dx.doi.org/10.1142/S012908350600099X

Kahoku Shimpo. (2011, July 7). Disposal of marine products ended. *Kahoku Shimpo newspaper.*

Kanai,Y., Tsujikawa, M., Yunoki, M., Nishiyama, S., Ikuta, K., & Hagiwara, K. (2009). Long-term shedding of hepatitis E virus in the feces of pigs infected naturally, born to sows with and without maternal antibodies. *Journal of Medical Virology, 82*(1), 69-76. http://dx.doi.org/10.1002/jmv.21647

Kessler, R. C., Barker, P. R., Colpe, L. J., Epstein, J. F., Gfroerer, J. C., Hiripi, E.,...Zaslavsky, A. M. (2003). Screening for serious mental illness in the general population. *Archives of General Psychiatry, 60*(2), 184-189. http://dx.doi.org/10.1001/archpsyc.60.2.184

Kishi, M., Nakaya, T., Nakamura, Y., Kakinuma, M., Takahashi, T. A., Sekiguchi, S., ... Ikuta, K. (1995). Prevalence of Borna disease virus RNA in peripheral blood mononuclear cells from blood donors. *Medical Microbiology and Immunology, 184*, 135-8. http://dx.doi.org/10.1007/BF00224350

Kojima, S., Kageyama, T., Fukushi, S., Hoshino, B., Shinohara, M., Uchida, K., ... Katayama, K. (2002). Genogroup-specific PCR primers for detection of Norwalk-like viruses. *Journal of Virological Methods, 100*, 107-114. http://dx.doi.org/10.1016/S0166-0934

Kouadio, K., Isidore, K., Aljunid, S., Kamigaki, T., Hammad, K., & Oshitani H. (2012). Infectious disease following natural disasters: prevention and control measures. *Expert Review of Anti-Infective Therapy, 10*(1), 95-104. http://dx.doi.org/10.1586/eri.11.155

Li, S., Wang, D., Zhang, C., Wei, X., Tian, K., Li, X., ... Yan, J. (2013). Source tracking of human leptospirosis: serotyping and genotyping of *Leptospira* isolated from rodents in the epidemic area of Guizhou province, China. *BMC Microbiology, 13*, 75. http://dx.doi.org/10.1186/1471-2180-13-75

Lu, L., Li, C., & Hagedorn, C. H. (2006). Phylogenetic analysis of global hepatitis E virus sequences: genetic diversity, subtypes and zoonosis. *Reviews in Medical Virology, 16*(1), 5-36. http://dx.doi.org/10.1002/rmv.482

Makita, K., Fèvre, E. M., Waiswa, C., Bronsvoort, M. D. C., Eisler, M. C., & Welburn, S. C. (2010). Population-dynamics focussed rapid rural mapping and characterization of the peri-urban interface of Kampala, Uganda. *Land Use Policy, 27*, 888-897. http://dx.doi.org/10.1016/j.landusepol.2009.12.003.

Mariner, J. C., & Paskin, R. (2000). Manual on Participatory Epidemiology. FAO, Agriculture and Consumer Protection Department, Rome, Italy.

Meerburg, B. G., Singleton, G. R., & Kijlstra, A. (2009). Rodent-borne diseases and their risks for public health. *Critical Reviews in Microbiology, 35*(3), 221-270. http://dx.doi.org/10.1080/10408410902989837

MIAC. (2011). Population, population dynamics and households number survey based on Basic Resident Register. *Ministry of Internal Affairs and Communications, Japan.*

MLITT. (2011). Survey report of the Tsunami damage caused by the Great East Japan Earthquake (Third Report). A survey report on the situation of evacuation from the Tsunami Press Release on December 26, 2011. Ministry of Land, Infrastructure, Transport and Tourism. Retrieved from http://www.mlit.go.jp/common/000186474.pdf

Mitsui, T., Tsukamoto, Y., Yamazaki, C., Masuko, K., Tsuda, F., Takahashi, M., ... Okamoto, H. (2004). Prevalence of hepatitis E virus infection among hemodialysis patients in Japan: evidence for infection with a genotype 3 HEV by blood transfusion. *Journal of Medical Virology, 74*(4), 563-572. http://dx.doi.org/10.1002/jmv.20215

Ministry of Environment. (1991). The standard level of Designated Hazardous Substances in soil (Table 3), The Soil Contamination Countermeasures Act, *Japan Ministry of Environment.*

Miyagi Prefecture. (2012). A guide book of livelihood support for the disaster victims. Retrieved from http://www.pref.miyagi.jp/uploaded/attachment/123693.pdf

NCNP. (2013, Sep 20). 2010 database of K6 from Comprehensive Survey of Living Conditions. National Information Center of Disaster Mental Health, National Center for Neurology and Psychiatry (NCNP). Retrieved from http://saigai-kokoro.ncnp.go.jp/document/medical.html

NPA. (2013). Damage situation and police countermeasures associated with 2011 Tohoku district off the Pacific Ocean Earthquake, 10 April 2013. Emergency Disaster Countermeasures Headquarters, *National Police Agency (NPA) of Japan.*

PMJHC. (2011, April 1). *Statement by the Prime Minister Naoto Kan. Prime Minister of Japan and His Cabinet (PMJHC).* Retrieve from http://www.kantei.go.jp/jp/kan/statement/201104/01kaiken.html

Ranjan, R. K., Ramanathan, A., Singh, G., & Chidambaram, S. (2008). Assessment of metal enrichments in tsunamigenic sediments of Pichavaram mangroves, southeast coast of India. *Environmental Monitoring and Assessment, 147*, 389-411. http://dx.doi.org/10.1007/s10661-007-0128-y

Reconstruction Agency. (2013). Summary of the 2013 Preliminary Budget. Retrieved from http://www.reconstruction.go.jp/topics/20130329_25zanteiyosangaiyou.pdf

Sack, D. A., Sack, R. B., Naire, G. B., & Siddique, A. K. (2004). Cholera. *Lancet, 363*(9404), 223-233. http://dx.doi.org/10.1016/S0140-6736(03)15328-7

Sera, E., & Yanagisawa, T. (1992). The Takizawa PIXE facility combined with a baby cyclotron for position nuclear medicine. *International Journal of PIXE, 2*(1), 47-55. http://dx.doi.org/10.1142/S0129083592000051

Sera, K., Goto, S., Takahashi, C., Saitoh, Y., & Yamauchi, K. (2013). Effects of heavy elements in the sludge conveyed by the 2011 Tsunami on human health and the recovery of the marine ecosystem. *Proceedings, the 13th International Conference on Particle Induced X-ray emission*, Gramado, Brazil, March 3-8, 2013.

Shibahara, S. (2011). The 2011 Tohoku Earthquake and Devastating Tsunami. *Tohoku Journal of Experimental Medicine, 223*, 305-307. http://dx.doi.org/10.1620/tjem.223.305

Staeheli, P., Sauder, C., Hausmann, J., Ehrensperger, F., & Schwemmle, M. (2000). Epidemiology of Borna disease virus. *Journal of General Virology, 81*, 2123-35.

Statistics Bureau. (2011). East Japan Great Earthquake and public statistics. The 32nd Statistics symposium. *Statistics Bureau, Director-General for Policy Planning and Statistical Research and Training Institute, Japan.*

Suito, H., Nishimura, T., Ozawa, S., Kobayashi, T., Tobita, M., Imakiire, T., … Kawamoto, S. (2011). Coseismic deformation and fault model of the Pacific cost of Tohoku Earthquake, based on GEONET (in Japanese). *Bulletin of the GSI, 122*, 29-37.

Sverdrup, H. U., Johnson, M. W., & Fleming, R. H. (1961). *The oceans: their physics, chemistry, and general biology* (Modern Asia edition), Prentice-Hall.

Ueda, S., Hanzawa, K., Shibata, M., & Suzuki, S. (2012). High prevalence of deep vein thrombosis in Tsunami-flooded shelters established after the Great East-Japan Earthquake. *Tohoku Journal of Experimental Medicine, 227*, 199-202. http://dx.doi.org/10.1620/tjem.227.199

Vinetz, J. M. (2001). Leptospirosis. *Current Opinion in Infectious Diseases, 14*(5), 527-538. http://dx.doi.org/10.1097/00001432-200110000-00005

Vose, D. (2008). *Risk Analysis – A quantitative guide* (3rd Ed.). John Wiley & Sons, Ltd. Chichester, UK.

WHO. (2005). Epidemic-prone disease surveillance and response after the tsunami in Aceh Province, Indonesia. *Weekly Epidemiological Record, 80*, 157-164.

Wong, D. C., Purcell, R. H., Sreenivasan, M. A., Prasad, S. R., & Pavri, K. M. (1980). Epidemic and endemic hepatitis in India: evidence for a non-A, non-B hepatitis virus aetiology. *Lancet, 2*(8200), 876-879. http://dx.doi.org/10.1016/S0140-6736(80)92045-0

Yamai, S., Okitsu, T., & Katsube, Y. (1998). Detection of *Vibrio cholerae* from river water. (in Japanese) *Journal of the Japanese Association for Infectious Diseases, 70*(12), 1234-1241.

Yoshii, T., Imamura, M., Matsuyama, M., Koshimura, S., Matsuoka, M., Mas, E., & Jimenez, C. (2013). Salinity in soils and Tsunami deposits in areas affected by the 2010 Chile and 2011 Japan Tsunamis. *Pure and Applied Geophysics, 170*(6-8), 1047-66. http://dx.doi.org/10.1007/s00024-012-0530-4

The Impact of Transport Infrastructure Modernisations on Acoustic Climate on the Example of the City of Szczecin (Poland) Intersections Redevelopment Effects

Katarzyna Sygit[1], Witold Kołłątaj[2], Marian Sygit[1,3], Barbara Kołłątaj[4], Ryszard Kolmer[5], Renata Opiela[5] & Paweł Zienkiewicz[1]

[1] Department of Physical Education and Health Promotion, University of Szczecin, Poland

[2] Department of Paediatric Endocrinology and Diabetology, Medical University of Lublin, Poland

[3] Department of Physical Education and Health Education, University of Szczecin; Institute of Rural Health, Lublin, Poland

[4] Department of Epidemiology, Medical University of Lublin, Poland

[5] Voivodship Sanitary and Epidemiological Station in Szczecin, Poland

Correspondence: Katarzyna Sygit, Department of Physical Education and Health Promotion, University of Szczecin, Al. Piastów 40 b, bl.6 Szczecin 71-065, Poland. E-mail: ksygit@poczta.onet.pl

Abstract

The source of most noise worldwide is mainly caused by machines and transportation means, including motor vehicles such as cars, buses, trains, aircrafts and so on.

The excessive noise, called noise pollution, may harm the activity or balance of human or animal life. Noise pollution can cause annoyance, aggression and sleep disturbances. Chronic exposure to noise may cause noise-induced hearing loss, tinnitus and contribute to cardiovascular problems such as hypertension as well as increased incidence of coronary artery disease. Such may bring about deterioration in the wellbeing of people and increase the number of days of incapacity for work.

This paper is an attempt to analyze the impact of transport infrastructure modernisations on the noise pollution in the city of Szczecin. The main objective of this paper was to compare the level of traffic noise in the areas surrounding streets: *Powstańców Wielkopolskich, Mieszka I and Aleja Piastów Streets* and crossroads of the streets: *Taczaka-Łukasińskiego* as well as *Taczaka-Derdowskiego* before and after the modernizations.

The comparison of obtained results suggest that in some cases the modernization hasn't influenced on noise levels. In some, it improved the acoustic situation but hasn't reduced the noise to keep acceptable levels.

The results emphasizethe thesis that some accepted methods of streets and crosswords modernization are sometimes ineffective in the fight against noise pollution.

Conclusions: Modernization of intersections in Szczecin improved traffic flow but had a little impact on the noise levels. Modernisations that improve the traffic flow can cause even increment in noise pollution. It should be taken into consideration possible benefits of used methods of city traffic modernization related not only to traffic improvements but also to noise pollution reduction. We suggest computer aided stimulations and acoustic specialist advices prior to any restructures of city traffic. To minimize the noise pollution, comprehensive solutions are needed.

Keywords: noise pollution, pollution monitoring, city traffic, modernization, health

1. Introduction

Noise (acoustic noise) is any unwanted, unpleasant, annoying and even harmful sound.

It accompanies human beings all over the world. A certain level of background sound, depending on the time and place of human life is even necessary for the well-being and welfare of man. The complete silence is irritating and influences on human psyche in unwanted way, on the other hand random and loud sounds disturb people, can cause permanent irreversible hearing damage as well as manyother biological as well as psychical harmful effects.

The sounds that are too loud, unpleasant, unexpected, or undesired are called noise. The noise, regardless of its origin, the intensity and duration is a bothersome factor for humans and for the environment (Brzeźnicki, Bonczarowska, & Gromiec, 2009; Iwanek, Kobus, & Mitosek, 2007) including animals (Jaeger et al., 2008; Moura et al., 2008; Zhang, Chen, Gao, Pu, & Sun, 2008), plants (Watts, Chinn, & Godfrey, 1999) and buildings (Akdag, 2004; Naticchia & Carbonari, 2007).

Studies of noise exposure suggest some associations with hypertension and cardiovascular diseases (Kempen, 2011; Bluhm & Eriksson, 2011). There was described association of aircraft and road traffic noise with psychological symptoms such as depressiveness and nervousness and some psychiatric disorders (at much higher noise levels) (Stansfeld & Matheson, 2003).

It is very difficult to measure parameters of any noise that are described by the such words as "unpleasant" or "annoying" - they are subjective. To compare and define the noise, the parameters: sound level and frequency of sounds are used. Usually the noise if defined as sounds with the range of frequencies between 16 Hz and 16 000 Hz (Dz. U, 2001). Often noises are described only by the most characteristic parameter - sound level expressed in decibels (dB). The dB is a logarithmic unit used to describe a ratio. The threshold of hearing is assigned a sound level of 0 decibels (abbreviated 0 dB); this sound corresponds to an intensity of $1*10^{-12}$ W/m^2. If one sound is 10^n times more intense than a sound that corresponds to the threshold of hearing, then it has a sound level equal to 10 x n decibels. For example military jet take-off which is 10^{14} times greater than the threshold of hearing has the intensity of $1*10^2$ W/m^2 and the level: 10 x 14=140 dB. Threshold of pain means the level 130 dB, instant perforation of eardrum may be caused by the sound level close to 160 dB.

The noise is all around us, it accompanies us during our work and rest, it is in our houses, at streets, in means of communication, restaurants, gardens, parks, shops and so on. The structure of the noise that reaches us consists of many components including traffic noise, rail noise, industrial noise, aircraft noise as well domestic noise.

The most common types of noise that affect people living in modern cities are: traffic noise as well as industrial, residential and housing noises (Kucharski, 1996).

The most troublesome and the most common, especially in the urban environment, are noises coming from motor vehicles and other modern sources of locomotion and transportation. In advanced as well developed countries, roadway noise contributes a proportionately large share of the total societal noise pollution. In some of them (such as many European countries and the USA (Miedema, 2001) traffic noise contributes more to environmental noise exposure than any other noise sources.

Industrial noise refers to noise that is created in the factories. Such noise adversely affects not only the workers but also people living close to industrial plants.

Domestic noise (residential) refers to noise nuisance coming from the following sources: playing of amplified music, playing of musical instruments, loud TV and video, parties and barbecues, barking dogs, neighbours activity and possible their anti-social behaviours,ventilation systems, kitchen utensils, other home appliances as well as intruder alarms.

Community noise (also called environmental noise, residential noise) is defined by The World Health Organization as noise emitted from all sources except noise at the industrial workplace. This term includes: domestic noise and traffic noise (traffic noise is a combination of the noises produced by vehicle engines, exhaust, and tires).

The most troublesome and the most common, especially in the urban environment, is noise from motor vehicles (traffic noise, roadway nose).

Traffic noise has become a serious problem nowadays because of inadequate urban planning of the city in the past. Homes, schools, offices, hospitals, commercial business centres, and other community buildings have been routinely built close to the main roads of the municipality without buffer zones or adequate sound proofing - such neglect of the past make life in many cities really difficult. The problem has been compounded by increases in traffic volumes.

All over the civilized worlds, the traffic noise is the most annoying kind of noise.

In most developed countries, sound-pressure levels of road communications are high, ranging from 60-90 dB. It is accepted that sound levels in cities should not exceed the range between 60-70 dB (A), with suburban levels between 50-60 dB(A). The World Health Organisation has set guideline levels for annoyance at 55 dB(A) representing daytime levels.

In the 1999, in European Union countries, about 40% of the population were exposed to road traffic noise with an equivalent sound pressure level exceeding 55 dB (Berglund, Lindvall, & Schwela, 1999) during the daytime and 20% - to levels exceeding 65 dB, at night, more than 30% were exposed to equivalent sound pressure levels exceeding 55 dB (Berglund et al., 1999).

In 2009, the problem of noise pollution in residential areas concerned 30% of people in Romania and Cyprus, 52.8% in Germany, 25.3% in Netherlands 17.7% in Poland and 11.8% in Norvay as well as 12% in Esonia (Rybkowska & Schneider, 2011).

Noise was became irritating and increasing problem in big cities, especially in dwellings located close to busy streets, airports, railway linesand big factories.

In 2012, 24% of the population of Dublin region were exposed to undesirable night time sound levels and 5% were exposed to day time sound levels exceeding 70dB(A) (Noise Maps, 2012). In Paris (in 2007) 59% of the population were plagued by noise, more than 7% of the inhabitants of the French capital were exposed to noise levels exceeding 71 dB (traffic noise measured year-round between 6 a.m. and 10 p.m.) (Bruitparif, 2013).

The Environmental Noise Directive of the European Parliament and Council (EN Directive, 2002) suggests EU Member States to produce strategic noise maps in their main cities and in vicinity of transport infrastructures as well as and near industrial sites. Mentioned actions may enable making a diagnosis of noise pollution in Europe and give suggestions that can be implemented in terms traffic modernization planning and other actions called acoustical planning.

Awareness of the unfavourable health effects of noise pollution forces local authorities to take into consideration the influence of any modernizations of traffic routes on human environment. It is recommended to make efforts to control environmental noise in cities by using computational models for urban planning and traffic modernization. These models are considered as extremely helpful for environmental management and decision-making for solutions to potential environmental risks, including urban noise (Zannin & de Sant'Ana, 2011).

Currently many realized projects of modernizations of the streets and intersections are aided by experimental measurements and software simulations, especially in cases when the prediction cannot be performed in simple way (Zurita, Parrondo, Díaz, & Corrales, 2005; Guarnaccia, 2010a, 2010b).

2. Aim

The fight against noise pollution and its negative effects is a priority in the policy in the European Union, which resulted in the adoption in 2002 the Noise Directive (EN Directive, 2002). Requirements for acceptable noise in Poland are set out in the Regulations of the Minister of Environment of 29th July 2004 and 14th June 2007 on permissible levels of environmental noise (Dz. U., 2004; Minister of Environmental Protection, 2007). Because of the fact that in many Polish cities noise levels exceed the permissible noise limit values, it is the urgent need to make efforts to change the *status quo*.

To do this, the following means and procedures are necessary: actual noise levels maps, changes in priorities in traffic modernisations procedures and permanent supervision of influences of any changes in traffic characteristics (including fluctuations of traffic volume, vehicles movement velocity and flow of the traffic) on noise pollution.

Our paper deals with such aspect of noise pollution supervision.

The main objective of this study was to assess the impact of restructuring and modernization of communication systems (intersections) in Szczecin on traffic-related acoustic noise pollution.

For this purpose, the traffic noise levels were estimated - prior and after the modernisations of three intersections in Szczecin: 1) *Powstańców Wielkopolskich, Mieszka I and Aleja Piastów*, 2) *Łukasińskiego-Taczaka* and 3) *Taczaka-Derdowskiego*.

Such observations concerned intersections, typical elements of city traffic system, being sources of noise pollution.

All the intersections, mentioned above, have been the sources of noise exceeding the acceptable levels.

3. Material and Methods

The noise levels were analysed before and after the modernisation of three busy intersections in Szczecin (Figures 1-4).

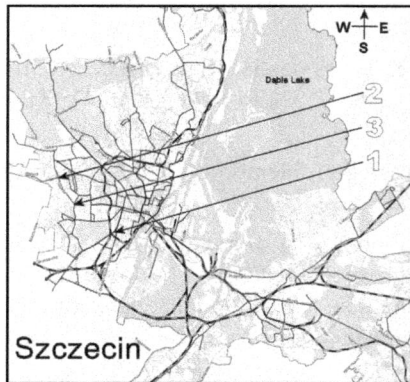

(1) the intersection: *Powstańców Wielkopolskich, Mieszka I and Aleja Piastów Streets*

(2) the intersection: *Łukasińskiego-Taczaka Streets*

(3) the intersection: *Taczaka-Derdowskiego Streets.*

Figure 1. Szczecin. Location of intersections where the noise measurement points have been settled

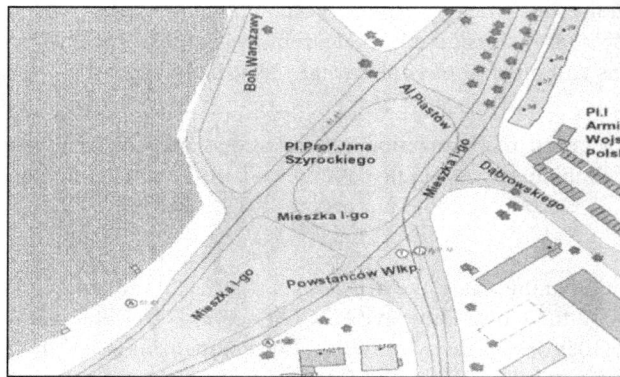

Figure 2. Noise measuring points (grey asterisks) around the intersection: *PowstańcówWielkopolskich, Mieszka I and Aleja Piastów.* Topographical map presenting the *status quo* before the modernisation

Figure 3. Szczecin. The intersection: *Łukasińskiego-Taczaka.* Topographical map presenting the *status quo* before the modernisation. Localisation of noise measurement points (grey stars)

Figure 4. Szczecin. The intersection: *Taczaka-Derdowskiego*. Topographical map presenting the *status quo* before the modernisation. Localisation of noise measurement points (grey stars)

Szczecin is the capital city of the West Pomeranian Voivodeship in Poland situated in the vicinity of the Baltic Sea, being the country's seventh-largest city and a major seaport in Poland. In 2011 its population was close to 408,000. The traffic noise problems described in Szczecin are typical also for other major cities in Poland.

The intersection *Powstańców Wielkopolskich, Mieszka I and Aleja Piastów* is located near the centre of Szczecin, in high-density residential area. It conducts a road traffic coming from the city centre of Szczecin to the western district of this city ('*Pomorzany*') inhabited by more or less 22 000 people and then to Germany (Motorvay A6).

The intersection *Taczaka-Derdowskiego* is located near a large housing estate '*Kalina*' and is the beginning of the bypass route of Szczecin and leads the road traffic to Police (city, 42 000 inhabitants). In the immediate vicinity of the intersection there is a residential area with single family housing.

The third intersection: *Łukasińskiego-Taczaka* is a part mentioned above bypass route of Szczecin leading the road traffic to Police. It is located near a residential area with single family housing, too.

The topographies of all three intersections are shown as the Figures 1-4.

All the intersection are equipped with traffic lights. All the measurements were made during the daytime, so at the time the road traffic was most intense.

The noise measurement was performed by the use of automatic sonometers SON - 50 and IM - 02 measuring equivalent noise levels. The characteristics of mentioned sonometers are as follows:

-   SON - 50 (Integrating SOUND LEVEL METER SON - 50; Producer: Sonopan) is an integrating sound level meter that measures levels of 3 types of signals: stable, unstable and pulse ones. The meter enables two independent measurement: RMS for measuring Leq, LMX, LMN, and second - to measure the peak. Accuracy class: 1; Measuring range: 15- 135 dB; range of RMS detector: 60 dB; dynamic characteristics: SLOW, FAST.

-   IM - 02(Integrating SOUND LEVEL METER IM-02; Producer: Sonopan) - an integrating sound level meter that measures levels of 3 types of signals: stable, unstable and pulse ones. Accuracy class: 2; Measuring range: 20-135 dB; range of RMS detector: 63 dB; frequency characteristics: A, C, Lin.

The noise measurements were realized in sampling points marked by asterisks on the attached maps (Figures 2-4). Measurements were taken at 1.0 m distance from curbs or elevations of residential buildings. The microphones were placed 1.2 ± 0.1 m above the ground surface.

The measured noise level was the result of noise levels coming various sources related to the use of vehicles; in the vicinity of measuring points there were no other (than traffic) sources of noise.

There were measured levels of only nonstable noise.

The measurements were provided during the following weather conditions: no precipitations, puffs of wind not exceeding 2m/s.

4. Results and Discussion

The results of measurements are shown as Graphs -Figures 5-13.

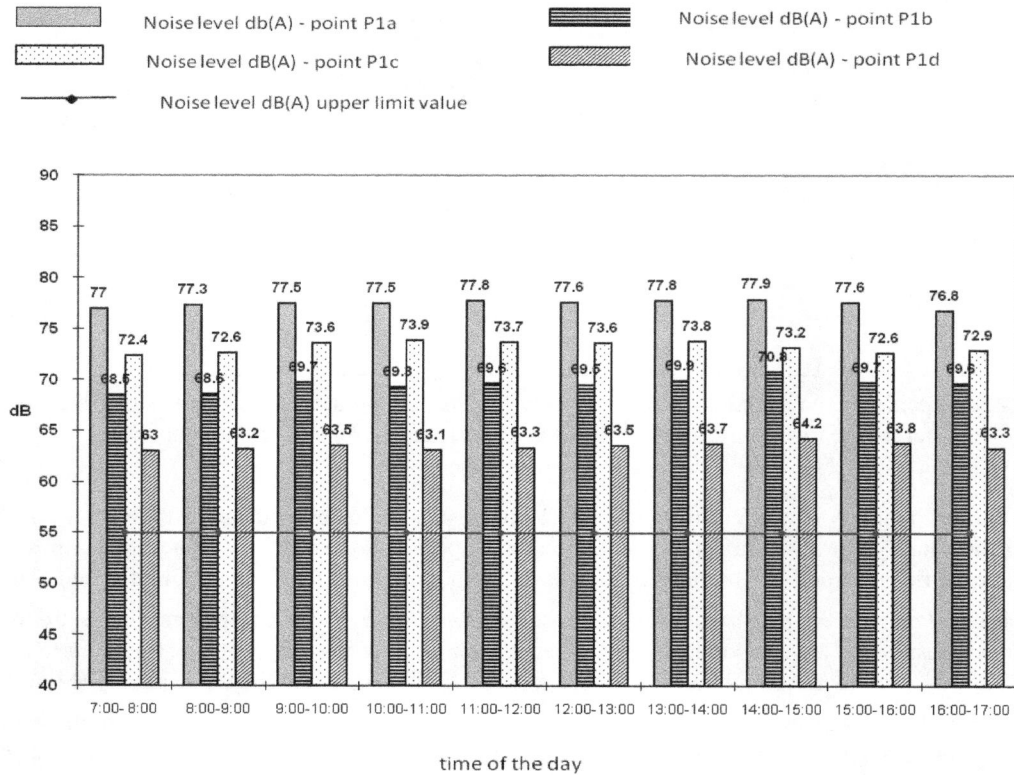

Figure 5. Noise level, depending on the time of day in the area of streets *Powstańców Wielkopolskich, Mieszka I and Aleja Piastów* measured before modernization the intersection (data from four measuring points: P1a, P1b, P1c, P1d)

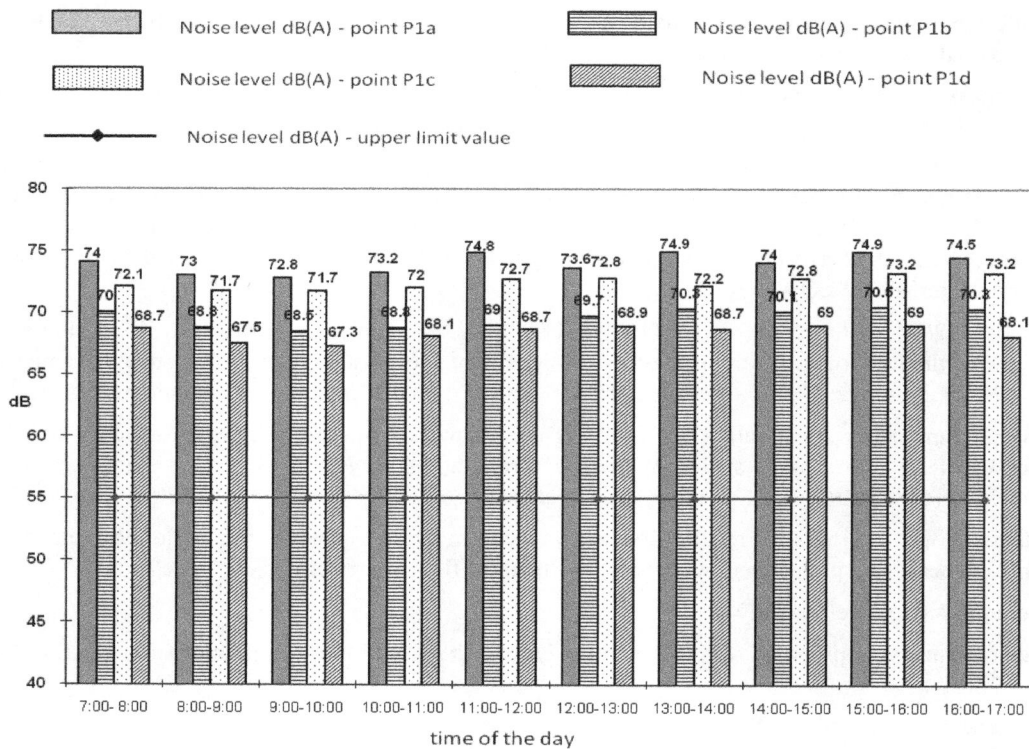

Figure 6. Noise level, depending on the time of day in the area of streets *Powstańców Wielkopolskich, Mieszka I and Aleja Piastów* measured after the intersection modernization (data from four measuring points: P1a-P1d)

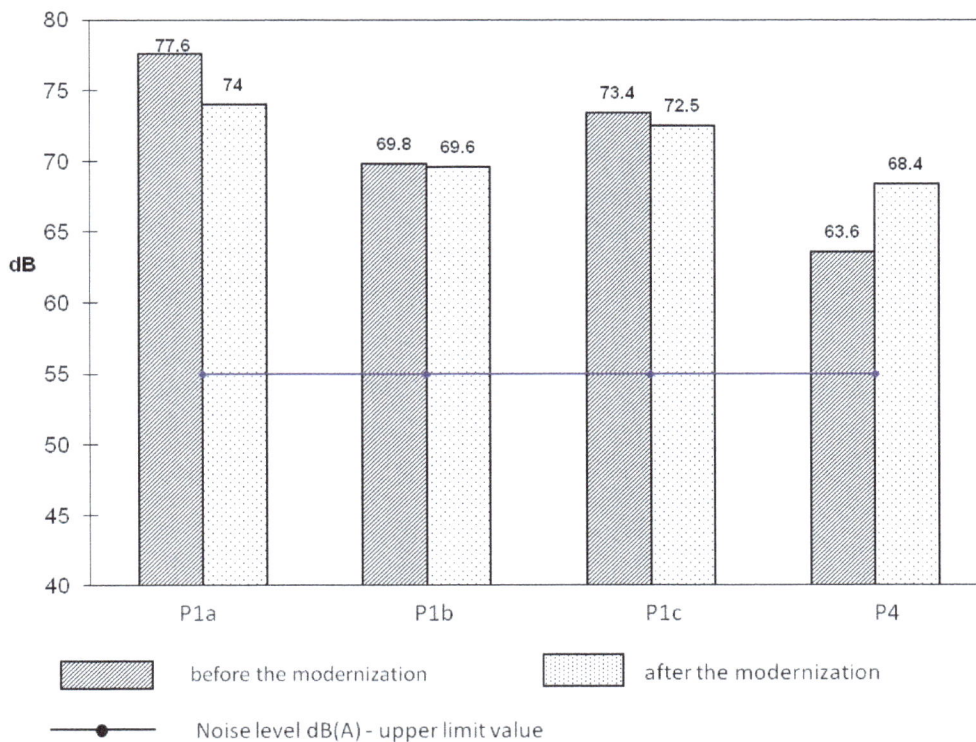

Figure 7. Average noise levels close to four measuring points (P1a-P1d) before and after the modernisation of the intersection: *Powstańców Wielkopolskich, Mieszka I and Aleja Piastów*

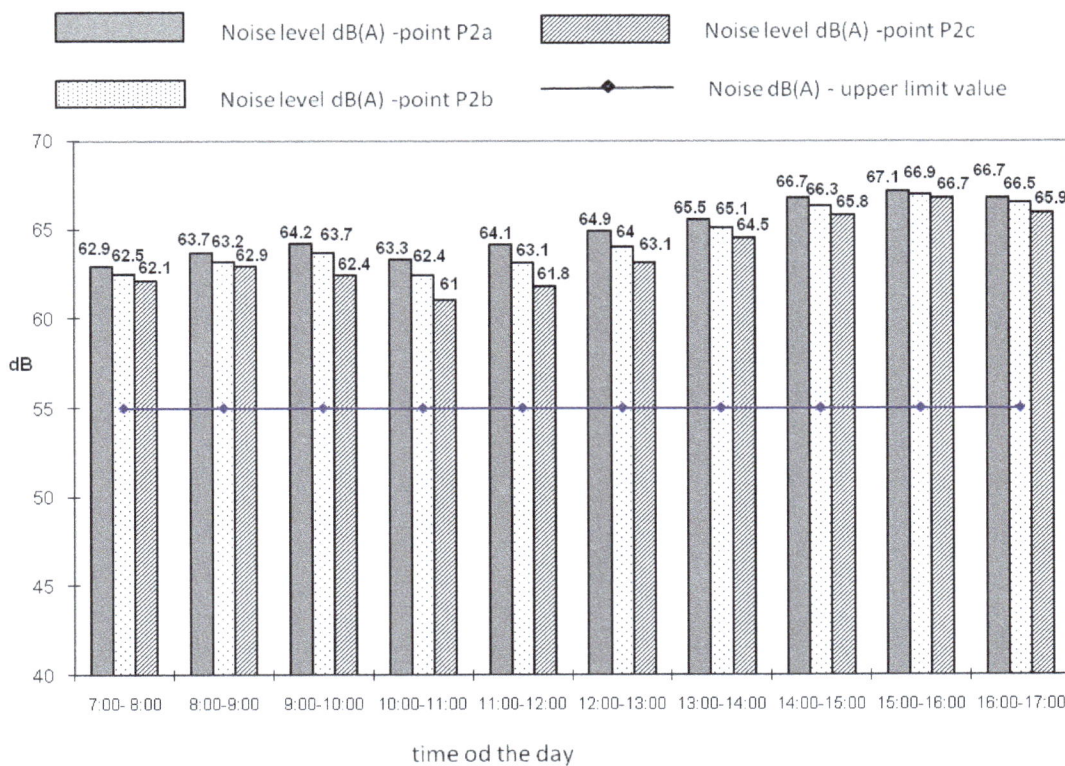

Figure 8. Noise level, depending on the time of day in the area of streets *Łukasińskiego-Taczaka* measured before the intersection modernization (data from three measuring points: P2a-P2c)

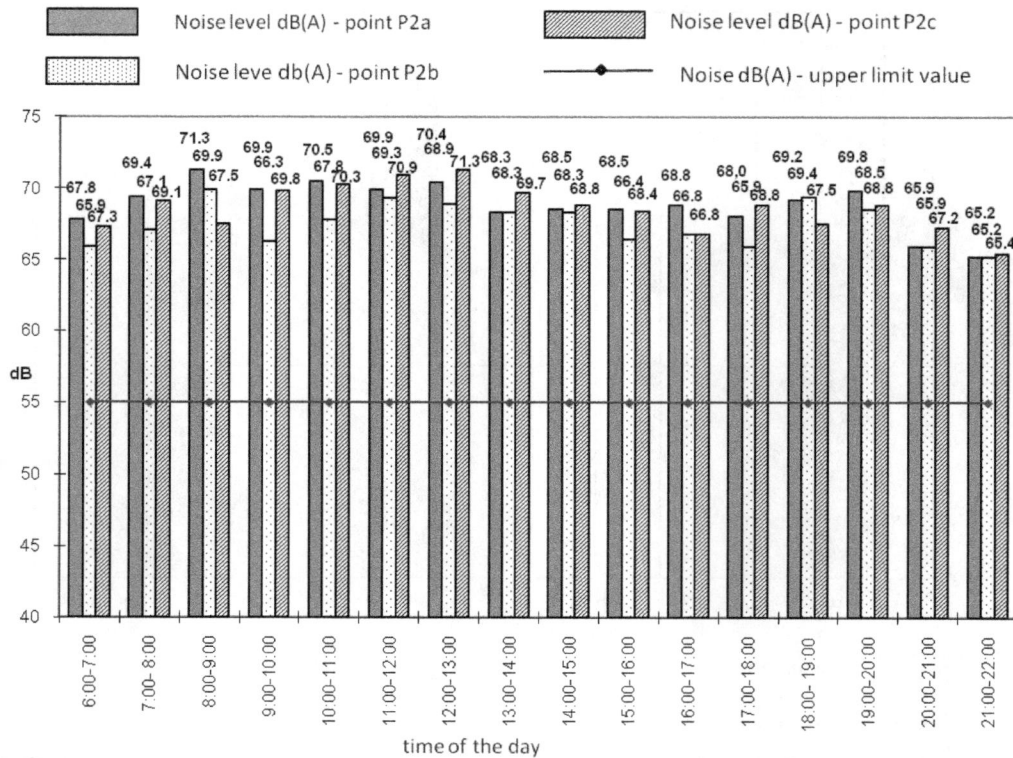

Figure 9. Noise level, depending on the time of day in the area of streets *Łukasińskiego-Taczaka* measured before the intersection modernization (data from three measuring points: P2a-P2c)

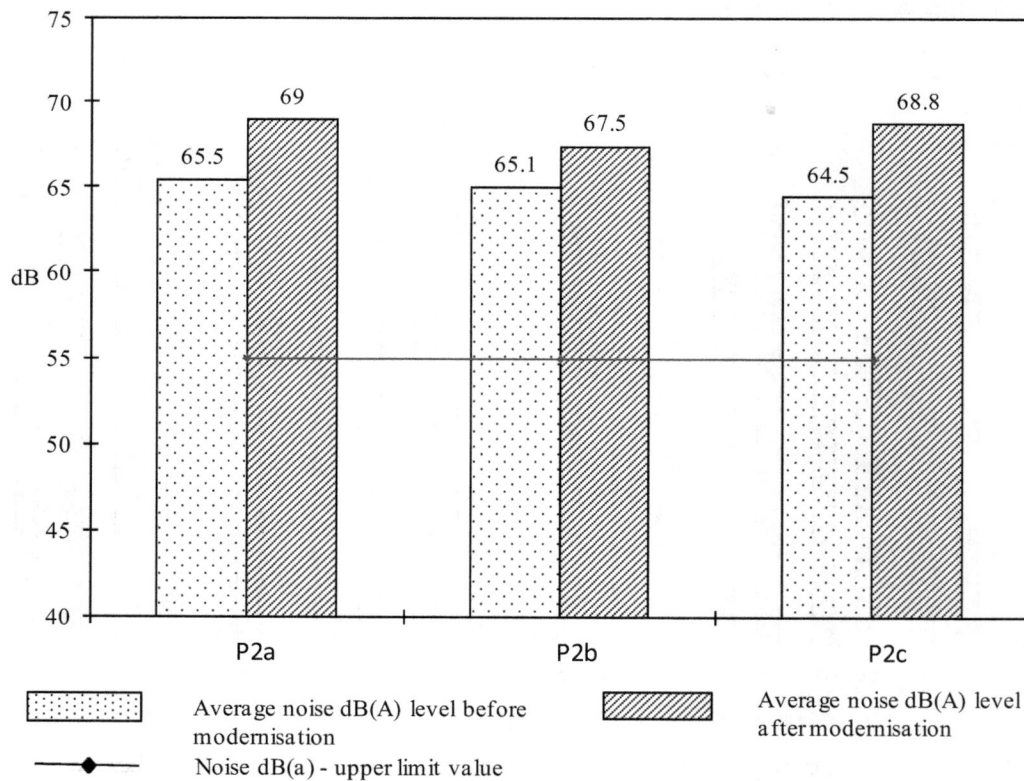

Figure 10. Average noise levels before and after modernization of the intersection: *Łukasińskiego-Taczaka* - data from the measurement pointsP2a-P2c

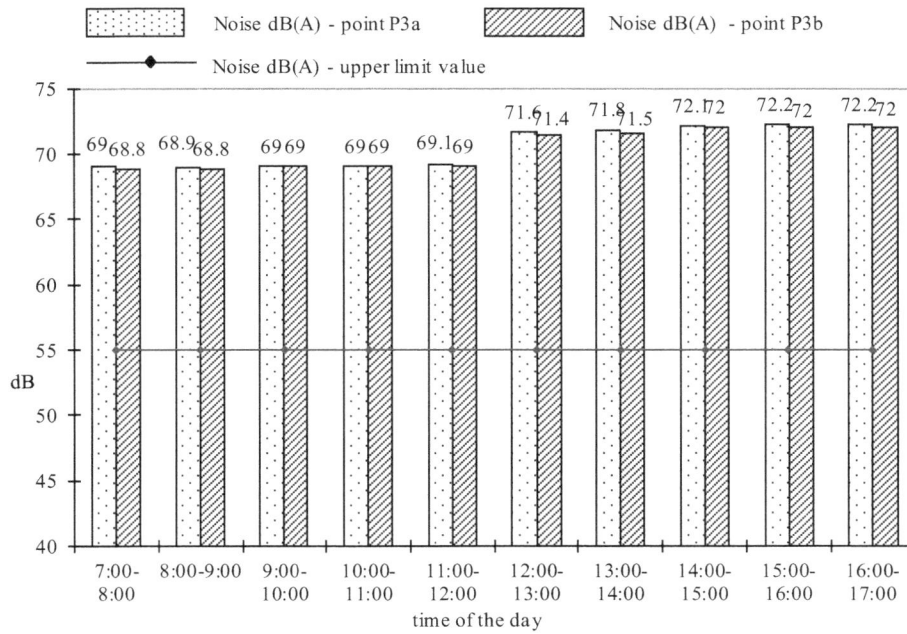

Figure 11. Noise level, depending on the time of day in the area of streets *Taczaka-Derdowskiego* - measured before the intersection modernization (data from two measuring points: P3a and -P3b)

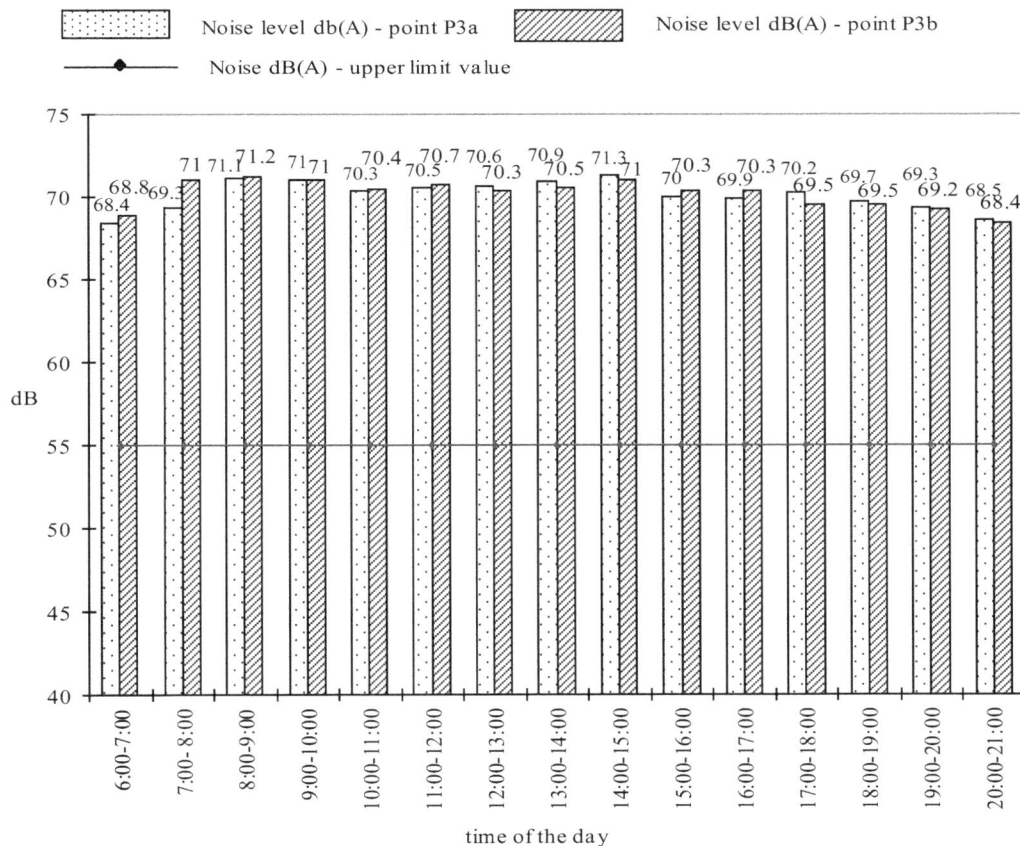

Figure 12. Noise level, depending on the time of day in the area of streets *Taczaka-Derdowskiego* measured after the intersection modernization (data from two measuring points: P3a and–P3b)

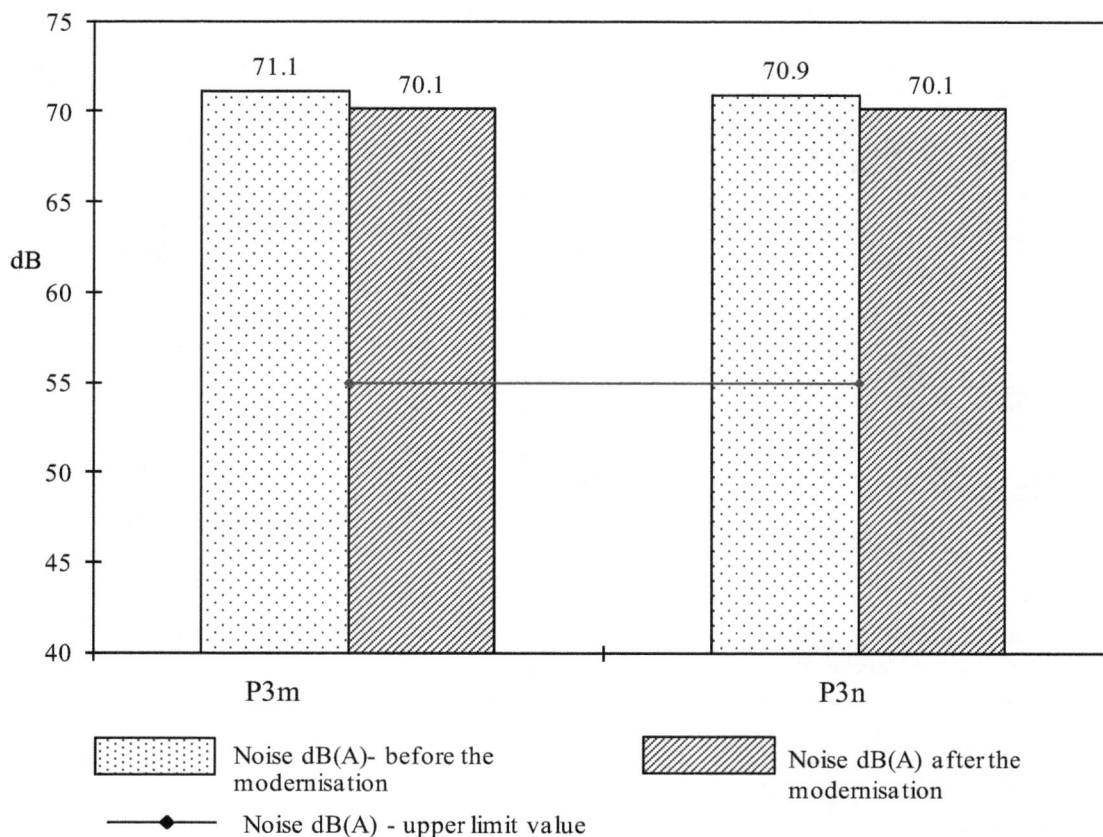

Figure 13. Average noise levels in two measuring points (P3m, P3n) exposed to highest noise levels. Data obtained before and after the modernisation of the intersection: *Taczaka-Derdowskiego*

Table 1. Permissible noise levelsin the environment according to the Regulation of the Minister of Environmental Protection, Natural Resources and Forestry (Dz. U, 2004)

Item in the Regulation	Areas	permissible noise levelsdB(A)	
		daytime	night
(1)	a) health-resorts b) hospital areas if located outside cities	45	40
(2)	a) single-family housing areas b) land buildings where children or young people live or stay for many hours c) areas of nursing homes d) hospitals in urban areas	50	40
(3)	a) multi-family housing and collective housing areas b) single-family housing areas with craft services c) recreational areas outside the city d) farms	55	45
(4)	downtown areas and cities with over100 thousands of residents with dense residential development and high concentration of administrative, commercial and service buildings	55	45

The results were compared to noise levels norms accepted by the Polish Ministry of the Environment in July of 2004 (Regulation of the Polish Minister of Environmental Protection, Natural Resources and Forestry on the permissible noise levels in the environment - 29.07.2004) (Dz. U., 2004) - Table 1. Location of measurement points (city > 100 000 inhabitants) and type of residential areas (high density housing) surrounded intersections gives the impression that the upper limits of permissible noise levels are defined by items 3 and 4 present in mentioned Regulation of the Polish Minister of Environmental Protection, Natural Resources and Forestry on the permissible noise levels in the environment (Table 1).

As is clear from the collected and shown as figures data (Figures 5-13), all the registered noise levels (both prior and after modernisations) exceed the upper limits of the applicable standards.

Comparison of differences in noise levels recorded before and after the modernizations of intersections made possible the evaluation of environmental profits obtained by carried out modifications.

The basic assumptions of modernization were: improvements in safety and traffic flow as well as in reducing the noise levels and then - improvement the acoustic conditions for people temporarily staying in the intersections areas, as well as living in the surrounding buildings.

The modernization of *Taczaka-Derdowskiego* intersection reduced the average noise level by 1 dB, the modernization of *Powstańców Wielkopolskich, Mieszka I and Aleja Piastów* intersection reduced the average noise level by 0-3.6 dB (different data in different measurement points), although in one measurement point there was noticed an increase in the noise level (about 5 dB). The modernization of *Łukasińskiego-Taczaka* intersection made the acoustic situation worse (an increase in the noise level by about 4 dB in almost all measurement points).

Therefore, in conclusion, it can be said that one of the most important goal (improvement in the acoustic conditions) has not been met. As a result, costly modernizations has given the effect: minimal improvement or even deterioration in the acoustic situation.

Noticed unfavourable effect can be explained in many different ways:

- the modernization of intersections improved the traffic flow, so the number of vehicles increased and (instead of some improvements) the noise volume increased (proportionally to increment in traffic density)
- the better traffic capacity influenced on drivers. Many of them have chosen better traffic routes (with improved intersections) instead of other streets jam-packed with vehicles
- modernisation of intersections improved traffic and enabled faster movement. Higher speed of vehicles means increment in noise volume.

The past decade is the period of sudden increment in traffic volume in all civilized countries. Such conclusion concerns Poland too. There is no doubt that the infrastructure of current and future large cities is a critical issue in our society. Streets and highways will remain critical transportation conduits, so their maintenance and improvement will remain an important challenge.

The rapid development of many civilised countries and increasing wealth of society led to a rapid increment in number of means of communication and transportation. Such increment - on one side and troubles with rebuilding communication routes (reconstruction are costly and time-consuming, in large cities is not enough space for tracing any new routes or broadening old ones) - on the other side, made that in many cities transportation is still ineffective, time-consuming and producinga lot of noise.

The modernisations, although, are necessary may brig only some, sometimes almost imperceptible effects - especially when it comes to the fight against noise. Such conclusion may be draw when one analyses effects of modernisation of mentioned three busy intersections in Szczecin.

The similar conclusion were drawn in other countries, where roads and cross-sections modernisations were considered as necessary but not sufficient efforts to achieve the objective of significantly reducing public exposure to traffic noise (Bing & Popp, 2011). One of the best solutions is: improving traffic flows without traffic growth (Rauterberg-Wulff, 2010) but typical modernisation increases traffic volume.

It seems to be necessary to look for other conceptions of fighting against noise pollution in big cities (Secretary of State for Transport UK, 2013).

Among them, nine (mentioned beneath) seem to be the most interesting (Jakovljevic, Paunovic, & Belojevic, 2009; Pilkington, 2000; Lee, 2013; Laoghaire, 2013; Secretary of State for Transport UK, 2013; Siemens, 2010; Węcławowicz-Bilska, 2012; Giuliano, O'Brien, Dablanc, & Holliday, 2013):

- ring roads
- underground roads/tunnels
- zones restricted to pedestrians only
- zones with public transportation means only
- zones with no transit traffic
- zones with no truck traffic
- zones with vehicles maximum speed limit lowered 50, 40 or even 30 km/h
- zones for electric or hybrid vehicles only
- intelligent systems controlling traffic lighting schemes (intelligent traffic regulation) to reduce the average speed and improve the flow of the traffic.

The most radical are concepts suggesting total ban on vehicles movement or reducing the traffic flow velocity to 30 km/h. Such radical ideas give the best results but disrupt normal city life and provoke many controversies over such points of view. In Europe, there are areas where the 30 km/h (19 mph) speed limit is in force. Such areas are present in some cities as Dublin (GB), Vienna (Austria) and Graz (Austria).

Mentioned regulations are forced both to reduce traffic noise and to improve safety for pedestrians (road traffic safety project started in Sweden in 1997 - "Vision Zero" (Fahlquist, 2006) - policy that requires that fatalities and serious injurious will be reduced to zero by 2020).

Mentioned modifications make some problems. Some drivers follow the lower speed limit while others ignore it, disrupting traffic and increasing the potential for collisions between slower and faster drivers. There are suggestions that speed limits that are inconsistent with driver expectations will not be kept (Skerritt, 2013). The reducing the traffic flow velocity has implications being in conflicts with the ideas of modern society - a large workers' mobility, flexible working hours and frequent changes of employment.

Reduced traffic flow makes trouble with worker's mobility and by influencing on the increase in the time spent on traveling to and from the work, reduces the amount of time that employees can spend on rest or devote for the families.

Almost all concepts of big cities have taken place in times when factories, offices and other places of employment were close to houses or residential areas inhabited by workers. So, the traffic was small and adequate to necessities and to the technical development was not so enjoying and the problem of noise was not so irritating.

Nowadays we have old concepts of cities (and their functions) and new concepts of life and working. They are simply not compatible.

It is an urgent need for changes in the functioning of societies, forcing reduced demand for the use of individual means of transportation (cars). It is quite possible that a change in the concept of work - work at home for remote companies and facilities (operating via the Internet) would be a reasonable way out of the current stalemate.

5. Conclusions

(1) Modernization of intersections in Szczecin improved traffic flow but had a little impact on the traffic noise levels.

(2) Modernisations that improve the traffic flow can cause even increment in noise pollution.

(3) It should be taken into consideration possible benefits of used methods of city traffic modernization related not only to traffic improvements but also to noise pollution reduction.

(4) We suggest computer aided stimulations and acoustic specialist advices prior to any restructures of city traffic.

(5) To minimize the noise pollution, comprehensive solutions are needed.

References

Akdag, N. Y. (2004). A simple method to determinate required Rtr values of building envelope components against road traffic noise. *Building and Environment, 39*(11), 1327-1332. http://dx.doi.org/10.1016/j.buildenv.2004.03.007

Berglund, B., Lindvall, T., & Schwela, D. H. (Eds.) (1999). *Guidelines For Community Noise.* Geneva: World Health Organization.

Bing, M., & Popp, Ch. (2011). Noise action planning in agglomerations Reduction potentials based on the example of Hamburg. *Environ Mental research of the Federal ministry of the environment, Nature conservation and nuclear safety.* Texte, 17/2011.

Bluhm, G., & Eriksson, C. (2011). Cardiovascular effects of environmental noise: research in Sweden. *Noise Health, 13*(52), 212-216. http://dx.doi.org/10.4103/1463-1741.80152

Bruitparif. (2013). *The noise observatory in Ile-de-France, the Paris Region.* A non-profit organisation established in 2004 by the Regional council, as requested by the environmental associations. Retrieved October 11, 2013, from http://www.bruitparif.fr/en

Brzeźnicki, S., Bonczarowska, M., & Gromiec, J. (2009). Occupational exposure limits for polycyclic aromatic hydrocarbons. Current legal status and proposed changes. *Medycyna Pracy, 60*(3), 179-185.

Dz. U. (2001). Ustawa z dnia 27 kwietnia 2001 r. Prawo ochrony środowiska. *Dziennik Ustaw.* Nr 62 poz. 627.

Dz. U. (2004). Rozporządzeniu Ministra Ochrony Środowiska z dnia 29.07.2004 r. w sprawie dopuszczalnych poziomów hałasu w środowisku. *Dziennik Ustaw* Nr 178 poz.1841.

EN Directive. (2002). *The Environmental Noise Directive of the European Parliament and Council. Directive 2002/49/EC of 25 June 2002.* Brussels, Belgium: European Parliament and Council.

Fahlquist, J. N. (2006). Responsibility ascriptions and Vision Zero. *Accid Anal Prev, 38*(6), 1113-1118. http://dx.doi.org/10.1016/j.aap.2006.04.020

Giuliano, G., O'Brien, T., Dablanc, L., & Holliday, K. (2013). Synthesis of Freight Research in Urban Transportation Planning. *Transportation Research Board 2013 Executive Committee.* NCRFP Report 23.

Guarnaccia, C. (2010a). Analysis of Traffic Noise in a Road Intersection Configuration. *Wseas Transactions On Systems, 8*(9), 865-874.

Guarnaccia, C. (2010b). New Perspectives in Road Traffic Noise Prediction. *Last Advances in Acoustics and Music, 8*(9), 255-260.

Iwanek, J., Kobus, D., & Mitosek, G. (2007). Wybrane problemy zanieczyszczenia powietrza w Polsce 2006 rok [w]; Zanieczyszczenie powietrza w Polsce w latach 2005-2006. *Biblioteka Monitoringu Środowiska GIOŚ.* Retrieved from http://www.gios.gov.pl

Jaeger, J. A. G., Bertiller, R., Schwick, C., Müller, K., Steinmeier, C., Ewald, K. C., & Ghazoul, J. (2008). Implementing Landscape Fragmentation as an Indicator in the Swiss Monitoring System of Sustainable Development (Monet). *J. Environ Manag, 88*(4), 737-751. http://dx.doi.org/10.1016/j.jenvman.2007.03.043

Jakovljevic, B., Paunovic, K., & Belojevic, G. (2009). Road-traffic noise and factors influencing noise annoyance in an urban population. *Environ Int., 35*(3), 552-556. http://dx.doi.org/10.1016/j.envint.2008.10.001

Kempen, Ev. (2011). Cardiovascular effects of environmental noise: research in The Netherlands. *Noise Health, 13*(52), 221-228. http://dx.doi.org/10.4103/1463-1741.80158

Kucharski, R. J. (1996). *Metody prognozowania hałasu komunikacyjnego, Biblioteka Państwowego Monitoringu Środowiska - PIOŚ – IOŚ.* Warszawa:Wydawnictwo SKON.

Laoghaire, D. (2013). Dublin Agglomeration Environmental Noise Action Plan December 2013 – November 2018. *Dún Laoghaire-Rathdown County Council.* Dublin.

Lee, D. (2013). Search for Quie. 2013 July. Civic Exchange. 23/F Chun Wo Commercial Centre, 23-29 Wing Wo Street, Central. Hong Kong. Retrieved September 11, 2013, from www.civic-exchange.org

Miedema, H. M. (2001). Annoyance from Transportation Noise: Relationships with Exposure MetricsDNL and DENL and Their Confidence Intervals. *Environmental Health Perspectives, 109*(4), 4019-4416. http://dx.doi.org/10.1289/ehp.01109409

Minister of Environmental Protection. (2007). Rozporządzenie Ministra Środowiska z dnia 14czerwca 2007 r. w sprawie dopuszczalnych poziomówhałasu w środowisku. *Dz.U.* 120 poz. 826.

Moura, D. J., Silva, W. T., Naas, I. A., Tolń, Y. A., Lima, K. A. O., & Vale, M. M. (2008). Real time computer stress monitoring of piglets using vocalization analysis. *Computers and Electronics in Agriculture, 64*(1), 11-18. http://dx.doi.org/10.1016/j.compag.2008.05.008

Naticchia, B., & Carbonari, A. (2007). Feasibility analysis of an active technology to improve acoustic comfort in buildings. *Building and Environment, 42*(7), 2785-2796. http://dx.doi.org/10.1016/j.buildenv.2006.07.040

Noise Maps. (2012). *Noise Maps, Report & Statistics. The Traffic Noise & Air Quality Unit. June 2012.* Dublin, Blk. 2, Floor 6. CivicOffices, Wood Quay, D8.

Pilkington, P. (2000). Reducing the speed limit to 20 mph in urban areas. Child deaths and injuries would be decreased. *BMJ, 320*(7243), 1160. http://dx.doi.org/10.1136/bmj.320.7243.1160

Rauterberg, W. A. (2010). Air Pollution & Noise Control in Berlin. Concepts and synergies. *Senate Department for Health, Environment and Consumer Protection, Unit III D,* M. Lutz.

Rybkowska, A., & Schneider, M. (2011). Population and social conditions. Statistics in focus. *Eurostat,* 4/2011.

Secretary of State for Transport UK. (2013). Action for RoadsA network for the 21st century. *The Stationery Office Limited on behalf of the Controller of Her Majesty's Stationery Office.* Crown copyright 2013.

Siemens. (2010). Adaptive network control Sitraffic Motion MX. The most intelligent answer to congestion and pollution. *Siemens AG Industry Sector Mobility Division Complete Transportation Intelligent Traffic Systems.* Hofmannstrasse 51, 81359 Munich. German.

Skerritt, J. (2013). Lower speed limit dangerous? City report says uneven compliance may bring new hazards to streets. *Winnipeg Free Press,* 01/8/2013.

Stansfeld, S. A., & Matheson M. P. (2003). Noise pollution: non-auditory effects on health. *Br Med Bull, 68*(1), 243-257. http://dx.doi.org/10.1093/bmb/ldg033

Watts, G., Chinn, L., & Godfrey, N. (1999). The effects of vegetation on the perception of traffic noise. *Applied Acoustics, 56*(1), 39-56. http://dx.doi.org/10.1016/S0003-682X(98)00019-X

Węcławowicz-Bilska, E. (2012). The city the future – trends, concepts, implementations. *Architecture, 1*(109), 323-342.

Zannin, P. H. T., & de Sant'Ana, D. Q. (2010). Noise mapping at different stages of a freeway redevelopment project - A case study in Brazil. *Applied Acoustics, 72*(8), 479-486. http://dx.doi.org/10.1016/j.apacoust.2010.09.014

Zhang, J., Chen, L., Gao, F., Pu, Q., & Sun, X. (2008). Noise exposure at young age impairs the auditory object exploration behavior of rats in adulthood. *Physiology & Behavior, 95*(1-2), 229-234. http://dx.doi.org/10.1016/j.physbeh.2008.06.005

Zurita, R., Parrondo, J., Díaz., Ó., & Corrales, J. A. (2005). *Numerical Simulation of the Noise Distribution due to Vehicle Traffic in Intersections of Urban Streets. Twelfth International Congress on Sound and Vibration. Lisbon 2005* (pp. 1-8).

11

A Preliminary Study on Genetic Variation of Arsenic Concentration in 32 Different Genotypes of Leafy Vegetable

Mathieu Nsenga Kumwimba[1], Xibai Zeng[1], Lingyu Bai[1] & Jinjin Wang[1]

[1] Institute of Environment and Sustainable Development in Agriculture, Chinese Academy of Agricultural Sciences / Key Laboratory of Agro-Environment, Ministry of Agriculture, Beijing, China

Correspondence: Zeng Xibai, Chinese Academy of Agricultural Sciences, Beijing 100081, China. E-mail: zengxb@ieda.org.cn

Abstract

Leafy vegetables are a food crop with higher protein and are also important source of minerals which are essential for good health. Due to the large consumption, it is necessary to decrease the arsenic (As) concentration in leafy vegetable to avoid the potential risk to human health. The current study is aimed at assessing arsenic (As) accumulation ability and identification of cultivars with less As concentration that could be grown in As contaminated farmland for food safety. A set of thirty two leafy vegetable cultivars from 5 species were compared in hydroponics for 2 weeks having moderate level of 0- control and 6 mg As L^{-1}. At harvest, plants were sampled and analyzed for As concentration. Significant genotypic variations were observed in the shoots As concentration, translocation and bioaccumulation factors revealing more than 8 and 25 times cultivar differences in shoot As concentration, and in translocation factors respectively. This result revealed that As concentration in shoot was in part governed by the greater ability of root-shoot translocation. Cultivar Sijibaiye (SJBY) had the lowest shoot As concentration while the highest was detected in Dayekongxincai (DYKXC). The average As concentration in roots were found to be ten to twenty times higher than those observed in shoots, indicating that there is restricted transport of As from the root system to the shoot of cultivars. Therefore, it has been suggested that there is possibility to lower the As concentration in leafy vegetables by selecting and breeding cultivars with less As concentration that can be safely grown in contaminated soils with the slight and moderate levels of As for safe consumption.

Keywords: arsenic (V), leafy vegetable, hydroponic screening, accumulation, cultivar variation

1. Introduction

Industrial activities such as mining, the disposal of industrial and municipal wastes in agricultural lands, excess use of fertilizers and pesticides have contaminated the large areas of agricultural soils with arsenic (As) in many countries (McGrath et al.2001). These pollutants are especially harmful where vegetables are grown in contaminated soils, which not only reduce the plant growth but also cause health hazards to the human beings as well as animals. Islam et al. (2007) reported that crop species are regulated by genetic basis for heavy metal transport and accumulation. However, the toxic effects of As may cause high susceptibility in some cultivars than others. The genotypic variation in the different cultivars for arsenic concentration, as a strategy to reduce the movement of As from contaminated soils to edible parts of crop plants has been receiving close attention (Norton et al., 2012). Researchers have developed the arsenic stress toxicity tolerant rice cultivars through locally-grown rice cultivars in Bangladesh, India and China (Norton et al., 2009). This will give a long-term effective and economical means of reducing As contamination in vegetables. Moreover, the fundamental requirement for breeding low As-accumulation cultivars is to know the genotypic differences and the mechanisms governing As accumulation in crops. Zia et al. (2011) found substantial genotypic variation in rice As accumulation, suggesting that genetic factors determined differences in As accumulation. Consequently, genetic variation in plants for uptake and translocation of As provides the potentiality to develop the cultivars with reduced dietary exposure to arsenic which either assimilates less As or restrict As translocation (Meharg et al., 2002). As the arsenic cannot be removed from the contaminated soils, therefore, there is a need to develop locally adapted leafy vegetable cultivars which have less arsenic accumulation. The uptake and translocation of As in plants vary greatly not only among plant species but also among cultivars within the same species, thus, an effective way to face this problem is growing of more As-excluder vegetable cultivars. The objectives of this study were (1) to screen and identify the cultivars with less As concentrations that can be grown in As contaminated farmland for food safety, (2) and also to obtain information on the basis of their genetic variation in the shoots As concentration.

2. Materials and methods

2.1 Plant Material and Preparation of Seedlings

Seeds of 32 cultivars of leafy vegetable belonging to 5 species including Romaine (Lactuce Sativa var. longifolia), Lettuce (Lactuce Sativa), Celery (Apium graveolens L.), and water spinach (Ipomea aquatica) (Table 1) were purchased from Chinese Academy of Agricultural Sciences, Beijing P. R. China and were cultivated at Institute of Environment and Sustainable Development in Agriculture, Beijing, China in 2012. The seeds were surface sterilized with 30% H_2O_2 for 15 min, rinsed five times with distilled water were sown in plates containing autoclaved vermiculite. On the 7th day, at third leaf stage, the healthy and uniform sized seedlings were carefully removed from vermiculite, and washed with distilled water then transferred to hydroponic bottles each containing 1500 ml of full-strength of modified Hoagland and Arnon with pH adjusted to 6.3 by using NaOH or HCl for 10 days. Nutrient solution was renewed twice a week.

2.2 Hydroponic Screening

On the 21st day of transplant to the hydroponic bottles, the As in the form of As(V) $Na_3AsO_4.12H_2O$ was applied at the concentration of 0- control and 6 mg As L^{-1} as the treatment in the basic nutrient solution for 14 days. There were four replicates for each cultivar and each replicate comprises four plants. The As treated nutrient solution was renewed once a week. Plants were grown in hydroponic bottles for 35 days, and then harvested. The shoots and roots of all harvested plants were separated carefully, washed with water, and then rinsed several times using deionizised water to remove adhering particles. These samples of roots and shoots were gently packed in envelopes. The concentration of As in cultivars was determined using Atomic Fluorescence Spectrometry (AFS) following the procedure described by Smith et al. (2009). Statistical analysis was performed by using SPSS software version 16.0 and Sigma Plot 10.0.

2.3 Kinetics of As(V) uptake

After 4 weeks of growth before the uptake experiment, the seedlings were transferred to the deionized water for 5 days, and then uniform seedlings from TXYLS and SJBY were selected washed from root system and transferred to plastic bottles containing 1200 ml of As uptake solution. Each bottle was wrapped with opaque plastic membrane. Eight different concentrations of arsenic (0, 0.5, 1, 2, 4, 6, 8, and 10 mg L^{-1}) were used to study the uptake kinetics of arsenic. Each treatment was replicated three times. Each 1200-mL container had 4 plants. The pH of the nutrient solution was maintained at 6.3. At each time interval (0, 0.5, 1, 2, 4, 6, 8, 10, 12, 22, and 24 h), sample of 1 ml was withdrawn from each uptake solution to measure depletion in As, as a result of its uptake by plants, the depletion period was for 24h. Water losses through transpiration and absorption were compensated by additions of deionized water at hourly intervals. After 24h of uptake experiment, the plants were harvested and quickly rinsed with distilled water. The plants were separated into roots and shoots, blotted dry with paper tissue and dried at 65°C for 72 h, and the dry weights were recorded. Arsenic concentrations were determined as previously described.

2.4 Calculations of As Uptake Rates and Kinetic Parameters

The amount of uptake rate (V [mg As g^{-1} shoot d.wt.hr^{-1}]) was calculated from the depletion of As in the uptake solution. As uptake by the plant per unit time per gram dry weight was calculated according to the following formula: (Ajaelu et al., 2011).

$$V = \frac{(Vi \times Ci) - (Vf \times Cf)}{t \times Bdwt} \tag{1}$$

Where Vi, Ci Vf, Cf and B d.wt are the volumes (V) and quantities in mg/L of substrate in solution at the start and the end of uptake experiment, and t is the time in hours and B dwt is the biomass dry weight of the plants, respectively. The reduction in the volume was very small, so this was not taken into account in the calculation.

The Michaelis-Menten Equation is:

$$V_0 = \frac{V_{max}.[S]}{K_m + [S]} \tag{2}$$

Where

V_0:   is the initial velocity of the reaction at substrate concentrations (S)

[S]:   is the substrate concentration

Vmax: is the maximum uptake rate ($mgAsg^{-1}$ shoot $d.wt.hr^{-1}$) achieved at which the enzyme is saturated with substrate

Km: is the Michaelis-Menten constant

The parameters Km and Vmax characterize the ability of plant to absorb nutrients from their soil environment [8]. The equation describes the relationship between the uptake rate of the nutrient and the nutrient concentration. Uptake rate reaches a constant or saturated rate (Vmax) at high ambient concentration. Lineweaver &Burk (1998)] pointed out that equation (2) becomes linear in form upon taking the reciprocal of both sides of the equation to calculate the key parameters in the Michaelis-Menten equation:

$$V_0 = \frac{V_{max}.[S]}{K_m+[S]} \rightarrow \left(\frac{1}{V}\right) = \left(\frac{K_m}{V_{max}}\right) \cdot \left(\frac{1}{S}\right) + \left(\frac{1}{V_{max}}\right)$$

i.e. (y= m x+ c), where (y=1/V and x= 1/[s])

In order to obtain the kinetic uptake parameters representing the maximum uptake rate (Vmax) and the half saturation constant (Km), by plotting 1/V against 1/[S] will give a straight line graph having a slope of Km/Vmax and a y- intercept on the ordinate at 1/Vmax. From this plot, the Km and Vmax values of arsenic uptake were determined by linear regression analysis, thus giving a convenient method for obtaining both Vmax and Km.

2.5 Statistical Analysis

All data were performed using One-way analysis of variance (ANOVA), using SPSS software version 16.0 and Sigma Plot 10.0. Significant differences among data were evaluated with the least significant difference (LSD) test at a 5% probability level. The tolerance index (TI) was expressed on the basis of root and shoot biomass, calculated as the following:

TI =100 x [biomass] As /[biomass]control

The translocation factor (TF), bioaccumulation factors (BCF) and As accumulation in shoot were calculated as follows:

TF = [As] shoot / [As] root

BCF = [As] shoot /[As]nutrient culture

As accumulation in shoot= [biomass] shoot x [As] shoot

Table 1. Name, abbreviation, origin, and seed provider of 32 leafy vegetable cultivars used for hydroponic experiment

No.	Cultivar name	Abbreviation	Origin	Seed provider
	Celery	**Apium graveolens L.**		
1	Mei qin	MQ	US	IVF, CAAS
2	Meiguo baili xiqin	MGBLXQ	US	IVF, CAAS
3	Boli cui shi qin	BLCSQ	China	IVF, CAAS
4	Meiguo wentu la xi qin	MGWLTXQ	US	IVF, CAAS
5	Texuan si ji xi qin	TXSJXQ	US	IVF, CAAS
6	Daye qin	DYQ	China	IVF, CAAS
7	Shengjie bai qin	SJBQ	China	IVF, CAAS
8	Siji xiao xiang qin	SJXXQ	Hongkong	IVF, CAAS
9	Wentula qin	WTLQ	US	IVF, CAAS
10	Riben xiao xiang qin	RBXXQ	Japanese	IVF, CAAS
	Lettuce	**Lactuce Sativa**		
11	Texuanyanling sun	TXYLS	China	BGSRDC
12	Sijibaiye	SJBY	China	BGSRDC
13	Huayesun jing yong woju	HYSJYWJ	China	BGSRDC
14	Yeyong woju	YYWJ	China	BGSRDC
15	Guayuanyiyesun	GYYYS	China	BGSRDC
	Romaine	**Lactuce Sativa,var. longifolia**		
16	Helanziye shengcai	HLZYSC	Netherlands	BGLATI
17	Nanhan zixiushengcai	NHZXSC	South Korea	BGLATI
18	Xingyun ruan wei	XYRW	China	BGLATI
19	Dasushengcai	DSSC	US	BGLATI
20	Jieqiushengcai	JQSC	China	BGLATI
21	Ziluolan	ZLL	China	BGLATI
	Water Spinach	**Ipomea aquatica**		
22	Chunguo kongxincai2	CGKXC2	China	BSVSPC
23	Dayekongxincai	DYKXC	China	BSVSPC
24	Chunguo kongxincai 1	CGKXC1	China	BSVSPC
25	Jingyanye kongxincai	JYYKXC	China	BSVSPC
26	Taiguo kongxincai	TGKXC	Thailand	BSVSPC
	Amaranth	**Amaranthus mangostanus L.**		
27	Huahongxianxcai	HHXC	China	BSBC
28	Yuanyebaixiancai	YYBXC	China	BSBC
29	Taiwanluxiancai	TWLXC	Taiwan	BSBC
30	Baiyexiancai	BYXC	China	BSBC
31	Tedayehuahongxiancai	TDYHHXC	China	BSBC
32	Jingxuanhongxiancai	JXHXC	China	BSBC

IVF, CAAS: The Institute of vegetable and flowers, Chinese Academy of Agricultural Sciences

BGSRDC: Beijing Gardening Seeds of Research and Development Center

BGLATI: Beijing Golden Land Agricultural Technology Institute

BSVSPC: Beijing Special Vegetable Seed and Plant Co.

BSBC: Beijing seed and breed center

3. Results

3.1 Genotypic Variations in Shoot As Concentration and Accumulation

As it can be seen in table 2, very wide genotypic variations in the average shoot As concentrations and accumulations among 10 cultivars of celery, 6 of romaine, 5 of lettuce, 6 of amaranth and 5 of water spinach, which were tested under As treatment (6 mg L^{-1}). Out of the 32 cultivars, As concentrations ranged from 14.07 to 112.16 mg kg^{-1} dry weight (8-fold variation). The average As concentrations in root was ten to twenty- two times higher than that in shoot (root > shoot) as an indicator for translocation ability from root to shoot. Figure 1 shows the variations among vegetable species in terms of As concentrations ranging from 14.065 to 20.555, 16.918 to 27.030, 21.418 to 47.778, 29.54 to 87.51 and 91.470 to 112.160 mg kg^{-1} dry weight for Lettuce, Romaine, Amaranth, Celery and Water spinach, respectively. Lettuce was significantly lower than romaine, amaranth, celery and water spinach. There were significant genotypic differences ($P < 0.05$) which accounted for 12.54% of the variation among them. The main results were shown as follows: there was a greater difference among species and different cultivars and were also significantly different in the ability to accumulate As. SJBY had the lowest shoot As concentration (14.065 mg kg^{-1} dry weight), followed by GYYYS, HYSJYWJ, XYRW, JQSC, YYWJ, etc. While the highest shoot As concentration were detected in DYKXC (112.160 mg kg^{-1} dry weight), followed by JYYKXC, CGKXC 2, CGKXC 1. It was so interesting that five species could be placed into 3 groups and can be differentiated corresponding to: (a) high, (b) intermediate and (c) low arsenic. Arsenic concentrations in shoot of 5 species were found to follow the trend: Lettuce <Romaine< Amaranth< Celery< Spinach (Figure 1). Shaibur et al. (2009) reported that the specialty of water-spinach is that it accumulated higher concentrations of As in shoot without showing visible toxicity symptoms. This was very similar to what we found in this study.

3.2 Arsenic accumulation

The data presented in Table 2 shows the amounts of arsenic (As) accumulated in the shoot of 32 cultivars. The values are expressed in µg-1 pot dry weight each calculated from the formula: (dry weight × As concentration) and they show significant different (p <0.05), as mentioned in Table 2 ranging from 61.79 to 188.47, 30.25 to 55.1, 26.03, 92.37 to 173.21 to 51.64, 16.75 to 71.76 µg^{-1} pot dry weight for celery, lettuce, romaine, water spinach and amaranth, respectively. Arsenic accumulated to the greatest extent in RBXXQ, followed by MGBLXQ, JYYKXC, CGKXC 1, CGKXC 2, etc. while the lowest shoot As accumulation was found in the TDYHHXC.

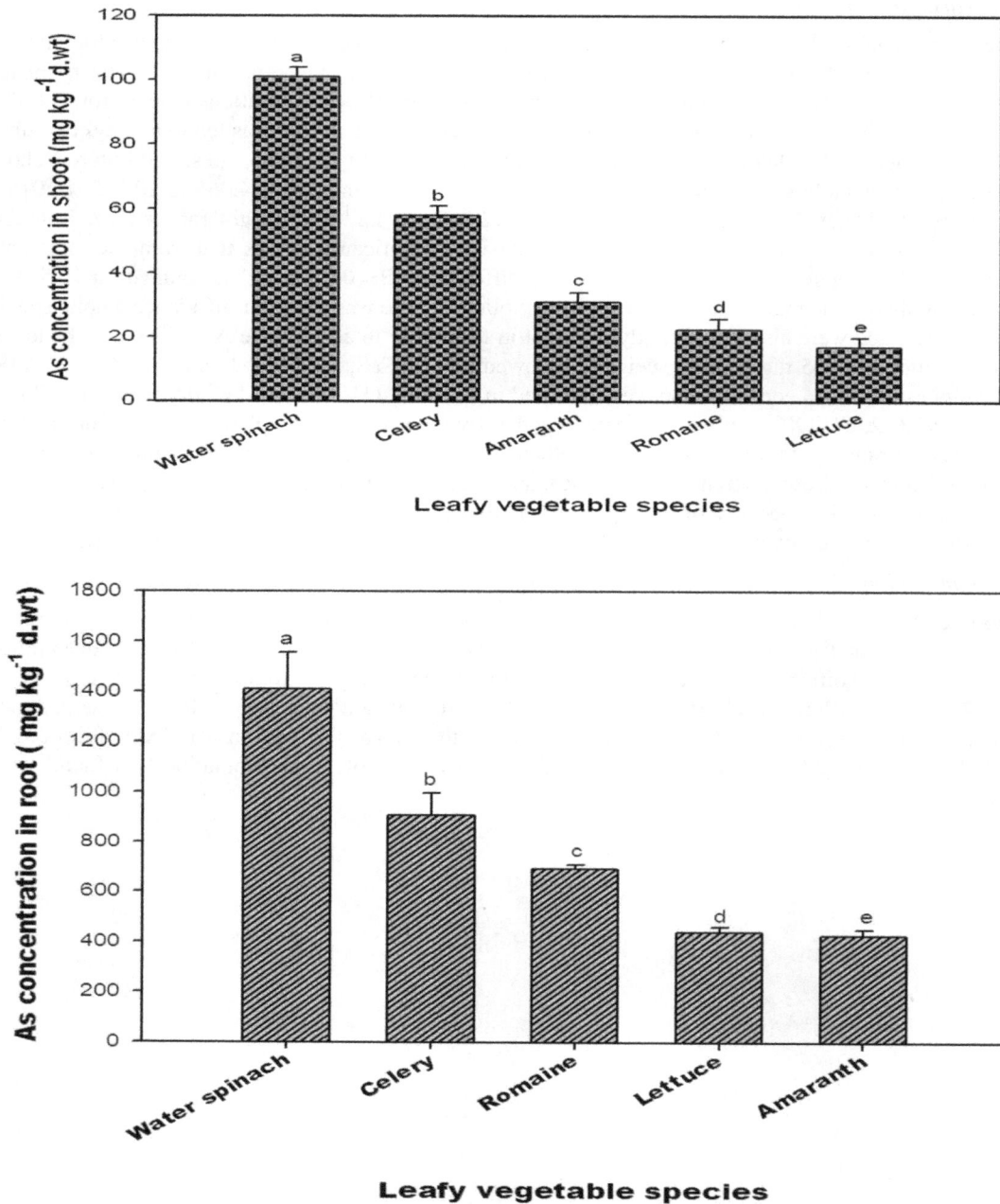

Figure 1. As concentration (mg kg^{-1} d.wt) in shoot and root of 5 group of species

Values are mean ± SD of the cultivars within the species. Different letters indicate significant differences at the 0.05 level based on LSD test.

Table 2. As concentration (mg/kg DW) and accumulation (μg^{-1} pot dry weight) of thirty two leafy vegetable cultivars.

Cultivar	As concentration	As accumulation
Celery		
Mei qin	56.51 ±1.70 ef	61.79±2.66 fghij
Meiguo baili xiqin	87.51 ±2.22 c	179.75±4.51 a
Boli cui shi qin	42.68±2.82 efgh	93.90±5.79 ef
Meiguo wentu la xi qin	33.74±1.50 ghijk	82.81±2.94 efg
Texuan si ji xi qin	61.00±8.88 de	93.92±14.98 ef
Daye qin	29.54±1.69 ghijk	25.56±1.51 jk
Shengjie bai qin	38.14±3.21fghij	49.47±4.13 ghijk
Siji xiao xiang qin	77.19±2.73 cd	90.06±4.34 ef
Wentula qin	77.14±2.37cd	79.82±2.65 efgh
Riben xiao xiang qin	77.17±0.74cd	188.47±12.93 a
Lettuce		
Texuanyanling sun	20.56±0.83 ijk	55.11±2.49 ghij
Sijibaiye	14.07±0.21 k	30.25±4.39 jk
Huayesun jing yong woju	16.27±0.71 k	45.07±3.06 hijk
Yeyong woju	17.67±0.63 jk	47.97±4.86 ghijk
Guayuanyiyesun	16.22±0.50 k	40.62±1.21 ijk
Romaine		
Helanziye shengcai	27.03±0.79 hijk	42.40±0.98 ijk
Nanhan zixiushengcai	20.47±1.35 ijk	31.27±2.44 jk
Xingyun ruan wei	16.92±0.84 jk	26.03±2.67 jk
Dasushengcai	25.62±0.30 hijk	30.46±1.37 jk
Jieqiushengcai	17.74±0.71 jk	27.23±1.38 jk
Ziluolan	25.91±0.27 hijk	51.64±1.86 ghijk
Water Spinach		
Chunguo kongxincai2	96.42±2.11 abc	139.50±10.09 cd
Dayekongxincai	112.16±16.45 a	92.37±14.63 ef
Chunguo kongxincai 1	94.44±16.8 abc	147.84±42.06 bc
Jingyanye kongxincai	110.39±19.33 ab	173.21±26.65 ab
Taiguo kongxincai	91.47±10.37 bc	109.53±12.14 de
Amaranth		
Huahongxianxcai	47.78±6.76 efg	71.76±11.93 fghi
Yuanyebaixiancai	22.63±2.23 hijk	25.69±2.50 jk
Taiwanluxiancai	40.51±3.91fghi	31.11±3.23 jk
Taiyexiancai	30.59±2.92 ghijk	30.31±5.41 jk
Tedayehuahongxiancai	22.36±1.67 hijk	16.75±1.27 k
Jingxuanhongxiancai	21.42±1.62 ijk	25.15±2.32 jk

Values (mean±S.E. n=4) with different letters in the same column are significantly different according to the duncan's test (P<0.05)

3.3 Genetic Variations in Shoot Biomass Trait Under As Stress

By growing 32 cultivars of leafy vegetable in the greenhouse, the responses of the selected cultivars differed and showed significantly differences for vegetable species ($P < 0.05$) and display various plant morphological differences that are predicted to be the cause of variation in biomass production. The shoot biomass (FW) of tested cultivars varied between 6.010- 47.315 (average 26.655) g pot^{-1} for celery cultivars, 46.340-60.505 (average 53.213) g pot^{-1} for lettuce, 34.610-64.350 (average 48.279) g pot^{-1} for romaine, 18.583-31.498 (average 22.29) g pot^{-1} for water spinach and 8.353-15.590 (average 10.5) g pot^{-1} for amaranth under 6 mg L^{-1} (Figure 2). This suggests that shoot biomass variation due to genetic differences among plant species and with cultivars was much greater. The maximum reduction in shoot biomass was observed in DYQ (48.2%), followed by BYXC (45%), YYBXC (39%), JXHXC (26.8%), MGBLXQ (26.5%), TGKXC (15.8%), YYBXC (15.7), TDYHHXC (12.5%), TXYLS (11%), RBXXQ (10.9%), JYYKXC (10%), TWLXC (9.8%), DYKXC (8.3%), YYWJ (8.8%), HYSJYWJ (7.6%), WTLQ (6.7%), SJXXQ (5.1%), NHZXSC (5.2%), SJBY(4.5%) (Table 3).

The tolerance index (TI) in terms of the biomass ratio of As treatment to control was high in all cultivars except DYQ, YYBXC and BYXC. By contrast, 69% of tested cultivars in 6 mg L-1 As showed a tolerance index higher than 90% (Table 3). The result suggests that shoot biomass variation due to genetic differences among cultivars was greater. MQ and CGKXC 2 cultivars had the highest shoots averaged over the level of arsenic used. Therefore, the effects of As on leafy vegetable growth and development appear to be cultivar dependent.

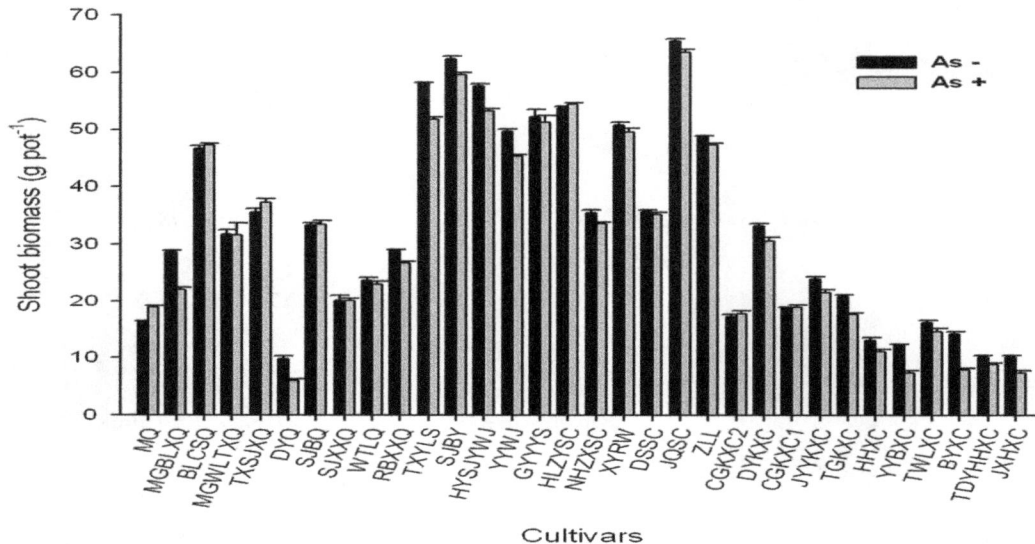

Figure 2. Shoot biomass of 32 cultivars exposed to 6mg L^{-1} As for 14 days

Data are expressed as means±S.E. of four replicates. p<0.05

Figure 3. The growth of leafy vegetable response to As stress

Table 3. Tolerance index (TI) based on shoot biomass of 32 different leafy vegetable cultivars exposed to 6 mg L^{-1} As for 14 days.

Cultivar	Tolerance index
Celery	
Mei qin	1.10±0.05 a
Meiguo baili xiqin	0.74±0.03 k
Boli cui shi qin	1.00±0.01 bcde
Meiguo wentu la xi qin	0.97±0.14 bcdefgh
Texuan si ji xi qin	1.02±0.05 bc
Daye qin	0.52±0.08 m
Shengjie bai qin	0.98±0.02 bcdefg
Siji xiao xiang qin	0.95±0.04 cdefghi
Wentula qin	0.93±0.03 defghi
Riben xiao xiang qin	0.89±0.02 ghij
Lettuce	
Texuanyanling sun	0.89±0.02 hij
Sijibaiye	0.96±0.02 bcdefghi
Huayesun jing yong woju	0.92±0.02 efghij
Yeyong woju	0.92±0.01 efghij
Guayuanyiyesun	0.98±0.02 bcdef
Romaine	
Helanziye shengcai	1.01±0.02 bcd
Nanhan zixiushengcai	0.95±0.04 cdefghi
Xingyun ruan wei	0.98±0.02 bcdef
Dasushengcai	0.99±0.02 bcde
Jieqiushengcai	0.97±0.02 bcdefgh
Ziluolan	0.97±0.01 bcdefgh

Water spinach

Chunguo kongxincai2	1.04±0.06 ab
Dayekongxincai	0.92±0.05 efghij
Chunguo kongxincai 1	1.01±0.05 bcd
Jingyanye kongxincai	0.90±0.06 fghij
Taiguo kongxincai	0.84±0.04 j
Amaranth	
Huahongxianxcai	0.85±0.04 j
Yuanyebaixiancai	0.61±0.07 l
Taiwanluxiancai	0.90±0.08 fghij
Baiyexiancai	0.56±0.04 ln
Tedayehuahongxiancai	0.88±0.05 ij
Jingxuanhongxiancai	0.74±0.10 k

Values (mean±S.E., n=4 with different letters in the same column are significantly different according to the Duncan's test (P<0.05)

3.4 Genetic Variations in Root-to-Shoot Transport and Nutrient Solution-to-Plant

3.4.1 Translocation Factor (TF)

The TF can be used to evaluate the ability of the plant to translocate arsenic from the root to shoot. As shown in table (4), the average TF differed greatly among cultivars (p <0.05). The root-shoot translocation factor ranging from 0.010 to 0.251. The lowest TF was found in DYQ while RBXXQ showed the highest TF, similar to As accumulation. The results revealed twenty five fold genotypic variation in root- shoot translocation ability. In addition, among 32 cultivars, the TFs of 12 cultivars were not significantly different. The translocation factor (TF) was calculated by dividing the As concentration in the shoot by its concentration in the root.

3.4.2 Bioaccumulation factor (BCF)

The quotient of the As concentration in shoot to that in nutrient solution can be used to evaluate the ability of plants to accumulate heavy metals. Table 4 shows the average BCF of 32 cultivars under As treatment (6 mg L-1), ranging from 2.34 to 2.95, 2.82 to 4.51, 4.92 to 12.83, 3.573 to 7.965, 15.25 to 18.695 for lettuce, romaine, celery, amaranth and water spinach, respectively. The lowest BCF was found in lettuce (SJBY cultivar) and the highest BC was detected in water spinach (DYKXC cultivar), which was up 18.695.

Table 4. Average TF and BCF of 32 selected leafy vegetable genotypes exposed to 6 mg L^{-1} As

Genotypes	BCF	TF
Celery		
Mei qin	9.42 ef	0.13 bc
Meiguo baili xiqin	14.58 c	0.15 ab
Boli cui shi qin	7.12 efgh	0.10 cde
Meiguo wentu la xi qin	5.62 ghijk	0.02 nm
Texuan si ji xi qin	10.17 de	0.08 efgh
Daye qin	4.92 ghijk	0.01 m
Shengjie bai qin	6.36 fghij	0.05 ghijkln
Siji xiao xiang qin	12.86 cd	0.09 def
Wentula qin	12.85 cd	0.16 ab
Riben xiao xiang qin	12.87 cd	0.17 a
Lettuce		
Texuanyanling sun	3.43 ijk	0.08 efg
Sijibaiye	2.34 k	0.03 klnm
Huayesun jing yong woju	2.71 k	0.04 jklnm
Yeyong woju	2.95 jk	0.04 jklnm
Guayuanyiyesun	2.71 k	0.03 klnm
Romaine		
Helanziye shengcai	4.51 hijk	0.03 klnm
Nanhan zixiushengcai	3.42 ijk	0.03 lnm
Xingyun ruan wei	2.82 jk	0.02 nm
Dasushengcai	4.27 hijk	0.05 ghijkln
Jieqiushengcai	2.96 jk	0.04 ijklnm
Ziluolan	4.32 hijk	0.04 jklnm
Water Spinach		
Chunguo kongxincai2	16.07 abc	0.07 fghij
Dayekongxincai	18.70 a	0.09 def
Chunguo kongxincai 1	15.74 abc	0.05 hijkln
Jingyanye kongxincai	18.40 ab	0.13 bc
Taiguo kongxincai	15.25 bc	0.07 efghi
Amaranth		
Huahongxianxcai	7.97 efg	0.13bc
Yuanyebaixiancai	3.77 hijk	0.05 ghijkln
Taiwanluxiancai	6.75 fghi	0.12 cd
Baiyexiancai	5.10 ghijk	0.06 fghijk
Tedayehuahongxiancai	3.73 hijk	0.06 fghijkl
Jingxuanhongxiancai	3.57 ijk	0.04 ijklnm

Different letters in a colomn indicate significant differences between the cultivars at the 0.05 level

3.5 Depletion of As concentration in nutrient solution
The uptake time is one of the most important factors affecting the uptake of heavy metals by plants. Short-term, time-dependent arsenic depletion from hydroponic solution was measured at a series of different time intervals

(after 0, 0.5, 1, 2, 4, 6, 8, 10, 12, 22 and 24 h) and various concentration of arsenic. As shown in Figure 4, the depletion of As in TXYLS was higher for at least 12 hours and then starts to decrease rapidly with the passage of time until 24 hours, while SJBY was higher at the first 10 hours then starts decreasing until 24 hours. It can be explained that arsenic was absorbed by cultivars. Both the cultivars and treatments influenced concentrations of As in nutrient solution. At the end of the depletion (24 h), SJBY and TXYLS have up taken 23.09 % and 28.40 % of arsenic from nutrient solution, respectively (Figure 4). Interaction between treatments and cultivars influenced depletion of As concentration.

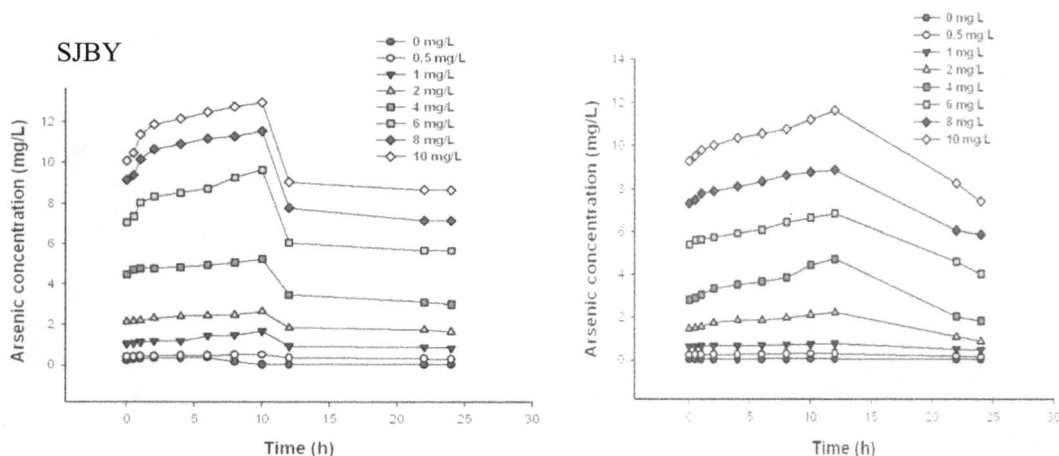

Figure 4. Depletion of As concentration (mg L^{-1}) with time (h) in the cultivars with high and low-As Accumulation (TXYLS and SJBY)

3.6 Short-term Arsenic Uptake Kinetics

The data for kinetic parameters which were derived from a linear regression on Lineweaver-Burk plot transformations are shown in table 5 and indicated that the uptake characteristics were different between cultivars. values of Km and Vmax were found to be much lower in SJBY (26.721 mg L^{-1} and 0.7737 mgAsg-1 shoot d.wt.hr^{-1}) than that of TXYLS (81.922 mg L^{-1} and 1.8488 mgAsg^{-1} shoot d.wt.hr^{-1}), indicating SJBY had the lowest uptake ability than TXYLS plants. Furthermore, TXYLS had the highest Km value than SJBY, confirming that uptake in the plant cell plasma membrane of TXYLS has a higher affinity for As than SJBY plants. The Vmax/Km values (Table 5) indicated that the arsenic uptake ability of cultivars followed TXYLS > SJBY. Therefore, we could conclude that TXYLS's cell wall had a relatively strong arsenic binding capacity, while SJBY had relatively lower.

Table 5. Comparison in the low-As-accumulating cultivar 'SJBY' and the high-As-accumulating cultivar 'TXYLS' in kinetic parameters (Km, Vmax and Km/Vmax) for concentration-dependent uptake of As into intact plants immediately after 24 h of As exposure

Cultivar	V_{max} (mg As g-1 shoot d.wt.hr^{-1})	K_m (mg/L)	K_m/V_{max}	R^2
TXYLS	1.848	81.922	44.312	1
SJBY	0.773	26.721	34.537	0.999

4. Discussion

4.1 Genetic Variations in As Concentration among Cultivars

The uptake of heavy metal/metalloid in plants differs greatly, not only among plant species but also among cultivars within the same species. In this study, the results showed that with uniform arsenic treatment (6 mg L^{-1}), wide genetic variations of As concentration was found among 32 tested leafy vegetable cultivars, revealing 6-8 fold variation ranged from 14.07 to 112.16 mg kg^{-1} d.wt in shoot As concentration (Table 2). This finding is consistent with earlier observations by Delowar et al. (2005) and Williams et al. (2006). The lower As concentrations were observed in the shoots of all cultivars and increased in the roots (Figure 1), this could be due to the inability of plant to translocate As beyond the roots (Smith et al., 2009). These results reveal that genetic variability for As accumulation exists within a species/cultivar. It can be concluded that this large genotypic variation in shoot As concentrations shows that there are opportunities to select for low As cultivars As and to breed cultivars that can be used in contaminated agricultural land.

In this study, three groups of species could be differentiated, species with low, intermediate, and high shoot As concentration (Figure 1). The reasons for differences in As concentration in these species might be greater ability of As uptake, translocation ability and high biomass. For example, in this study water spinach and celery species had the ability to accumulate more arsenic with average concentrations of 100.97 mg kg^{-1} and 58 mg kg^{-1} in their shoots. Our results are consistent with others studies who have indicated that leafy vegetable spinach contained higher concentrations of As in shoots (Shaibur et al., 2009). Moreover, people may intake much amount of As without knowing the presence of high As in the vegetable. However, among 5 group of species, lettuce had the lowest average shoot As concentration (16.95 mg kg^{-1} dw) followed by romaine (22.28 mg kg^{-1}) and amaranth (30.87 mg kg^{-1}), similar trend was observed with other studies, Warren et al. (2003) detected 6.8 and 17.8 mg kg^{-1} dw in lettuce grown in soil that received As. In addition, Munoz et al. (2002) also reported a lower range of As concentrations in lettuce (0.68 to 4.5 mg kg^{-1} dw.). Therefore, high As concentrations in the root and low translocation to the shoot in these group species may suggest that leafy vegetable cultivars had a high tolerance to As and it mainly accumulates As in the root. On the other hand, water spinach group species had the highest average shoot As concentration (100.97 mg kg^{-1}) followed by celery (58.059 mg kg^{-1}) when compared to other group species, which visibly demonstrate that the translocation of As was higher from the root to the shoot, which suggest that these species should be examined for phytoremediation potential.

In this study, the results indicated that cultivar was the key factor controlling the As transport into shoot and may lead to a large variation in leafy vegetable cultivars. Zia et al. (2011) found that As concentration in rice grain was affected by soil type (69% and 80%), followed by genotype (9% and 10%), among 38 rice varieties. Our results demonstrated that, leafy vegetable uptake might depend on vegetable cultivar and the rate of arsenic concentration under hydroponics.

4.2 Growth Response of Vegetable to As Toxicity

Under the same arsenic level, all cultivars seemed healthy, green and still alive and without showing any toxic symptoms (Figure 3), except for visible growth reductions. For most of the tested cultivars (69%) showed a tolerance index higher than 90%; this result indicates that leafy vegetable had considerable tolerance to As. In this study, the shoot biomass and TI were used as indicators to evaluate As toxicity to leafy vegetable. The results showed that almost all cultivars produced greater or similar shoot biomass except DYQ, YYBXC and BYXC, were able to cope up with 6 mg L^{-1} As stress with the maximum reduction less than 50% (Table 3). This dose (6 mg L^{-1}) is much higher than normal As level in the Chinese soils and in heavily contaminated soils. Therefore, it is possible to grow leafy vegetable in As contaminated agricultural land. Consequently, farmers may not get sufficient warning about the toxic concentration of As in leafy vegetable based on yield change alone.

4.3 The Important Factors Affecting As Uptake by Cultivars

In this study, the transport of As from nutrient solution to root and shoot of the cultivars, different acquisition abilities of root for As, variation in the capability to absord As and accumulation abilty were the main important factors which resulted in cultivar difference. The results revealed that all the cultivars had TFs less than 1, (17 fold variation) and (8 fold) in BCF, indicating that As was limitedly transported into the shoot. Therefore, these leafy vegetable cultivars can be considered as a As excluder (Baker and Whiting, 2002), comprising avoidance of metal uptake and restriction of metal transport to the shoots. This is consistent with the results reported by Huang et al. (2005) using lettuce and when compared with other plants such as Pteris vittata (Francesconi et al.,2002), the As concentrations were high in the shoot (aboveground) than in root, also the TFs were greater than 1.0. Wei et al. (2005) found that the exclusion of metals from shoot has been regarded as a metal tolerant strategy. Metals, once taken up by roots, can either be stored in the roots or exported to the shoot.

4.4 Identification of Genotypes with Low-As Uptake

Hydroponic identification has the advantage of rapidness, environmental control, and repetitiveness, which is optimal for screening a large number of cultivars. However the identification in field provides direct evaluation, but it is time-consuming and costly (She et al., 2011). Hydroponic has already proved useful for screening interesting properties in plants (i.e., arsenic tolerance or accumulation), and the result can be confidently extrapolated to field conditions (Moreno et al., 2010). An important aspect of this study was to establish a screening and identification system for cultivars with the ability to accumulate less As in shoot that would be applicable on uniform arsenic-contaminated farmland. It should be noted that up to now, there is no scientific rules for hydroponics screening of cultivars like leafy vegetable that could be applied to identify suitable cultivars. To our knowledge, the number of hydroponic tests performed to study low As accumulation in shoot among different leafy vegetable species/cultivars is rather small and most of studies have been investigated on As hyperaccumulation. Results are therefore hardly comparable between different experiments and different species.

Screening of cultivars to be grown on contaminated farmland involves several criteria which may vary, but generally include consideration of tolerance and accumulating ability of the cultivar in response to the specific metal (Liu et al., 2009). Based on the previous literature and our research, we proposed three rules that were selected as a criterion. (1) Tolerance property, (2) As translocation factor and (3) lower As concentration in their shoots. It is well-known that some plants can grow in heavy metal contaminated soils without accumulating significant amounts. These types of plants are called excluders. Besides, Baker and whiting (2002), suggested that excluders can be characterized by TF less than 1.0, whereas in accumulators TF is more than 1.0. Therefore, due to the different growing environments, the same rules use in this study may have different effectiveness in field and pot identifications. Based on the average shoot As concentration among 5 group of species, water spinach appears to be relatively high accumulator of As and thus it could be argued that this species could probably not be suitable to be grown on farmland with elevated As level. Celery also exhibited relatively high accumulation concentration and should probably be avoided on As-contaminated soils. On the other hand, lettuce and romaine species appear to accumulate less As concentration and so would be suitable species to grow. Amaranth had intermediate As concentration, but the significant differences among cultivars and the arsenic uptake ability based on kinetic parameters indicate that SJBY is the most suitable for minimizing As intake on As contaminated soils. When shoot/root quotient (TF) was taken into account, according to Baker and Whiting (2002) who suggested that excluders are characterized with TF <1.0, as showed in table 3-5, all the cultivars had TFs <1.0 and thus 12 cultivars were significantly lower than other cultivars. However there were selected as suitable cultivars. we also tried to confirm regarding to their tolerance to As stress; most of the tested cultivars (69%) were tolerant showed a tolerance index higher than 90%, thus, only two cultivars was identified to high TI compared to others and when BCF was further considered, 17 cultivars appeared to have significantly lower BCF than other species. In general, if the tolerance to As stress, arsenic concentration, TF, BF are all considered in the study, lettuce species (e.g SJBY) was the most appropriate cultivar for planting in arsenic contaminated farmland.

5. Conclusion

A hydroponics screening methodology was used in this study to screen and identify cultivars for low-As concentration ability in order to be grown on arsenic contaminated farmland. The results obtained exibit the existence of genotypic variations in the shoot As concentration among 32 cultivars under arsenic treatment (6 mg/L). Distinctive differences were also identified when comparing species to another. Lettuce and romaine species tending to be low accumulators, amaranth species tending to be moderate accumulator, and water spinach being high accumulator. The TFs of all cultivars were lower than 1.0. This detailed study was effective in providing complete information regarding the variation of arsenic concentration and accumulation in different cultivars based on the results from hydroponics.

It can be also concluded that our results could aid rapid development of leafy vegetable genotypes with decrease As accumulation in shoot by selection and bredding techniques while great care is needed for using lettuce cultivars. The selected cultivars had considerable tolerance to As according to their shoot biomass, except in DYQ and BYXC.Thus these cultivars can be cultivated in arsenic contaminated areas without any major risk of significant, considering the shoot biomass alone. Based on the hydroponic experiment alone, SJBY could be finally identified as suitable cultivar to be grown on arsenic contaminated farmland with low arsenic contaminated soil.

These results suggested that the uptake, translocation and accumulation of As in the cultivars may be strongly linked to genetics and associated to As concentrations in the plant, cultivars differences and the retention of As in the roots. Therefore, it seemed worthwhile to carry out similar screening tests with a larger number of cultivars of other species. In addition, further research comparing the different leafy vegetables under hydroponics, pot condition and field condition would be of great interest.

Acknowledgements

This study is a part of a research project. Thanks should be given to Dr Shuming and Dr Li LIANFANG for their assistance in the laboratory experiments, data analysis and manuscript preparation. This study was financially supported by the National Scientific and Technological Program of the "12th Five-Year" Plan of China (Grant No. 2012BAD14B02), and the National Natural Science Foundation of China (Grant No.41171255).

References

Ajaelu, C. J., Ibironke, Oluwafunke, L., Adedeji, V., & Olafisoye, O. (2011). Equilibrium and Kinetic Studies of the Biosorption of Heavy Metal (Cadmium) on Cassia siamea Bark. *American-Eurasian Journal of Scientific Research, 6*, 123-130.

Baker, A. J. M., & Whiting, S. N. (2002). In search of the Holy Grail - a further step in understanding metal hyperaccumulation. *New Phytologist, 155*, 1-7. http://dx.doi.org/10.1046/j.1469-8137.2002.00449_1.x

Delowar, H. K. M., Yoshida, I., & Harada, M. (2005). Growth and uptake of arsenic by rice irrigated with As-contaminated water. *Journal of Food Agriculture and Environment, 3*, 287- 291.

Francesconi, K., Visoottivieth, P., Sridokchan, W., & Goessler, W. (2002). Arsenic species in an arsenic hyper accumulating fern, Pityrogramma calomelanos: a potential phytoremediation of arsenic-contaminated soils. *Science of the Total Environment, 284*, 27-35. http://dx.doi.org/10.1016/S0048-9697(01)00854-3

Glass, A. D. M. (1989). *Plant nutrition: An introduction to current concepts.* Boston, MA, USA: Jones and Bartlett Publishers. p. 234.

Huang, B., Kuo, S., & Bembenek, R. (2005). Availability to Lettuce of Arsenic and Lead from Trace Element Fertilizers in Soil. *Water Air and Soil Pollution, 164*, 223-239. http://dx.doi.org/10.1007/s11270-005-3023-6

Islam, E., Yang, X., He, Z., & Mahmood, Q. (2007). Assessing potential dietary toxicity of heavy metals in selected vegetables and food crops. *Journal of Zheijang University Science B, 8*, 1-13. http://dx.doi.org/10.1631/jzus.2007.B0001

Lineweaver, H., & Burk, D. (1934). The determination of enzyme dissociation constants. *American Chemical Society, 56*, 658-666. http://dx.doi.org/10.1021/ja01318a036

Liu, W. T., Zhou, Q. X., & Jing, A. (2009). Variations in cadmium accumulation among Chinese cabbage cultivars and screening for Cd- safe cultivars. *Journal of Hazardous Materials, 8*, 14-147.

McGrath, S. P., Zhao, F. J., & Lombi, E. (2001). Plant and rhizosphere process involved in phytoremediation of metal-contaminated soils. *Plant and Soil, 232*, 207-214. http://dx.doi.org/10.1023/A:1010358708525

Meharg, A. A., & Hartley, W. J. (2002). Arsenic uptake and metabolism in arsenic resistant and nonresistant plant species. *New Phytologist, 154*, 29-43. http://dx.doi.org/10.1046/j.1469-8137.2002.00363.x

Moreno, J. E., Esteban, E., Fresno, T., Egea, C. L., & Penalosa, J. M. (2010). Hydroponics as a valid tool to assess arsenic availability in mine soils. *Chemosphere, 79*, 513-517. http://dx.doi.org/10.1016/j.chemosphere.2010.02.034

Munŏz, O., Diaz, O. P., Leyton, I., Nuñez, N., Devesa, V., & Súñer, M. A. (2002). Vegetables collected in the cultivated Andean area of northern Chile: Total and inorganic arsenic content in raw vegetables. *Journal of Agricultural and Food Chemistry, 50*, 642-647. http://dx.doi.org/10.1021/jf011027k

Norton, G. J., Pinson, S. R. M., Alexander, J., Mckay, S., Hansen, H., & Duan, G. L. (2012). Variation in grain arsenic assessed in a diverse panel of rice (Oryza sativa) grown in multiple sites. *New Phytologist, 193*, 650-664. http://dx.doi.org/10.1111/j.1469-8137.2011.03983.x

Norton, G. J., Duan, G. L., Dasgupta, T., Islam, M. R., Lei, M.,Zhu, Y. G., et al. (2009). Environmental and genetic control of arsenic accumulation and speciation in rice grain: Comparing a range of common cultivars grown in contaminated sites across Bangladesh, China, and India. Environmental Science & Technology, 43, 8381-8386. http://dx.doi.org/10.1021/es901844q

Shaibur, M. R., & Kawai, S. (2009). Effect of arsenic on visible symptom and arsenic concentration in hydroponic Japanese mustard spinach. *Environmental and Experimental Botany, 67*, 65-70. http://dx.doi.org/10.1016/j.envexpbot.2009.06.001

She, W., Jieyu, C., Xing, H. C., Wei, Y. L., Huang, M., & Li, W. K. (2011). Tolerance to Cadmium in Ramie (Boehmeria nivea Genotypes and Its Evaluations Indicators). *Acta Agriculturae Sinica, 424*(37), 348-354.

Smith, E., Juhasz, A. L., & Weber, J. (2009). Arsenic uptake and speciation in vegetables grown under greenhouse conditions. *Environmental Geochemistry and Health, 31*, 125-132. http://dx.doi.org/10.1007/s10653-008-9242-1

Warren, G. P., Alloway, B. J., & Lepp, N. W. (2003). Field trial to assess the uptake of arsenic by vegetables from contaminated soil and soil remediation with iron oxides. *Science of the Total Environment, 311*, 19-33. http://dx.doi.org/10.1016/S0048-9697(03)00096-2

Wei, S. H., Zhou, Q. X., & Wang, X. (2005). Identification of weed plants excluding the absorption of heavy metals. *Environment International, 31*, 829-834. http://dx.doi.org/10.1016/j.envint.2005.05.045

Williams, P. N., Islam, M. R., Raab, A., Hossain, S. A., & Meharg, A. A. (2006). Increase in rice grain arsenic for regions of Bangladesh irrigating paddies with elevated arsenic in groundwater.

Zia, U. A., Golam, M. P., Hugh, G. J., Susan, R. M., Wricha, T. I., Mohammed, S., & John, M. D. (2011). Genotype and environment effects on rice (Oryza sativa L.) grain arsenic concentration in Bangladesh. *Plant and Soil, 338*, 367-382. http://dx.doi.org/10.1007/s11104-010-0551-7

12

Vulnerability and Climate Change Perceptions: A Case Study in Brazilian Biomes

Teresa da Silva-Rosa[1], Michelle Bonatti[2], Andrea Vanini[3] & Catia Zuffo[4]

[1] Vila Velha University, Center of Socioenvironmental and Urban Studies -- NEUS/UVV, Brazil

[2] University Buenos Aires -UBA, Argentina

[3] Technical Office for Management and Environmental Education at the Fiocruz Atlantic Forest Campus, Brazil

[4] Federal University at Rondônia – UNIR, Brazil

Correspondence: Teresa da Silva Rosa, Núcleo de Estudo Urbanos e Socioambientais/NEUS, Programa de Pós Graduação em Sociologia Política/PPGSP - UVV-ES, Rua Comissário José Dantas de Melo, 21, Boavista – Vila Velha, Espírito Santo, Brazil. E-mail: tsrosaprof@yahoo.com.br

Abstract

Based on the assumption that vulnerability is socially constructed, and may thus change according to transformations in human action, it appears necessary to consider the issue at the core of studies on the social aspects of Climate Change (CC), risk level and disaster prevention. The social nature of vulnerability is determined by elements such as poverty, inequality, exclusion and access to sanitation, water, food and education among other factors.

In 2010, a study was undertaken in the ambit of the project "Climate Change, Social Inequalities and Vulnerable Populations" about the perception of and an assessment of vulnerability. Information was gathered through interviews using structured questionnaires administered in communities in three Brazilian biomes: the Atlantic Forest, the Amazon and the Semi-Arid and Cerrado region. This paper discusses the result of three case studies: one in Rondonia in the Amazon, and two in the Atlantic Forest biome, one in Rio de Janeiro and another in Santa Catarina.

Two points in common among the communities should be highlighted. (1) None of the communities have plans or actions to adapt to natural climate variability, much less to CC. This makes the communities more vulnerable and unprepared to act with protective or reactive preventive measures. (2) Thus, it is important to develop a plan with actions that address the situation of "organized irresponsibility", understood as a network of mechanisms that treat environmental problems as normal, or regard them as being of governmental responsibility alone.

Keywords: climate change, social organization, vulnerability

1. Introduction

The paper is based on research that was the result of a collective endeavor by social actors from distinct backgrounds who are involved in Brazilian environmental governance, including academia, organized civil society, and government. The aim is to shed light on the understanding of climate change (CC) found in a variety of Brazilian communities. The study emerged from activities of the national network known as the Committee of Entities in the Fight on Hunger and for Life (National COEP) (Note 1) undertaken among economically under privileged communities throughout Brazil, with the common purpose of developing actions against poverty.

As a member of the Brazilian Forum for Climate Change (*Fórum Brasileiro de Mudanças Climáticas/FBMC*), COEP raised the issue of social inequality at this forum, propitiating the creation of the working group "Climate Change, Poverty and Inequality", as of April 2009. One of the major activities (Note 2) of this working group is the study "Climate Change, Social Inequalities and Vulnerable Populations in Brazil: Building on Skills" along with two subprojects, one of which addresses corporations, and the other vulnerable populations. This paper is concerned with the latter, and was coordinated by the Reference Centre for Food and Nutrition Security (*Centro de Referência em Segurança Alimentar e Nutricional/CERESAN*) at the Federal Rural University at Rio de Janeiro (UFRRJ), and the Centre for Urban and Socio-environmental Studies (*Núcleo de Estudos Urbanos e*

Socioambientais/NEUS) at the University of Vila Velha (UVV-ES). A number of researchers with different academic backgrounds from two Brazilian universities (Note 3), and one research institution (Note 4), located in different biomes (Note 5) and close to communities presenting specific socio-economic scenarios were involved in the study. The research was made possible by funding from the Ministry of Science, Technology and Innovation (MCTI) through the National Council for Scientific and Technological Development (CNPq), and by research scholarships for scientific initiation from the National Foundation for the Development of Private Higher Education (FUNADESP).

The research was organized around three interconnected themes (Note 6), and this paper focuses on case studies about the vulnerability and capacity for adaptation of selected populations from distinct biomes. One is in the Amazon, where the focus is on a riverine community located within a project for settling previously landless families, which is part of the federal government agrarian reform program in Rondônia state. The two other communities are located in the peripheries of the metropolitan regions of Rio de Janeiro and Florianópolis, in the Atlantic Forest biome.

In Brazil, intense weather events can have impacts on the populations and economies of big cities, small towns and rural areas that create states of emergency or calamity. Losses in agricultural yield, infra-structure damage and the number of victims from such disasters in recent decades prompt us to pursue a broader understanding of these phenomena, and more effective actions by the communities and government to mitigate the negative effects by more thorough preparation to protect against these events.

The approach adopted here presupposes that the climate issue, similarly to other environmental matters, creates an opportunity to review the current basis of economic development, particularly the use of natural resources. The form of this use is made more apparent by the approach that exposes the many degrees of vulnerability of the populations, in particular those that are already in a situation of risk as a result of their social and economic exclusion, especially in countries with late development such as Brazil. In other words, we begin from the premise that the vulnerabilities suffered by the populations are historically constructed by a development model that creates socio-economic inequalities (Valencio et al., 2009; Mattos & Da-Silva-Rosa, 2011; Da-Silva-Rosa & Mattos, 2012), placing the excluded populations in situations of risk or vulnerability because they occupy regions such as floodable areas, hillsides, and river banks that for this reason are designated areas for permanent environmental preservation by Brazilian law.

In this context, it is important to consider the disproportional imposition of environmental risk on populations with less financial, political and informational resources. This situation has become known as environmental injustice (Acselrad, 2009). According to this concept, poor populations are driven to live in areas more subject to flooding, landslides and other environmental problems.

These risky situations exist before the occurrence of extreme weather events, which aggravate the existing vulnerabilities. This makes it inappropriate to speak of "natural" disasters, because in fact they are usually socio-environmental problems that are made evident by natural events. It is worth noting that these events do not only affect economically disadvantaged populations, as seen in the disaster that occurred in the mountainous region of Rio de Janeiro state, in January 2011 (Note 7), although the populations in historic situations of exclusion constitute the majority of those who suffer in the disasters. This prompts us to suggest that climate-related events are not the prime causes of so-called socio-environmental disasters, but rather reveal concrete situations of vulnerability.

In the aftermath of these events, it can be observed that government commonly declines to enforce environmental laws that prohibit the occupation of areas designated for permanent preservation; and also fails to implement public policies that would satisfy the needs of low income populations for education, healthcare, housing and sanitation. Moreover, it is clear that the affected communities are "unprepared" to react promptly and in advance against imminent risks, revealing their inability to adapt to increasingly recurring situations.

This paper reflects on some key concepts related to the social dimension of climate change and vulnerability; it then presents and discusses the results of three case studies undertaken in different socio-environmental situations in Brazil.

This paper understands vulnerability to be the incapacity to deal with danger, and it can refer both to natural and social systems (Fussel & Klein, 2006; O'Riordan cited in Braga et al., 2006). With Regards to The Social System, Vulnerability Refers to the Fragility or Precariousness of the Living conditions of a group, which hinders a population's capacity to react to or confront the impacts of an event, whether because of its unpredictability or its intensity. Vulnerability, as highlighted by O'Riordan (cited in Braga et al., 2006) is "...a consequence of a combination of economic, social, environmental and political processes" (p. 2), which exemplifies the level of complexity that must be acknowledged when tackling such issues. Vulnerability may also be understood as being an "...intrinsic condition to...the receptor system, which in interaction with the magnitude of the event..., defines adverse effects..." (Valencio, 2011, p. 9).

The categorization "socially and environmentally vulnerable populations" indicates the fact that these populations face a dual exposure (O'Brien & Leichenko, 2000). After all, not only are they living in a condition of social exclusion and deprived of basic needs (due to unjust development models), but they are also placed in a situation of environmental vulnerability, exactly because of the socio-economic exclusion and injustice to which they are submitted, considering that they are pushed to occupying areas of risk or of environmental degradation (Alves, 2006). There is, therefore, a tendency towards a spatial overlapping of socio-economic and environmental problems (Alves, id.). In addition to biophysical factors (such as declivity, vegetation and soil impermeability), social factors contribute to vulnerability, as pointed out by Cardona (2003). Thus, a situation of deprivation, which generally speaking characterizes poorer populations, fragile infra-structure, and precarious public services are some contributing factors. Brazilian studies that analyze vulnerability in urban areas describe these variables as socio-economic and socio-demographic factors (Mello et al., 2010; Alves et al., 2010; Gamba, 2010).

Furthermore, other research shows a relationship between vulnerability and risk. Jacobi (1995) suggests a relationship between precarious or nonexistent public services (such as sewage and garbage collection) and greater exposure to environmental risks. The exposure to risk of disease (such as Leptospira epidemics, typically associated to floods) is directly related to extreme weather conditions, as suggested by Confalonieri (2003). This reveals the interdependency between risk from disaster and the vulnerable structural characteristics of a community. Given that the notion of double exposure is inherent to the concept of socio-environmental vulnerability, a situation of double exposure was one of the criteria for the selection of the social groups that took part in this study.

For the reasons presented, we suggest that a state of vulnerability makes a population highly prone to greater suffering from the unprecedented weather events that have recently occurred in Brazil (including intense rain, long-term droughts and tornados). The social concern for the damage and losses (the disasters) caused by natural or human phenomena are included in this affirmation, yet it is not clear that vulnerability is the cause of the disasters in contemporary times. After all, vulnerability, risk and disaster are closely related. According to Cardona (2003), the risk of disaster is composed of two factors: threat and vulnerability. For the author, threat corresponds to an external factor that may be forecasted, but is usually difficult to control, such as heavy rain, hurricanes or earthquakes. Vulnerability is an internal factor that represents the degree of susceptibility of a system or subject, to a threat. Vulnerability thus constitutes the condition of a threatened subject in his or her social system, and for that reason, it can be understood as a social state. Cardona also suggests that a decrease in the levels of threat and vulnerability may lead to a decrease of risk as a whole. Nevertheless, considering that vulnerability is a socially constructed state, it is therefore, susceptible to change by human agency.

This analysis suggests that addressing historically constructed vulnerability, and adopting measures to foster the capacity to face risks from extreme events should be the underlying concerns in the mitigation of disasters. We therefore agree with Marchezini (2009) who affirms that "...the nature of disasters must be sought in social organization, understanding them as a process associated with social vulnerability, the causes of which must be explained as structural problems, and duly contextualized." (p. 50). This implies the need to reconsider the idea that a disaster is an eminently natural event (and thus referred to as a natural disaster), and contextualize it historically as a socially constructed process resulting from exclusionary conditions and social inequalities caused by a given development model (Valencio et al., 2009). Thus, within the context of CC, the term natural disaster has been replaced by the expression social or socio-environmental disaster.

Populations living in the Third World are considered the most vulnerable, and especially children, women and the elderly (Cardona, id.; BRASIL/PNMC, 2009), even if vulnerability is not restricted to these populations (O'Brien & Leichenko, 2000) as evident in the disaster of January 2011 in the mountainous region of Rio de Janeiro (Brazil). On that note, three aspects related to vulnerability are highlighted by the work of Moser (1998) and Cardona (2003): (i) the exposure of populations to climate events such as droughts, floods, landslides or cyclones; (ii) the susceptibility of these groups to risky events; and (iii) the resilience or adaptive capacity of vulnerable populations to withstand those events and disasters.

Nevertheless, the discussion of vulnerability and conditions of deprivation, or poverty, reveals that the adaptation measures must consequently prioritize improving conditions by deploying strategies to meet basic needs to minimize situations of risk, and consequently disasters. It is worth clarifying that adaptation is a process and concept that is central to this study. For this reason it is necessary to discuss this term to better understand the reasons that led researchers to the field with a questionnaire based on this concept.

These adaptation measures follow similar guidelines as those in the human rights field as suggested by Sachs (2003; 2008). In the context of CC, the notion of adaptation emerged with greater strength after CoP13, in Bali (2007), as a consequence of the Fourth Evaluation Report by the IPCC. The late arrival of adaptation as a concern is noteworthy, when compared to mitigation, but it is nevertheless a growing concern, as noted by Klein et al. (2003), while they are both the sustaining pillars of the social dimension of CC.

The underlying purpose of adaptation measures is, in other words, to decrease the probability of the occurrence of disasters, by mitigating vulnerable situations, risk exposure, or insecurity caused by deprivation. This involves increasing the adaptation capacity of vulnerable populations through a change of values and behavior by enhancing awareness among these populations about more sustainable and just ways of life.

The Fourth Report by the IPCC describes adaptation as the capacity to adjust to change, or to react to impacts from climate and non-climate events and to mitigate harm, while taking advantage of opportunities spawned by the respective adversities. The report attempts to link the capacity for adaptation to social and economic development, which, in our view, reinforces the need to establish more sustainable and just ways of life fostered by changes in values and behavior.

On this note, adaptation should be understood not as a simple matter of adjustment as inferred in the definition by the IPCC, but as a skills-building process needed to prepare populations to face impacts resulting from weather variations. This approach differs from one that focuses on adjusting to a given event, since it occurs prior to the occurrence of the extreme event. The process requires evaluating the probable scenarios that could arise from climate or weather conditions and assess the vulnerabilities of communities, such as those that occupy hillsides and unstable land, or areas susceptible to floods.

In this line of reasoning, the capacity to adapt is related to resilience, which is the capacity some systems have to react or resist impacts, and requires greater flexibility from the elements of the impacted system. According to Homer-Dixon (2009), the resilience of both social and complex systems is a property that, with proper strengthening, can avoid disasters, but which requires a cultural change based on the strengthening of systemic connections, while encompassing all the components of a social group.

This process therefore requires a preliminary diagnosis of the community that will participate in the action, to identify the respective vulnerabilities. Afterwards, and with the due participation of the community members, those involved can search for ways to strengthen characteristics intrinsic to existing adaptation strategies and overcome the vulnerabilities identified. In both cases, the aim should be for the community to achieve ecological sustainability, social justice, and a capacity for resilience. This brings us to two points that must be highlighted. Firstly, resilience cannot be strengthened without prior identification of vulnerabilities (Giddens, 2009). Secondly, one must acknowledge the local aspect of the adaptation measures to be undertaken. The latter point illuminates the close relationship of the cultural, social, economic and ecological characteristics of the communities in question (IPCC, 2007).

The local aspect both characterizes adaptation and differentiates it from mitigation, as evidenced by the National Plan for Climate Change (Plano Nacional de Mudanças Climáticas - BRASIL, 2008) which states that "Adaptation actions – contrary to those of mitigation, whose results are reflected globally – are normally perceived in the place where they are carried out, granting adaptation a high degree of specificity" (p. 102). Although the scales of adaptation actions may be distinct (national, municipal, or even pertaining to a particular watershed), the local aspect actually influences the elaboration and choice of measures, reinforcing the already mentioned aspect of the connection between development, adaptation and vulnerability. After all, adaptation measures attempt to minimize the vulnerability of communities while, according to the National Plan, they "…promote better living conditions including housing, food, healthcare, education, employment…[which means that the promotion of sustainable development]…is the most efficient way of fostering resilience in situations of climate change" (BRASIL, 2008, 102). The opportunity to take adaptation measures to improve many areas is thus proposed, favoring, for example, infrastructure less likely to emit greenhouse gases (GHG), without forgetting the need for community empowerment to better prepare it to react promptly when under stress, and when facing destabilization of social and environmental systems.

On that note, adaptation has been seen by the literature as a skill-building process for populations to better adapt to impacts, associated to Homer-Dixon's notion of resilience as the strengthening of systemic connectivity. This is above all a process of political construction of citizenship, which contributes to the empowerment of people while, in the case under analysis, fostering the mobilization of a community to react promptly against adversities.

This notion relates to what Lemos (2007) considers the double adaptation process: the identification of vulnerabilities, and the construction of preparedness plans, focusing on risk management. With regards to this management, the author identifies three main factors in the construction of resilience systems: adoption of good governance principles by those who elaborate management plans; the essential role of the political networks that orient that governance; and the importance of democratizing knowledge for decision making. These three points are closely related to the constructive intervention perspective of Sen (PNUD/RDH, 2008).

It seems apparent that this skill building process differs from public policies normally adopted in Brazil, which usually focus on emergency actions, and are not deemed efficient for a conscious mobilization of the population since they lack the empowerment aspect characterized by independent, participatory and responsible actions. The role of public policies in the context of CC should be geared towards stimulating pro-activity, rather than focusing on reaction (Giddens, 2009). This brings us to two distinct approaches for the skills building process related to adaptation: induced adaptation, and spontaneous adaptation. The former is that induced or promoted through public policies by means of a skills building process related to adaptation. It is the perspective known as the institutional approach (UNFCCC, 2006). The latter is the autonomous approach (ibid), in other words, it is the one that is promoted by the community often through existing strategies, and for that reason it is historically consolidated.

2. Area Descriptions

In this section three case studies will be presented, and their specificities highlighted. First, each case study is contextualized, and an assessment of the vulnerabilities is presented. The perceptions of the resident families (the key-actors) are introduced and finally a chronology of the local actions based on the identified outcomes. It must be highlighted that in the presentations below, we sought to maintain the particular approaches of the authors responsible for each case study.

2.1 FIOCRUZ CAMPUS – Atlantic Forest (Rio de Janeiro State)

The region of the communities studied is an Atlantic Forest fragment in Rio de Janeiro state that had been occupied by large sugar-cane plantations that were transformed to coffee plantations, until the land was transferred to the government. A hospital complex was constructed in these regions to treat mental health problems, tuberculosis and leprosy. Therefore, part of the population in these communities are descendants of former slaves, farm workers, and employees who worked at the hospitals and who received permission to live there, as well as patients who did not have families and were encouraged to marry and form families.

Four urban communities located inside the FIOCRUZ campus in a buffer zone of the Pedra Branca State Park, were studied: Sampaio Correa, Caminho da Cachoeira, Viana do Castelo and Faixa Azul, with a total of 96 families. The Fiocruz campus of the Atlantic Forest (CFMA) is located on land that encompassed a group of psychiatric hospitals known as the Juliano Moreira colony, in Jacarepaguá, and encompasses an area of 5,097,150.24 m². Fiocruz is a government-affiliated health research organization.

The mostly low-income population that resides in these irregular urban settlements feel generally insecure about their permanence, given that it is an extremely environmentally fragile place, particularly due to the floods that are commonly hazardous to health. The main vulnerability factors among the communities under analysis are poor infrastructure, in terms of garbage collection, water supply, and basic sanitation. These shortcomings contribute to the propagation of disease in the region, and can be intensified by extreme weather events. This territory is therefore characterized by situations of socio-environmental vulnerabilities.

Thus, the objective of this case study was to contribute towards the construction of a shared process of knowledge production about the territory in question, to empower the populations to better face the challenges brought about by extreme weather events in the four communities located in the campus. The communities were selected based on socio-environmental vulnerability factors (Tables 2 and 3 attached), and environmental challenges already faced, focusing on those that may be related to the effects on the environment and health, and those caused by climate change.

A total of 36 families were interviewed (procedure 4), five interviews with local actors (procedure 3) and 12 participants of the focus group (procedure 2).

2.2 AMAZON - the Gleba Aliança Case Study (Rondonia State)

There have been many experiments in the Amazon region with policies related to population and economic development, which, nevertheless, have failed to integrate communities to the local biodiversity. With the growth of global apprehension about climate change, concern over deforestation in the Amazon has also increased, since according to Marengo (2006), the impacts resulting from deforestation can lead to a rise in temperature, rainfall evaporation and surface drainage.

This characterization results from the case study undertaken at Gleba Aliança, a rural region 30 km northeast of the center of Porto Velho, capital of Rondônia state (ZUFFO et al., 2010).

Regarding the methodology (procedure 3), the local actors interviewed included five women, and one man: who was a fishing community leader, and president of the Association of Small Farmers of Gleba Aliança - AGRILANÇA; and four government employees, with distinct backgrounds (a university lecturer and environmentalist, a journalist, an economist, and a researcher with a background in social communication). To identify the perceptions and vulnerabilities about climate change, a sample questionnaire was given to 47 riverine dwellers (Table 1): traditional populations, small farmers and cattle raisers, all residents in the five communities of Gleba Aliança.

According to the Agrarian Reform Information System (SIPRA) produced by the National Institute of Colonization and Agrarian Reform (INCRA), Gleba Aliança, became federal property in 1990 (MAGALHÃES, 2005). It is divided into three micro-regions with characteristics specific to its occupation process, namely: one portion is a Settlement Project (Projeto de Assentamento - PA) Aliança, and two others are spontaneous occupations. The latter are areas of land tenure regularization, one of which is located on solid ground, while the other is on flooded plains on the shores of Cujubim Grande lake and the Candeias, Jamari and Madeira rivers.

During the field activities in June and July 2010, it was observed that the most present segment, among the traditional population of the Amazon biome under study are the riverine communities, concentrated especially in the locality referred to as "Agrovila" (which is located in the Gleba Aliança).

2.3 TAPERA DA BASE - Atlantic Forest (Santa Catarina State)

Climate adversities have significantly affected Santa Catarina state throughout its history, mostly as a result of some specific features in the region, namely: high rainfalls and heavy storms contrasted by droughts that can last many months. According to the "Natural Disasters Caused by Climate Adversities in the state of Santa Catarina Assessment (1980-2000)", this is one of the states where there has been an increase in the occurrence of storms that lead to declarations of a state of emergency or public calamity (Herrmann et al., 2001). One of the communities where these events take place is Tapera da Base.

Located 27 km south of the center of Florianópolis, Santa Catarina state, in 2010, Tapera da Base had a population of around 12 thousand people. Part of the municipality of Florianópolis, the community is situated between an ocean bay and a mangrove, and there is an Air Force Base between the community and the center of the city. The village has had an ongoing process of occupation, encompassing modest to very precarious housing.

Most of the houses are built on terrain with high water levels that are strongly affected by the tides. About half are on lots less than two meters above sea level, which makes them vulnerable to flooding. Near the mangrove, clandestine subdivisions and other lots are occupied on landfill composed of a wide range of materials that are usually inadequate to sustain tidal dynamics and human health. Thus, a combination of factors such as high tides, lack of storm sewerage, the soil quality and topography of the region and impermeable structures contribute towards the occurrence of high impact floods, by hindering the drainage of water from intense rainfalls. For that reason, rainfalls are seen as a daily risk by the population. Constant flooding in the mangrove region affects almost half of the inhabitants in the community of Tapera da Base.

To assess the living conditions of some of these populations, the State Secretariat for Urban Development and the Environment of Santa Catarina (SDM/SC) applied the Local Human Development Index (IDHL) to 88 neighborhoods in Florianópolis (Cesa, 2008). According to the CESA (2008) index, although living conditions improved between 1991 and 2000, they are not uniform throughout the city. Tapera da Base came in 84th place among the 88 neighborhoods analyzed.

Independent of possible problems related to climate change, the population of Tapera da Base is vulnerable to common and daily weather alterations. This vulnerability is determined by the difficulties of the location, and by the lack of public policies, as well as by the nature of the social reproduction of particular groups in the community. This is particularly the case of the collectors of berbigão (the common cockle) in the region, and of the residents of more flood-prone streets, as indicated by the interviews with the families.

Considering the extension and quantity of Tapera's population, it was decided to focus the interviews on the regions with the most vulnerable household's. Therefore, the interviews were conducted with 16 families from two segments of the community who were indicated by the community council as the most vulnerable to climate dynamics, more specifically 8 families of cockle collectors and 8 families living on Rua da Juca (Table 1). The vulnerability of these segments is related to the increasing difficulty for the families to survive through shellfish collection and by the frequent flooding of their homes, respectively.

Table 1. Total monthly family income of those interviewed percent by state and total

Range	Rio de Janeiro	Rondonia	Santa Catarina
Up to US$ 691.56 (R$ 1.500)	75%	87%	100%
US$ 691.56 (R$1.500)- US$ 1,383.12 (R$ 3.000)	22%	9%	0%
More than US$ 1,383.12 (R$ 3.000)	3%	4%	0%

3. Methods

The choice of communities for the case studies was based, on one hand on the criteria of socio-spatial, environmental and cultural diversity in the country. The case study sites are located in areas with high frequencies of extreme climate events, which can make their populations more vulnerable. The selection of case studies and interviewers was designed to promote dialog with people who are directly affected by extreme climate events, particularly flooding and drought. Populations located in areas where activities had previously been undertaken by researchers and institutions belonging to the National Mobilization Network COEP, were given prioirty. Furthermore, there were two other preponderant considerations: the various Brazilian biomes, and the rural and urban populations in the large metropolitan centers in the country.

The research focused on a study of the communities, their territory and culture, as suggested by the literature, exploring the populations' perceptions and recognition of climate phenomena. The focus on the local dimension is also justified by the interest in investigating the capacity of each community to engage with the institutions present in its territory, and of its members to organize themselves as citizens at the local level. Thus, the research undertaken in 2010 entailed four procedures common to all the case studies:

(1) Identification of social actors present in the communities (local governmental institutions, non-governmental organizations among others) and their programs;

(2) Configuration of focal groups for discussion of pre-established themes;

(3) Interviews with the most relevant local actors; and

(4) Presentation of standard questionnaires to the families that made up the sample.

Table 2. Number of families interviewed in rural and urban areas by state

Rural or Urban	Rio de Janeiro	Rondonia	Santa Catarina	Total
Rural	0	47	0	47
Urban	36	0	16	52
Total	36	47	16	99

In addition to the specificity of each one of these procedures, the number of informants in the study also sought to respect the social characteristics of the areas studied. Two criteria were adopted to define the number of people interviewed, or who would be given the questionnaire that is part of the fourth methodological procedure above. The first was that more than 15% of the sample in each community should be from vulnerable households. Thus, the number of respondents was considered sufficient when it was possible to identify a pattern or common terms found in the material, allowing the establishment of generalizations. The second is that the respondents were present when the questionnaires were issued. Thus, the composition of the sample is not the same for the three case studies, except the fact that they were people living in a vulnerable area or in a historically vulnerable situation.

Five key questions orientated the case studies based on the work of Smit et al. (cited in Fussel & Klein, 2004), and on the document by UNDP/GEF (Note 8):

(1) *Adapt to what?* – Climate variations (projected patterns of temperature and rainfall), threats and vulnerabilities, risks and answers suggested by the resulting information.

(2) *Who will be affected?* – Socioeconomic conditions of vulnerable communities and groups in specific risk situations.

(3) *What will be affected?* – Threats and impacts that will probably have to be faced by the vulnerable group, and possible causes that justify those vulnerabilities. Some areas of impact identified in the literature as vulnerable, or seen as being in situations of risk (biodiversity, water, agriculture/food, health, housing), were selected for the research.

(4) *How to adapt?* – Measures to reduce vulnerabilities, or risk management already in place in the community, other adaptation measures that could possibly be adopted by the community based on examples; opportunities and difficulties to implement the reduction measures, or risk management identified; identification of actors/partners who may become involved.

(5) *How adequate is the adaptation proposal?* – It is necessary to know the contextual situation of the area where probable impacts and adaptation strategies will be implemented.

4. Results and Discussion

Considering the different social contexts found in each case study, the results will be presented and discussed according to geographic region. The paper will only discuss the data related to the key questions in the study.

In the case of the *Fiocruz Campus*, the interviewees that belonged to the family groups reported that temperature variations coincided with a higher incidence of mosquitoes that carry disease. They also noted that the environment is degrading due to construction in coastal areas, deforestation and pollution, including the establishment of a quarry nearby.

The social actors interviewed associated climate variations with human activities, even if they did not see themselves as being responsible either for improving or changing the scenario.

It can be observed that the interviewees in the family group have no adaptation plans for the climate events taking place in these communities. Some of the families interviewed reside in areas where Brazilian environmental law prohibits construction, known as Areas of Permanent Preservation (APPS), because they are close to water, a fact that increases their vulnerability to increases in the volume of rainfall. Nevertheless, there was no mention of how to foster adaptation to the recurring floods in this region, which were deemed common by the residents. Heavy rainfall also provokes landslides on hilly sites, which can threaten housing.

Since the communities in the study are located in the buffer zone of the Pedra Branca State Park, they are also susceptible to high levels of humidity, and are close to disease vectors and reservoirs. The high humidity affects houses with poor ventilation and aggravates respiratory problems and allergies resulting from fungi and parasitic mites. In the broader scope of this study, the populations were also monitored to gauge the control of these diseases over the years, so that they do not get worse, in the case of possible extreme climate events.

Based on the results above, some propositions for an action plan can be indicated. As a strategy for risk mitigation in future action plans of the communities under study, it is worth mentioning the enactment of a plan to regulate land tenure coordinated by FIOCRUZ, including the reallocation of families in areas of risk to nearby areas, in accord with current legislation. New houses will be built on lots of around 400 m², while residents who remain in their residences will participate in a residential improvement program. Moreover, a contingency plan is being developed, with the participation of FIOCRUZ, the state department of Civil Defense, and the community. Lastly, in addition to the above initiatives, the populations were included in the federal works project known as PAC - to receive sanitation and street paving, and improved access to public transportation, which was precarious at the time of this research.

In addition to these projects, it is also important to focus on the reduction of deforestation and forest fires in the area, the disposal of waste in appropriate locations, a more efficient garbage collection system and an ecological restoration project. These kinds of mitigating strategies offer direct and indirect benefits, such as: reducing the risk of landslides, maintenance of the local micro climate, a decrease in the proliferation of disease-transmitting vectors, reduction of waterway sedimentation and subsequent reduction of flood risks.

Evaluating the issues of "Adapt to what" or "What will be affected" the aspects most cited were: deforestation, floods and invasion of land for real estate speculation. About "What would be a suitable local action plan," the relocation of houses to areas not subject to flooding and landslides should be considered as well as the elimination of the use of fire to clear pasture and burn garbage. More recently, the community became part of the rain alert plan promoted by the Rio de Janeiro state civil defense agency, which will alert residents by cell phone about the arrival of heavy rain.

In the *Amazon case*, the significance of the principal changes in local conditions observed by those interviewed in the Gleba Aliança in Rondônia cannot be overemphasized. From the most to the least important they include: changes in volume and season of rainfall, changes in temperature, interruption of water supply, and changes in agriculture, small-scale cattle rearing, and extractivism (collection of natural products such as Brazil nuts, açaí fruit and vegetable oils - copaiba and andiroba). Since there are different population groups (riverine populations, small farmers, and cattle raisers) in Gleba Aliança, the study understood that the characteristics and degree of climate vulnerability are different for each group. The information gathered in the interviews point to a variety of factors regarding the areas suffering impacts selected by the study and climate-change related events.

It was perceived that the time of residence in the location is reflected in the type of construction, its suitability and conservation as well as in family size. In terms of health issues and basic sanitation, the situation is a bit precarious and concerning vectors of tropical diseases, a high rate of malaria was found. In relation to water, in the months of June and August, due to the drop in rainfall, a good portion of the igarapés (forest waterways) suffer considerable reductions in flow, coming to dry out, mainly in places where deforestation was very intense, harming agricultural activities. Moreover, due to fragile soil quality, there has been a gradual change in the areas used for planting and animal raising, which is expanding strongly in the region.

Predatory fishing has threatened the subsistence of the riverine residents (fishermen) and driven them to organize associations that strengthen the fight against predatory fishing and promote the integration between residents of the riverine communities. It was also noted that deforestation to make space for agricultural activities, leads to a gradual reduction in the local biodiversity, especially of the flora and fauna typical to dry land.

The tools applied in this study and their results, including those from the field trips, indicate that the climate change impacts that most need attention, considering the special needs of the Amazon biome are biodiversity, water and agriculture. Based on the results obtained, the research team suggested the following elements for a future plan of action:

- expand the cooperation and exchange of experiences between different areas in the Amazon;

- minimize environmental degradation and guarantee the maintenance of protected areas;

- foster changes in perception, and the involvement of people interested in a shared vision of the what kind life conditions should be achieved in the future, including an awareness of sustainability;

- promote agro-forestry systems as a sustainable alternative for Rondônia;

- broaden lines of credit available to sustainable entrepreneurs, particularly from official banks;

- foster the industrialization of primary resources in activities that according to Bartholo Jr. & Bursztyn (1999) have social and environmental repercussions, enhancing the value of agro-industrial enterprises with a biotechnological focus, as well as agro-forestry systems that link extractivism and agriculture.

These suggestions invite the understanding that while in some aspects there are similar or related processes in different parts of the Amazon, it is difficult to develop homogeneous criteria for fostering the integration of the Amazon in the context of national economic development. The formulation of a scientifically adequate and specific policy, which respects and enhances traditional populations, while integrating other sectors of the community to promote the sustainable use of the Amazon biome, must be given priority.

According to the social actors and other participants in the study, the main adaptations needed for the region related to adverse climate change are related to the fluctuation and uneven distribution of rainfall, as well as temperature extremes. These factors affect the most vulnerable and needy communities as was clear in the responses given to the question "Adapt to what?". In relation to the issue "What will be affected" those interviewed affirmed that decreased water availability, harming agricultural production and increasing the incidence of various diseases, will be the most prominent items. Educational campaigns (15%) and rational use of natural resources (19%) were mentioned as actions for adaptation, although 32% of respondents did not know how to react to the consequences of future climate events. Even though the community did not have a specific adaptation plan spontaneous adaptation was observed to the degree that, to reduce the adverse effects, farmers for example sought hardier animal and crop species. The suitable local action agenda that was suggested by the researchers who conducted this study should be enacted through a participatory planning workshop with the identification of needs and emergencies aimed at the realization of effective actions in the communities of the Gleba Aliança.

From the *case study at Tapera* it can be inferred, broadly speaking, that its residents make little association between global climate dynamics and daily phenomena. The answers provided by the focal group with the social actors seemed to suggest some degree of knowledge about climate change and possible global implications. Nevertheless, they did not differentiate global climate change from more mundane meteorological variations. The answers invariably focused on issues (Note 10) related to the current precarious living conditions, or to the government's persistant lack of acknowledgement of the community, rather than to possible climate dynamics.

In the issues of "Adapt to what?" or "What will be affected?", the answers pointed to aspects such as: lack of infra-structure (basic sanitation, paved streets, leisure areas, access to education, transportation, and healthcare); occupation of the mangrove by residents; large numbers of abandoned animals; and a high rate of marginal social behavior, particularly related to drug trafficking. With regards to the issue "What would be an adequate local action plan?", the most commented were the following: tree planting; waste recycling; distribution of sunscreen lotion; monitoring of spaces by the public sector; and adoption of maritime transportation to connect the communities and the city.

When addressing the consequences of environmental transformation, but not necessarily connected to climate change, the perception of the decreasing quantity of the mollusk (*Anomalocardia brasiliana*) or cockles, known locally as *berbigão*, was a common remark. This reduction in the mollusk population is normally attributed to the effects of real estate development in the mangrove area, and to disrespect, even by the mollusk gatherers themselves, of the periods when capturing this species is prohibited.

With regards to the issues most related to climate dynamics, the social actors most commonly commented that the risk of floods (81%) is the most present impact in the community. In fact, floods affect a large portion of the residents, and are especially related to a higher incidence of diseases and health complications – such as flu, asthma, allergies, migraines, skin infections, worms, diarrhea, or other ailments resulting from lack of treated water.

When addressing the capacity of the community to deal with possible risks from extreme climate events, 69% of interviewees admit not being able to prepare or react to the consequences of climate change, while 19% did not answer this question. Considering this response and the precarious infrastructure in the community, the researchers evaluated the community's capacity for reaction as fragile. In Table 3 is showed what events most affect their community, interviewees responded intense rain (88%), floods (81%) and extreme temperatures (44%).

When asked to address adaptation to climate change, the social actors consulted point to factors such as education, sanitation, and social services as the most promising strategies, even if they do not specify which actions. In other words, they point to basic sectors that serve the primary needs of any population.

When attempting to identify initiatives for an action plan, the social actors pointed to the need to undertake activities with people involved with recycling, as well as the realization of infrastructure projects to provide the community with appropriate places for garbage disposal. Another aspect given similar emphasis is the evident need to treat waste

As mitigating measures for dealing with risk, the interviewees pointed to the evident need for improvements in infrastructure, including the elements that affect hydrological dynamics, the implications of which include recurring floods, and a permanent condition of poor health among the population. Nevertheless, the social actors in Tapera da Base do not associate a possible increase in their vulnerability to climate dynamics.

Table 3. Comparison different studies case about events that occur in the region where the families interviewed live, percent by state and total

Climate Events *	Rio de Janeiro	Rondonia	Santa Catarina
More intense rains	75%	34%	88%
High tides, inundations or floods	58%	21%	81%
Drought	14%	68%	0%
More frequent lack of rain	22%	38%	0%
Changes in seasonal flooding of forest	0%	11%	0%
More intense heat	72%	83%	44%
More intense cold	53%	9%	44%
Deforestation	17%	38%	31%
Fire or burnings (accidental or purposeful)	69%	30%	19%
Pests	53%	26%	38%
Loss of planting area	6%	17%	0%

5. Conclusion

Based on a theoretical discussion of socio-environmental vulnerability, development and extreme climate phenomena, this paper presents the outcomes of three case studies undertaken in specific regions of Brazil to evaluate the perception of populations deemed already vulnerable, in terms of their knowledge and recognition of climate phenomenon and variations that have affected them. Distinct methodological procedures were used to gather data from the different local actors and populations.

The research was based on the understanding that the climate issue provides an opportunity to reconsider the bases of economic development, characterized as employing specific models for the use of natural resources that can place different populations in situations of risk. These situations become more evident according to the different levels of vulnerability of these populations, in particular those already exposed to social and economic exclusion in countries of late development such as Brazil. Thus, this work began with the premise that a population's vulnerabilities are historically constructed and result from a development model that creates socioeconomic inequalities. When a population is excluded from the development process, it is placed in a situation of risk and often comes to occupy areas designated by law as sites for permanent environmental preservation, as illustrated in the cases of this study, which encompasses populations who occupy mangroves, hillsides and coastal areas.

The threat posed by climate change to the natural bases that sustain life for these communities, and for the planet as a whole, highlights the need to place the ethical and social aspects of this phenomenon on the political agenda. A human rights perspective is thus encompassed, since it is based on the principle that future generations also have a right to the environment.

Properly addressing these conditions of vulnerability, and the impacts of climate events that expose these conditions, involves constructing capacities for resilience that question the understanding of development from a purely economic dimension. In other words, this process must contribute towards the empowerment of communities to facilitate their resilience. Furthermore, the framing of this research tried to contribute to the development of action plans aimed at mitigating vulnerabilities and risks according to the peculiarities of the territories.

The case study of the Fiocruz Campus community observed that all the communities have long suffered from precarious access a route, which has the residents transportation and movements, and have also lacked access to basic public services and infrastructure such as sanitation and health care. These are problems that can be aggravated by extreme climate events, with subsequent implications for living conditions and health. Meanwhile, it is important to help residents in this community become pro-active, because only this way can they push for the changes needed to decrease the vulnerabilities of families facing climate change, and diseases resulting from these events.

The case study of the Gleba Aliança community clearly demonstrates the great variation of the traditional communities (riverine fishermen) and those with which they co-exist in the surrounding area. Those other communities are constituted mostly of migrants from different regions of Brazil who conduct other types of economic activities (planting and/or cattle rearing), and who have distinct ways of feeling and understanding the interaction with the environment and the effects of climate variations on their lives and means of production. It is necessary to consider these peculiarities to avoid failure when proposing public policies and mitigating actions for the Amazon biome.

The case study of the Tapera da Base community found that the vulnerability of the community is not merely evidence of probable climate threats. The climate transformations understood to be threatening reveal the community's vulnerability to the potential risk of flooding, and other risks such as disease. Therefore, the need for investment in basic actions, constitutionally the responsibility of the state (education, healthcare and infrastructure), is of utmost importance for reducing the community's vulnerability.

For this community, the growing risks associated with threats posed by difficulties common to daily life are still greater than the growing risks related to climate threats, about which residents have little, if any, information. It is worth recalling that more than 80% of the group interviewed has no knowledge that significant climate changes are taking place. Even when they are made aware of these risks, the residents in the community of Tapera da Base do not reflect upon the need to adapt to possible severe implications of global climate change.

The main priority for the residents of Tapera, and sometimes the only one, is to adapt to the world they live in, given its precarious current state. This manner of channeling attention to socio-economic difficulties to overcome immediate difficulties in the urban environment is both a source of strength for the community, considering that it develops its capacity to prioritize essential aspects, and of growing vulnerability, since it hinders the reflection on the residents' future needs.

Another issue to be highlighted is that the vulnerability of the inhabitants of the communities studied is part of a historic context that naturalizes the risks faced by the economically disadvantaged communities and isolates local social subjects from decision-making processes. This situation is sustained by historic social exclusion, especially in terms of a lack of access to resources, education and basic healthcare. Moreover, the historically structured living conditions of these subjects are subject to different threats, creating a situation in which survival requires adaptation. That is, climate change is only one of the existing threats to which they are historically exposed (others include disease, lack of basic sanitation, various forms of pollution and deforestation).

It is in this sense that a parallel naturalization occurs of the problematic contexts by the part of responsible public and private institutions. The phenomenon of "organized irresponsibility" (Beck, 2006), takes place in this situation, in which to "disguise" the inabilities of the political, scientific and legal spheres to address current risks, institutions act symbolically, creating the sense of a normality and control that are not effective for reducing problems.

Thus, it is important to develop action plans that involve civil society, NGOs and municipal, state and federal agencies to confront environmental disasters and prevent fatalities when they occur.

Considering the three cases, we were able to observe that to foster adaptation of community to climate change, it is necessary to solicit actions that are in tune with the way the system operates at the local level, establishing partnerships with civil defense agencies and municipal governments. It is important to promote changes in the way that a community operates, which can only be accomplished trough dialog among its components that addresses and disseminates information about the main emergency actions in this context of climate change. The goal is to support conditions that can allow community residents to become active social actors in their life trajectory. It is therefore necessary to provide them conditions to interpret the phenomena and its possible implications, and that these and other individuals can have some control of their realities, under the threat of disasters caused or intensified by climate changes.

Future studies should focus on accompanying the establishment of local action plans to confront risks such as flooding and landslides that are intensified by climate change and to verify the participation of the local community, who are the principal interested parties. The local population must be involved in the construction of these plans because they are the ones who know the territory and can contribute to the efficiency of these actions. These studies should be frequently re-evaluated to analyze the effectiveness of the agenda, seeking to mitigate the damage to the vulnerable populations, by increasing their capacity for reaction and readiness. In conjunction, public policies must be planned to build housing to relocate the affected population to safer places less subject to impacts from extreme weather events, yet close to their current homes. This is called for in the guidelines and objectives of the National Protection Policy PNPDC (Law 12.608, of 10/04/2012) and in the strategies for reducing risks and disasters proposed by Hyogo in the Disaster Reduction Plan 2005-2015 (UN, 2005). In other words, the actions must be adopted by the various government sectors and should be integrated into various public policies as a transversal theme, to reduce the risks and disasters involving the vulnerable populations.

Acknowledgements

Our acknowledgments to the respondents who agreed to cooperate with these studies, the Brazilian National Council for Scientific and Technological Development (CNPq), a Fundacao de Amparo a Pesquisa do Espírito Santo/[Research Support Foundation of Espirito Santo] (FAPES) and to the National Foundation for the Development of Private Higher Education/ FUNADESP for their research grants.

References

Acselrad, H. (2009). *O que é justiça ambiental*. Rio de Janeiro: Garamond.

Alves, H., Mello, A., D'Antona, A., & Carmo, R. (2010). Vulnerabilidade socioambiental nos municípios do litoral paulista no contexto das mudanças climáticas. Retrieved from http://www.abep.nepo.unicamp.br/encontro2010/docs_pdf/tema_3/abep2010_2503.pdf

Alves, H. P. F. (2006). Vulnerabilidade socioambiental na metrópole paulistana: uma análise sociodemográfica das situações de sobreposição espacial de problemas e riscos sociais e ambientais. Rev. bras. estud. popul. v.23 n.1 São Paulo jan./jun. http://dx.doi.org/10.1590/S0102-30982006000100004

Bartholo Jr., R. S., & Bursztyn, M. (1999). *Amazônia sustentável: uma estratégia de desenvolvimento para Rondônia 2020*. Brasília: IBAMA.

Beck, U. (2006). *Living in the World Risk Society*. British Journal of Sociology Centennial Professor. London School of Economics and Political Science. February 2006.

Braga, T., Oliveira, E., & Givisiz, G. (2006). Avaliação de metodologias de mensuração de risco e vulnerabilidade social a desastres naturais associados a mudanças climáticas. Trabalho apresentado no XV encontro nacional de estudos populacionais. *São Paulo em Perspectiva, 20*(1), 81-95..

BRASIL. Plano Nacional e Mudanças Climáticas. (2008). Decreto nº 6.263 (21 / 11 / 2007), Brasília.

Cardona, O. (2003). La necessidad de repensar de manera holística los conceptos de vulnerabilidad y riesgo: una critica y una revisión necesaria para la gestión. Retrieved from http://www.desenredando.org/public/articulos/2003/rmhcvr/rmhcvr_may-08-2003.pdf

Da-Silva-Rosa, T., & Maluf, R. (2010). Populações vulnerabilizadas e o enfrentamento de eventos climáticas extremos: estratégias de adaptação e mitigação. *Boletim da Sociedade Brasileira de Economia Ecológica, 23/24*, 40-48.

Da-Silva-Rosa, T., & Mattos, R. (2012). Exclusion, Vulnerabilities and Climate Change. In LASA - Latin American Studies Association, 2012, San Francisco. XXX International Congress of the Latin American Studies Association, v. 1.

Füssel, H.-M. (2008). Adaptation to climate change: a new paradigm for action or Just old wine in new skins? International workshop: Prospects of safety and sustainability Science for our globe, Dec 4, Tokyo (Japan).

Füssel, H.-M., & Klein, R. J. T. (2004). Conceptual frameworks of adaptation to climate change and their applicability to human health, PIK Report No. 91, Potsdam, Germany, August.

Gamba, C. (2010). Avaliação da vulnerabilidade socioambiental dos distritos do município de São Paulo ao processo de escorregamento. Master's Dissertation, Fac. Filosofia, Letras e Ciências Humanas, USP, supervision by Wagner C. Ribeiro.

Giddens, A. (2009). *The politics of climate change*. Cambridge: Polity Press.

Herrmann, M. L. P. (2001). Levantamento dos desastres naturais ocorridos em Santa Catarina no período de 1980 a 2000. Florianópolis: IOESC, p. 89.

Klein, R. J. T., Schipperc, E. L. F., & Dessaid, S. (2003). Integrating mitigation and adaptation into climate and development policy: three research questions.

IPCC-Internacional painel for Climate Change. (2007). The Physical Science Basis. Contribution of Working Group I to the Fourth Assessment Report of the Intergovernmental Panel on Climate Change. In S. Solomon, D. Qin, M. Manning, Z. Chen, M. Marquis, K. B. Averyt, M. Tignor, & H. L. Miller (eds.). Cambridge University Press, Cambridge, United Kingdom and New York, NY, USA. pp. 996.

Jacobi, P. R. (1995). Moradores e meio ambiente na cidade de São Paulo. Cadernos CEDEC, São Paulo, n. 43.

Magalhães, J. M. (2005). Relatório de Consolidação do PA Aliança. Divisão de Desenvolvimento da Superintendência Regional do INCRA-RO. Porto Velho.

Maluf, R. S., & Da-Silva-Rosa, T. (2013). Mudanças climáticas, desigualdades sociais e populações vulneráveis no Brasil: construindo capacidades. Rio de Janeiro: CERESAN, 2011. Retrieved from http://www.coepbrasil.org.br/portal/Publico/apresentarArquivo.aspx?TP=1&ID=8fc52b23-e5bb-4039-96dd -2cacdee9e008&NOME=Relatorio%20Final%20da%20Pesquisa%20%28Volume%20I%29.pdf

Marchezini, V. (2009). DOS DESASTRES DA NATUREZA À NATUREZA DOS DESASTRES. In Sociologia dos desastres – construção, interfaces e perspectivas no Brasil/ organizado por Norma Valencio, Mariana Siena, Victor Marchezini e Juliano Costa Gonçalves. São Carlos :RiMa Editora.

Marengo, J. A. (2009). Mudanças climáticas globais e seus efeitos sobre a biodiversidade. Caracterização do clima atual e definições das alterações climáticas para o Território Brasileiro ao longo do século XXI – Brasília: MMA.

Mattos, R., & Da-Silva-Rosa, T. (2011). Reestruturação econômica e segregação sócioespacial: uma análise da Região da Grande Terra Vermelha. In I Seminário Nacional do Programa de Pós-Graduação em Ciências Sociais UFES, 2011, Vitória-ES. Anais do I Seminário Nacional do Programa de Pós-Graduação em Ciências Sociais UFES. Vitória, pp. 1-21.

Mello, A., D'Antona, A., Alves, H., & e Carmo, R. (2010). Análise da Vulnerabilidade Socioambiental nas Áreas Urbanas do Litoral Norte de São Paulo. V Encontro Nacional da Anppas 4th to 7th October 2010, Florianópolis, SC, Brazil.

Moser, C. (1998). The asset vulnerability framework: reassessing urban poverty reduction strategies. *World Development, New York, 26*(1).

Sachs, W. (2003). Environment and human rights.Wuppertal Paper 137, September.

Sachs, W. (2008). Climate change and human rights. *Development, 51*, 332-337. Doi 101057/dev.2008.35.

Valencio, N., Siena, Mariana; Marchezini, Victor e Costa, Juliano. Gonçalves (org.) (2009). Sociologia dos desastres – construção, interfaces e perspectivas no Brasil. São Carlos : RiMa Editora.

Valencio, N. (2011). A sociologia dos desastres: perspectivas para uma sociedade de direitos. Seminário estadual de emergências e desastres: estratégias latino-americanas de enfrentamento à questão. Vitoria, ES.

Zuffo, C. E. (2010). Relatório de estudo de caso: Rondônia. In Projeto de pesquisa "Mudanças climáticas desigualdades sociais e populações vulneráveis no Brasil: construindo capacidades", Porto Velho: COEP.

Notes

Note 1. COEP has engaged more than 1,100 organizations, thus giving rise to the COEP Network with approximately 20,000 people (the Mobilization Network) from more than 120 communities (the Communities Network).

Note 2. On this subject, COEP was involved with the Database on Climate Practices, Vulnerability and Adaptation (www.coepbrasil.org.br/projetosdeadaptacao) and with the presentation to the government of elements to support the elaboration of the National Plan for Adaptation of Human Impacts by Climate Change, based on discussions held in the working group with the participation of a wide number of social actors.

Note 3. The universities included: the Federal University at Santa Catarina (UFSC), led by Professor Luis D`Agostini and associate researcher Michelle Bonatti; the Federal University at Rondônia UNIR), led by Professors Cátia Zuffo and Joel Magalhães (INCRA).

Note 4. FIOCRUZ – the Oswaldo Cruz Foundation –Mata Atlântica Campus, led by Dr. Andrea Vanini.

Note 5. The study cases originally also included two other biomes that are not discussed in this paper: the Cerrado (tropical savannah biome), which involved a rural quilombola community in Mato Grosso do Sul state, led by Professors Dario Lima Filho and José Jesus Lopes of the Federal University at Mato Grosso do Sul (UFMS); and the Caatinga (xeric shrubland) which is home to family farmers in Pernambuco State, led by Professor Guilherme Soares (Federal Rural University at Pernambuco (UFRPE).

Note 6. The other two foci are: a survey of national and international reference documents, and national policies related to the theme and a synthesis of their implications for the case studies; and mapping and systematization of knowledge production in Brazil on the theme of climate change and social inequality.

Note 7. We are referring to the disaster occurred in the municipalities of Petropolis, Nova Friburgo and Teresópolis among others hit by intense rains. Generally classified as Brazil's major climatic tragedy, the event caused hundreds of deaths and left hundreds missing as well as infra structure losses. The National Policy on Protection and Civil Defense (Política Nacional de Proteção e Defesa Civil – PNPDEC, L. 12608, April 2012) is considered a direct consequence of this social disaster and mobilization.

Note 8. UNDP/GEF Technical Paper 7. Assessing and Enhancing adaptative capacity, in http://undp.prg/climatechange/adapt/apf.html

Note 9. PAC refers to the federal Growth Acceleration Program, which seeks to improve the living conditions of the local population.

Note 10. "Adapt to what?" Who will be affected? What Will be affected?, How to adapt oneself?" and What would be an adequate local action plan?

13

Comparison of Different Global Climate Models and Statistical Downscaling Methods to Forecast Temperature Changes in Fars Province of Iran

Mohamad Aflatooni[1] & Jafar Aghajanzadeh[1]

[1] Islamic Azad University, Shiraz Branch, Department of Water Resources, Shiraz, Iran

Correspondence: Mohamad Aflatooni, Islamic Azad University, Shiraz Branch, Department of Water Resources, Shiraz, Iran. E-mail: Moh01012500@yahoo.com

Abstract

In order to find a suitable climate model to forecast future temperature change in Fars province of Iran, three different Global Climate Models (GCMs); that is HADCM3 with scenarios A2 and B2, CCCMA-A2 and ECHOG with scenario A2a, were compared on coordinate point and whole area basis. GCM temperature variable was taken from Internet (http//www.cera-dkrz.de) and local measured minimum and maximum temperature were taken from 27 Synoptic Weather Stations (1989-2007) in Fars province and neighbouring areas. For downscaling GCMs, a variation of different regression models, namely; linear, second order, third order and multiple linear regression of stepwise type were tried in the form of 6 Methods using a detailed error analysis. In our study, the variables were minimum and maximum temperature and GCM model selection criteria were MSE and SS (Skill Score). The results showed that GCM model selection for the area depended on selection criteria and the kind of variable (being either minimum or maximum temperature). In most parts of the area, CCCMA-A2 was the best with the least error for minimum temperature and ECHOG-A2a for maximum temperature. Also, multiple linear regression of stepwise type, among other regression models, proved to be the best method of downscaling having the least error in all comparisons.

Six methods were then used to obtain temperature from 1950 to 2100. Results of the multiple linear regression of step wise type as the best method showed that the average monthly temperature in the control run (1995-2009) was 292.83 and for future period (2085-2099) was 297.95 degrees Kelvin showing temperature increase of 5.12 degrees for the next 90 years.

Keywords: GCM outputs, climate model, downscaling, error, minimum and maximum temperature, multiple regression, weighing technique, stepwise

1. Introduction

Recent use of fossil fuels, human life activity and technological developments have led to climate change on a world wide scale according to NRC (National Center for Atmospheric Research) and IPCC (Inter-governmental Panel on Climate Change) reports. Increase of green house gases has caused the temperature of the earth to sharply increase in recent decades and expected to increase in the coming future. This non-periodical increase can have different effects on climate of various parts of the world in different manner (David, Piercea, Barnetta, Benjamin, Santerb, & Glecklerb, 2009). Also, different climate change may have different effects on water resources (Beldring et al., 2006; Fowler et al., 2007; Hamlet et al., 2009; Misra et al., 2003; Wilby et al., 2006; Chen et al., 2003).

The main problems facing the researchers are how to downscale GCM outputs to consider the local effects and selection of suitable GCM model in any area to decrease the model errors involved (Jones et al., 1980; Hamlet et al., 2009; Hoar, 2008; Wilby et al., 2006). Due to large variability of GCM models and their outputs from different organizations throughout the world, care should be taken while selecting the models; one model may give good results in one area or point and the other one may give unacceptable errors in the same area considering the downscaling methods used. Thus the source of error can come from downscaling method on one hand and selection of the model itself on the other. Pros and cons of different GCM models and downscaling methods other than statistical are discussed in in various articles (Hoar & Nychka, 2008; Davis et al., 2009) and also by NRC and IPCC reports.

It is assumed that selecton of a GCM model variable on the fly for an area without a previous study on its suitability can cause eronious results. As an assumption in our study, there may be no specific GCM model for the south west of Iran and downscaling method is also of concern. The motivation, therefore, behind this research is two fold; first to find the specific GCM model for the area and second to find the suitable downscaling method for maximum and minimum temperature to adjust for local effects for the south west of Iran. In the latter case, different regression equations were tried to select a suitable downscaling method for the area.

2. Materials and Methods

2.1 Study Area and Selection of Common Interpolating Coordinates

The study area is located in south western Iran and extends in 50-55.375 degrees longitude and 26-33 degrees latitude. Figure 1 shows the area along with the major downloaded GCM points and local weather stations.

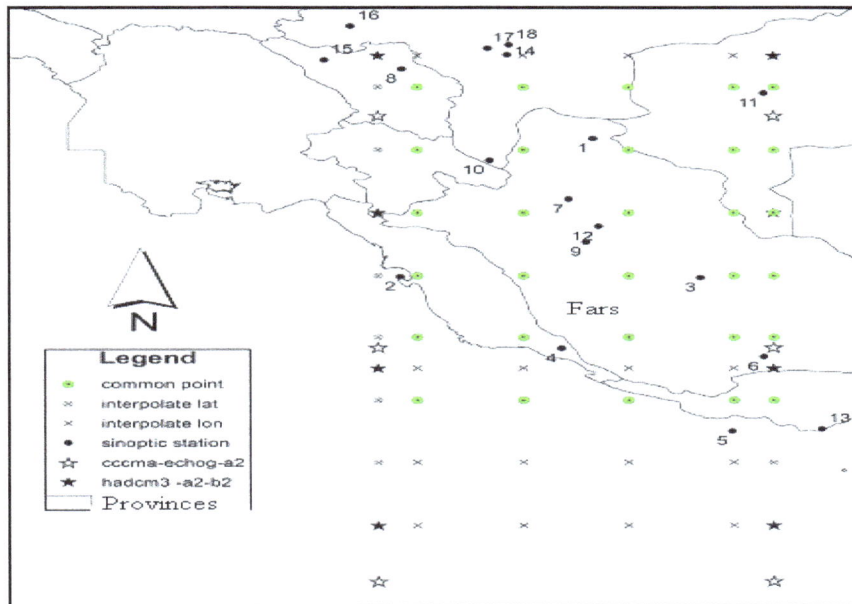

Figure 1. Graphical representation of the study area in south western Iran showing original and interpolated GCM locations along with Synoptic stations

Table 1 shows coordinates of different GCM models at which the data were downloaded.

Table 1. Coordinates of available downloaded GCM temperature data covering the area. Each box indicates a geographic coordinate point

GCM	Latitude-Longitude			
HADCM3 A2	32.5-50.625	30-50.625	27.5-50625	25-50.625
HADCM3 B2	32.5-54.375	30-54.375	27.5-54.375	25-54-375
ECHOG-A2	35.256-50.625	31.545-50.625	27.833-50.625	24.122-50.625
	35.256-54.375	31.545-54.375	27.833-54.375	24.122-54.375
ECHOG-B2	35.2556-50.625	31.5445-50.625	27.8334-50.625	24.1223-50.625
	35.2556-54.375	31.5445-54.375	27.8334-54.375	24.1223-54.375

Using these coordinates, the new coordinates common to all GCM models and measured data were constructed (Table 2).

Table 2. Coordinates of common points interpolated for each GCM model and measured data covering the area used for comparison purposes

Latitude-Longitude				
32 : 51	32 : 52	32 : 53	32 : 54	32 : 54/375
31 : 51	31 : 52	31 : 53	31 : 54	31 : 54/375
30 : 51	30 : 52	30 : 53	30 : 54	30 : 54/375
29 : 51	29 : 52	29 : 53	29 : 54	29 : 54/375
28 : 51	28 : 52	28 : 53	28 : 54	28 : 54/375
27 : 51	27 : 52	27 : 53	27 : 54	27 : 54/375

The common points (green)are also shown in Figure 1. The original GCM coordinates were linearly interpolated based on $1^o \times 1^o$ to get the common points. Temperature variable time series of three different global climate models; HADCM3 with scenarios A2 and B2, CCCMA-A2 and ECHOG with scenario A2 were taken from Internet (http//www.cera-dkrz.de). Resolution of the first model was $3.75^o \times 2.5^o$ ($367.5 \times 295\,Km^2$)and the other two were $3.75^o \times 3.711^o$ ($367.5 \times 438\,Km^2$). The temperature data were interpolated using the following relations (Aghajanzadeh, 2010):

$$TLa_N = \frac{(La_N - La_I) \times (TLa_F - TLa_I)}{(La_F - La_I)} + TLa_I \qquad (1)$$

$$TLo_N = \frac{(Lo_N - Lo_I) \times (TLo_F - TLo_I)}{(Lo_F - Lo_I)} + TLo_I \qquad (2)$$

where Lo and La are longitude and latitude, indices F, I and N correspond to coordinates of end, first and interpolated points, T is the temperature in respect to coordinates. Equations 1 and 2 are used to interpolate points for latitudes (columns) and longitudes (rows) respectively.

2.2 Measured Local Data and Interpolating Corresponding Points to GCM Outputs

Average monthly measured temperature data for 1989-2007 were taken from 27 weather stations (Iranian Synoptic Weather Organization) located within $50 - 55.375$ degrees longitude and $26 - 33$ degrees latitude of the study area. Only 18 stations which had common data in the period were selected. A detailed preprocessed time series data analysis consisting of finding lost data points using regression analysis, test of temporal data homogeneity using Double Mass analysis, test of stochastic nature of temperature data using Run Test technique were performed for further certainty purposes (Aghajanzadeh, 2010). There were no temporal outlier points in the data. Local temperature data were so determined to correspond with GCM data points using IDW (Inverse Distance Weighted) weighing method:

$$D_i = \sqrt{((Lo_N - Lo_S) \times 98)^2 + ((La_N - La_S) \times 118)^2} \qquad (3)$$

$$W_i = \frac{1 / D_i}{\sum_{i=1}^{n} 1 / D_i} \qquad (4)$$

Where indices S and N are, respectively, stations and interpolated points(i.e., common points), Di is the distance between S and N points, N is the number of stations within 1 degree (lat. & Lon.) of the interpolated GCM points, numbers 89 and 118 are equatorial distance (lat. and lon. respectively obtaind by the area map) in kilometer, Wi is the weight of each station. Therefore, for each GCM data point, the number of weather stations used in weighing method was between 1 to 6 each having a weight between 0 to 1. The weight of each station used in IDW method is given elsewhere in details (Aghajanzadeh, 2010). Finally, 27 common points out of 30 for which measured temperature data existed were used in comparing GCM models. The weights were applied to the time series of each station data and summed up according to Equation 4 so as the interpolated points to have a new time series corresponding to GCM point time series. In this way, number of temporal data points were

1800 (150 years) in 27 spatial locations, whereas, number of temporal measured data points were 228 (19 years; 1989-2007) in 27 spatial locations. This period is devided into tow periods; one is for calibration (1989-2005) and the other for validation (2006-2007). GCM and measured maximum and minimum mean monthly temperature were compared separately in the study.

2.3 Model Selection Criteria

Mean Squared Error, MSE and Skill Score, SS given in Equations 5 and 6, respectively, were the criteria for comparing measured and GCM data. However, to eliminate the effects of data unit and scattering in the error analysis (David et al., 2009), MSE was converted to SS (Skill Score) according to Equation 6:

$$MSE(m,o) = \frac{1}{N} \sum_{k=1}^{N} (m_k - o_k)^2 \qquad (5)$$

$$SS = 1 - \frac{MSE(m,o)}{MSE(\mu,o)} \; ; \; SS < 1 \qquad (6)$$

Where m is GCM data, μ is the mean observations and o is observed value, N is the number of observations and k is data index. It should be noted that whenever SS is closer to unity, it shows a better model capability. In case of zero SS, the model predicts temperature variable around mean observations. Percent error was calculated as follows:

$$\%Error = \frac{|\bar{o} - \bar{m}|}{\bar{o}} \qquad (7)$$

2.4 Downscaling Methods

A variation of linear, second order, third order and multiple linear regression equations were tried with 3 GCM models to define six downscaling Methods. These Methods are so defined to be referenced easily in the text.

Method 1-Raw GCM model data were first compared with local measured data at each point and depending on errors calculated, the best model was selected for that point. The selected model was then downscaled using linear, second and third order regression equations. The best regression model was selected with the highest correlation coefficient R^2.

Method 2- Three GCM models were directly downscaled separately at each point using linear, second and third order regression; the best regression model was selected with the highest R^2. Finally, all models for each point were compared with observations whichever had lowest error was selected for that point.

Method 3-Applying weights to the raw GCM outputs according to their respective errors and then downscaling the new time series according to the following equations (Aghajanzadeh, 2010; David et al., 2009):

$$W_i = \frac{e_i}{\sum_{i=1}^{4} e_i} \qquad (8)$$

$$e_i = \frac{MSE(\mu,o)}{MSE(m,o)} = \frac{1}{1-SS} \; ; \; i = 1,2,3,4 \qquad (9)$$

$$Nm = \sum_{i=1}^{4} W_i \times m_i \; ; \; i = 1,2,3,4 \qquad (10)$$

Where W_i is the weight of each GCM model for each point, Nm is the new and m_i is the four old time series data for each point (that is a total of 1800 values for 150 years at each point for new data). Equation 10 was used to convert the old to new time series of the selected GCM model. New time series data were then downscaled using linear, second and third order regression analysis. The best regression model was selected for each point.

Method 4-Applying weights to downscaled outputs S_i (instead of m_i in method 3) and then the new time series were downscaled again(double downscaling). Equations 8 to 10 were used accordingly as discussed in method 3.

The only difference is that Equation 11 is used instead of Equation 10 in which a new parameter S_i is introduced here. The methods 3 and 4 may be called weighing techniques.

$$Nm' = \sum_{i=1}^{4} W_i \times S_i \tag{11}$$

Method 5-Direct downscaling of outputs using multiple linear regression of the stepwise type in all GCM raw data using Equation 12:

$$DS = b_0 + \sum_{i=1}^{4} b_i \times m_i \tag{12}$$

Where b is the regression coefficient which could either be zero or non-zero. Four GCM models (i=1,2,3,4) were used in this method for each point. Each point, however, might need 1 to 4 GCM model to get the highest regression coefficient.

Method 6-The downscaled GCM data from Method 2 were downscaled again applying multiple linear regression of the stepwise type to already downscaled GCM data (double downscaling). Equation 13 is used for double downscaling:

$$DS' = b_0 + \sum_{i=1}^{4} b_i \times S_i \tag{13}$$

where S_i, downscaled data, were selected from Method 2. Briefly, for each point, 1 to 4 already downscaled GCM models were downscaled again using multiple linear regression of the stepwise type. Therefore, in methods 5 and 6, step wise multiple regression technique was used for downscaling. It should be noted that in all above mentioned downscaling methods, a point error analysis was first performed and was averaged over entire area to get a better picture of model selection.

3. Results and Discussion

3.1 Error Analysis

Error analysis for all coordinate points and entire area was performed and only typical results are shown here. The errors are based on MSE and SS appropriately. A typical point error analysis based on SS for CCCMA-A2 is given in Table 3 which shows that for each coordinate point certain error is obtained; therefore, different models may be selected for each point.

Table 3. Typical error analysis results based on SS for CCCMA A2 minimum temperature variable as compared with measured data (1989-2005)

Lat/Lon	51	52	53	54	54.375
32	0.8119	0.8341	0.8038	-0.1169	-0.2093
31	0.8542	0.8297	0.8048	0.1100	0.0365
30	0.8630	0.7900	0.7259		
29	-1.9358	0.8546	0.8099	0.7255	0.6943
28	-1.5594	-0.2144	0.2506	0.4739	0.4115
27		-0.0666	-2.0949	-2.4583	-2.4239

The range of errors is from -2.4583 to 0.8630 on SS basis. Range of errors, values of MSE and SS are all given in Table 4 for minimum and maximum temperature when comparing all raw GCM models.

Table 4. Comparison of GCM model temperature variable with different selection criteria (1989-2005)

Models	Min. MSE	Max. MSE	MSE over area	Min. SS	Max. SS	SS over area
			Minimum Temperature			
CCCSM3-A2	6.6799	130.4608	47.33	-2.4583	0.8630	0.02
ECHOG-A2a	16.0736	215.7153	109.87	-3.4230	0.7888	-1.2
HADCM3-A2	5.1999	196.1556	47.65	-3.0220	0.901	0.13
HADCM3-B2	4.8637	198.3750	47.62	-3.0675	0.9057	0.12
			Maximum Temperature			
CCCMA A2	17.3229	86.4460	35.93	-0.3158	0.8058	0.5
ECHOG-A2a	3.9219	63.0972	15.54	0.3157	0.9554	0.8
HADCM3-A2	6.4315	58.6877	28.84	0.2377	0.9282	0.61
HADCM3-B2	8.0748	57.2878	28.30	0.2519	0.9098	0.62

Minimum and maximum MSE and SS for minimum and maximum temperature are given in this table. The errors averaged on entire area are also given in this table. Based on mean MSE and SS over the entire area, CCCMA-A2 and HADCM3-A2 were the most suitable model for minimum temperature, respectively(shaded boxes in the Table 4). However, HADCM3 A2a was the most suitable based on the criteria mentioned. Table 5 shows appropriate model for points for minimum temperature which were used in downscaling Method 1.

Table 5. Models for each point having maximum SS for minimum temperature, Method 1, (1989-2005)

	51	52	53	54	54.375
32	CCCMA-A2	HADCM3-B2	HADCM3-B2	ECHOG-A2	ECHOG-A2
31	CCCMA-A2	CCCMA-A2	HADCM3-B2	ECHOG-A2	ECHOG-A2
30	CCCMA-A2	CCCMA-A2	HADCM3-A2	-	-
29	HADCM3-A2	CCCMA-A2	CCCMA-A2	CCCMA-A2	HADCM3-A2
28	HADCM3-B2	HADCM3-B2	CCCMA-A2	CCCMA-A2	CCCMA-A2
27	-	HADCM3-B2	HADCM3-B2	HADCM3-B2	HADCM3-B2

The empty boxes in this table are because no measured data were available at these points. Similar table was obtained for maximum temperature. Table 6 shows average weight of each GCM model over entire area for minimum and maximum temperature which indicates different model contribution to the area whether the model being raw or downscaled.

Table 6. Average weight of each GCM model for Method 3 and 4 (weighting Methods)

Models	W_i			
	Method 3		Method 4	
	Tmin	Tmax	Tmin	Tmax
CCCMA-A2	0.3402	0.1488	0.3093	0.0874
ECHOG-A2a	0.1525	0.4488	0.1164	0.4532
HADCM3-A2	0.2529	0.201	0.2879	0.2265
HADCM3-B2	0.2544	0.2014	0.2865	0.2329

Models CCCMA-A2 and ECHOG-A2a had more weight depending on downscaling method and minimum or maximum temperature. For example, comparing Method 3 and 4 and considering minimum temperature, CCCMA-A2 had nearly 34% and 30% weight, respectively. The errors for all raw GCM models are given in Table 7 for validation period (2006-2007).

Table 7. Error values of raw GCM models for entire area and validation period. Tmin and Tmax are minimum and maximum temperature

| Models | 2006–2007 | | | |
| | MSE | | SS | |
	Tmin	Tmax	Tmin	Tmax
CCCMA-A2	47.7853	38.1010	0.0121	0.4810
ECHOG-A2a	118.9524	14.1343	-1.3703	0.8156
HADCM3-A2	53.9809	31.0899	-0.0001	0.5799
HADCM3-B2	54.3870	28.8593	-0.0023	0.6130

The errors for calibration period (1989-2005) are given in Table 4 discussed previously. Errors for minimum and maximum temperatures along with the type of selection criteria are also given in these Tables. Table 7 shows that for validation period and minimum temperature, based on both criteria, CCCMA-A2 is the most suitable but ECHOG-A2a is the most suitable when predicting maximum temperature. The errors, therefore, depend on selection criteria and the GCM variable being minimum or maximum temperature. The point is that for minimum temperature with SS criteria, Tables 4 and 7 do not give the same exact results. Downscaling methods were also compared and the error values are given in Tables 8 and 9 for both calibration and validation period, respectively. In calibration period, the methods differ depending on MSE or SS, and minimum or maximum temperature. Method 5 is the most suitable for this period. In validation period, method 5 is preferred (Shaded area in Tables 8 and 9).

Table 8. Error values of different downscaling methods for calibration Period. Tmin and Tmax are minimum and maximum temperature (Degrees, K)

| Methods | 1989–2005 | | | |
| | MSE | | SS | |
	Tmin	Tmax	Tmin	Tmax
Method1	44.6683	38.5038	0.1729	0.4588
Method2	46.9830	37.9614	0.1303	0.4764
Method3	16.2622	39.5832	0.6750	0.4999
Method4	3.5566	3.3157	0.9266	0.9572
Method5	1.6674	3.0798	0.9681	0.9597
Method6	1.6987	3.2071	0.9675	0.9581

Table 9. Error values of different downscaling methods for validation period. Tmin and Tmax are minimum and maximum temperature (Degrees, K)

| Methods | 2006–2007 | | | |
| | MSE | | SS | |
	Tmin	Tmax	Tmin	Tmax
Method1	45.2790	39.5010	0.1637	0.4467
Method2	48.5031	38.5605	0.1019	0.4713
Method3	19.6730	39.0506	0.6107	0.5132
Method4	4.6136	4.0862	0.9050	0.9460
Method5	2.3335	3.8310	0.9546	0.9487
Method6	2.4495	3.9826	0.9526	0.9467

Percent errors for all raw and downscaled GCM models are summarized in Table 10 for both periods.

Table 10. Percent error for all GCM models and downscaling methods averaged over entire study area for two periods

| Model | % Error | | | |
| | 2006–2007 | | 1989–2005 | |
	Tmin	Tmax	Tmin	Tmax
CCCMA-A2	40.56%	16.79%	76.14%	7.27%
ECHOG-A2a	64.14%	10.62%	24.08%	10.40%
HADCM3-A2	22.66%	15.01%	24.40%	9.97%
HADCM3-B2	22.97%	14.56%	12.29%	10.60%
Methods				
Method 1	13.52%	12.83%	26.90%	9.32%
Method 2	18.63%	14.38%	16.04%	10.50%
Method 3	13.79%	15.91%	7.32%	4.21%
Method 4	11.36%	7.66%	6.43%	4.27%
Method 5	9.57%	7.48%	6.33%	4.36%
Method 6	9.72%	7.58%	20.65%	22.35%

Table 11. Priority of GCM models and downscaling methods for entire area and validation period (2006-2007)

| Models | MSE | | SS | |
	Tmin	Tmax	Tmin	Tmax
CCCMA-A2	6	7	7	8
ECHOG-A2a	10	4	10	4
HADCM3-A2	8	6	8	6
HADCM3-B2	9	5	9	5
Method1	5	10	5	9
Method2	7	8	6	10
Method3	4	9	4	7
Method4	3	3	3	3
Method5	1	1	1	1
Method6	2	2	2	2

This Table shows that method 5 has the lowest percent error compared to other methods. As far as the raw GCM model comparison is concerned, the GCM model selection are based on selection criteria(MSE or SS) and the type of variable (here minimum or maximum temperature) as expected (see Tables 4 and 7). Model selection priority is also given in Table 11 for validation period. This Table also emphasizes that downscaling method 5 has the first priority for the study area and priority of raw GCM data selection are based on selection criteria type (MSE or SS) and the GCM variable, minimum or maximum temperature. The priority of the GCM models and downscaling methods for calibration period gives the same results (Aghajanzadeh, 2010) (data not shown).

3.2 Graphical Model Comparison

Comparison of three raw GCMs using monthly average observed minimum and maximum temperature are given in Figures 2 and 3 respectively, for validation period.

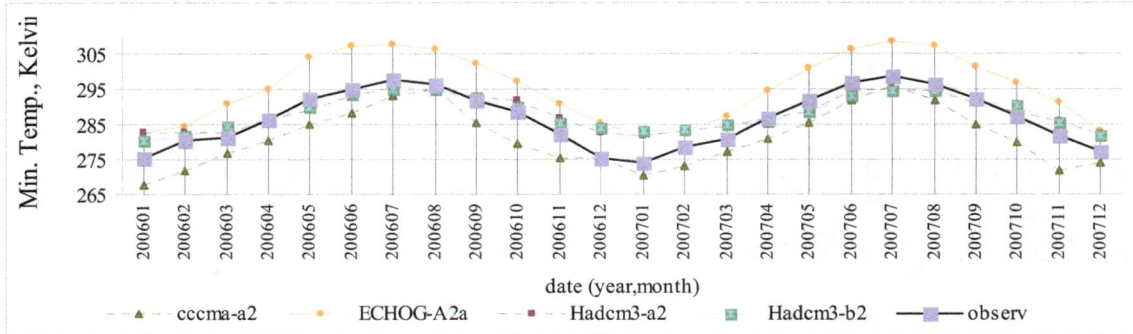

Figure 2. Comparison of different raw GCM models in validation period for average monthly minimum temperature (2006-2007)

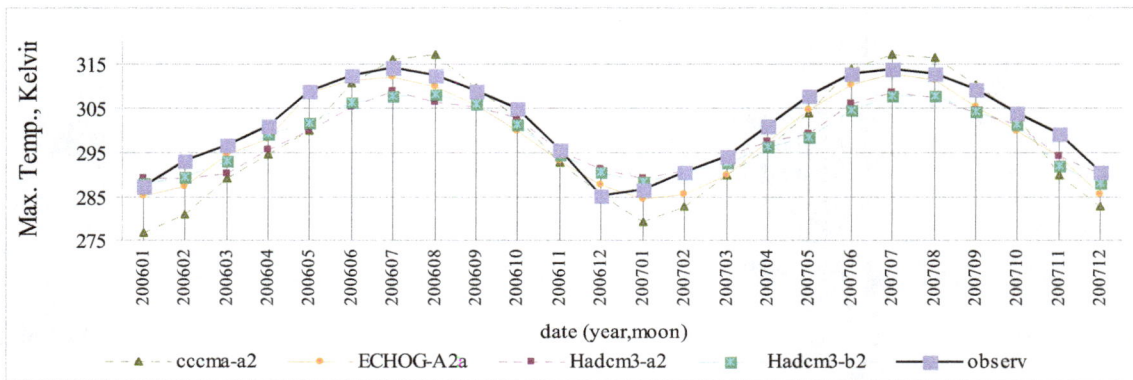

Figure 3. Comparison of different raw GCM models in validation period for average monthly maximum temperature (2006-2007)

Comparisons of six downscaling methods are given in Figures 4 and 5 for minimum and maximum temperature and validation period, respectively.

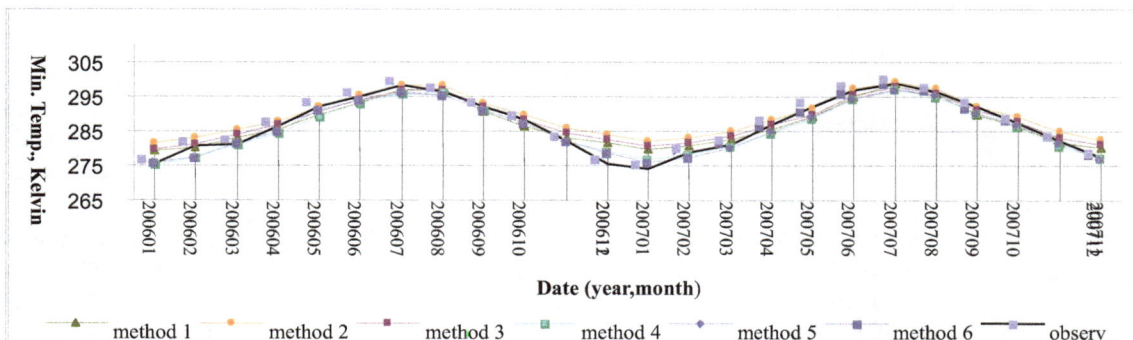

Figure 4. Comparison of different downscaling methods for average monthly minimum temperature (2006-2007)

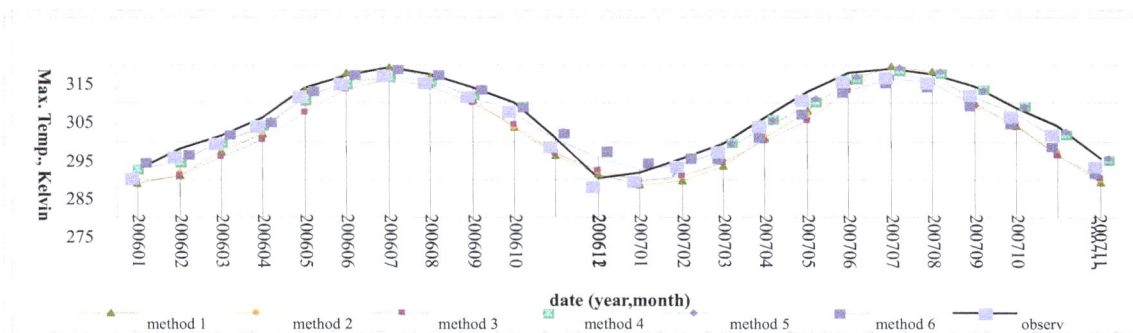

Figure 5. Comparison of different downscaling methods for average monthly maximum temperature (2006-2007)

Monthly average minimum and maximum temperatures are used to correspond to Figures 2 and 3. Graphical comparison of using raw and downscaled GCM models in the area indicates the need for downscaling before using the GCM models for local study. When no downscaling is done, the errors are high (about 26% for both calibration and validation period according to Equation 7) since the local effects such as terrain elevation and plant cover are not accounted for. Due to downscaling (i.e., using Method 5 and for validation period) these effects are considered and the errors are greatly diminished to about 9.57% and 7.48% for minimum and maximum temperature, respectively. Other downscaling methods, however, show a declining error trend somewhat different from the above compared to raw GCM models (Table 10).

Scatter diagrams comparing observed and estimated minimum and maximum temperature averaged over entire area for 1989-2005 and 2006-2007 periods were constructed for all GCM models and downscaling methods. Typical results for downscaling Methods 5 and 6 are given in Figures 6 and 7 for mean monthly maximum temperature, respectively.

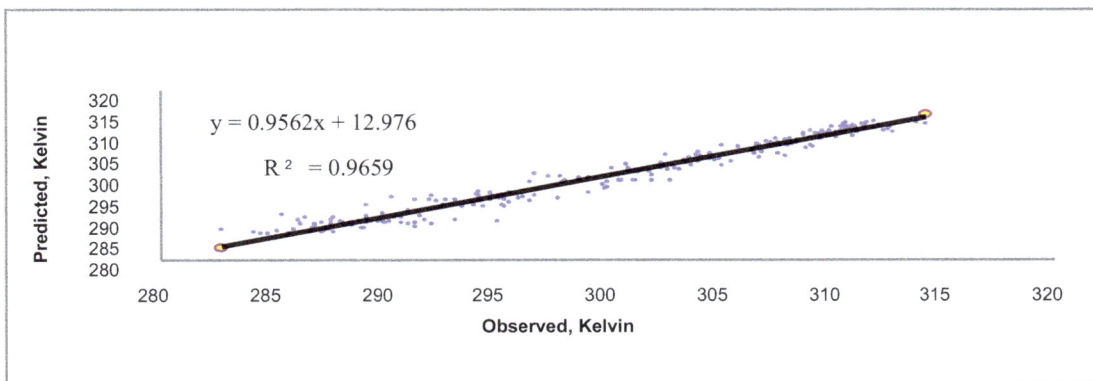

Figure 6. Scatter diagram for downscaling Method 5 for average monthly maximum temperature (1989-2005)

The regression equations and R^2 values for each diagram are shown in the Figures. Method 5 in Figure 6 as expected shows the best fit for maximum temperature with a R^2 value of 0.9659. This value for Method 6 is 0.9651. Also, the best fit for minimum temperature was ascertained for Methods 5 and 6 (Figures 8 and 9).

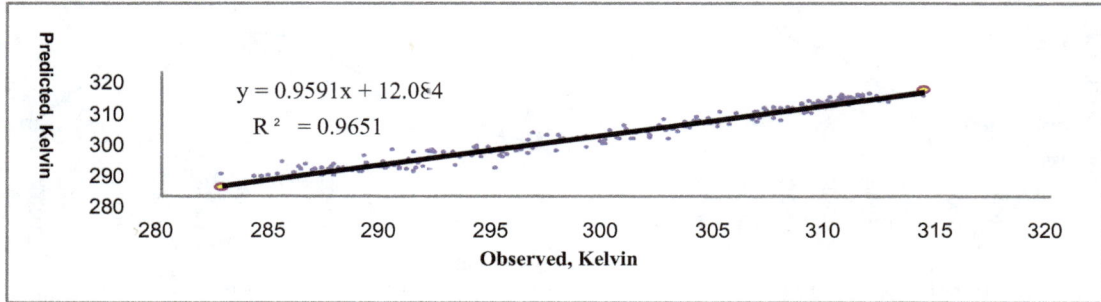

Figure 7. Scatter diagram for downscaling Method 6 for average monthly maximum temperature (1989-2005)

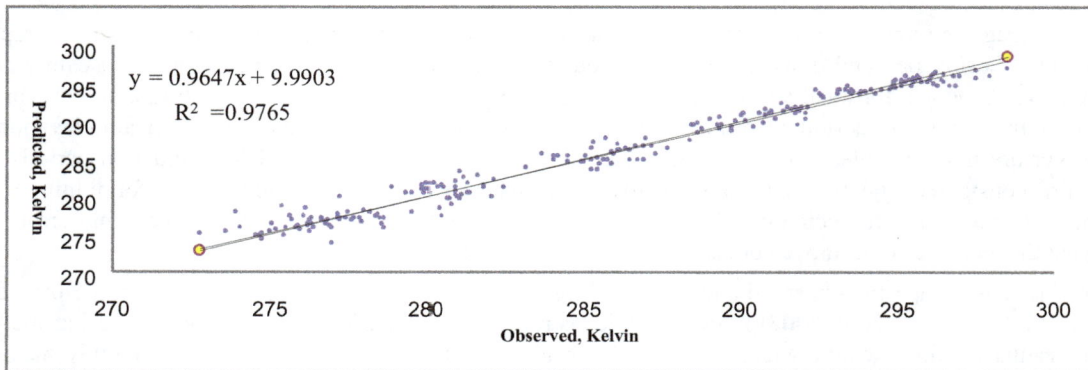

Figure 8. Scatter diagram for downscaling Method 5 for average monthly minimum temperature (1989-2005)

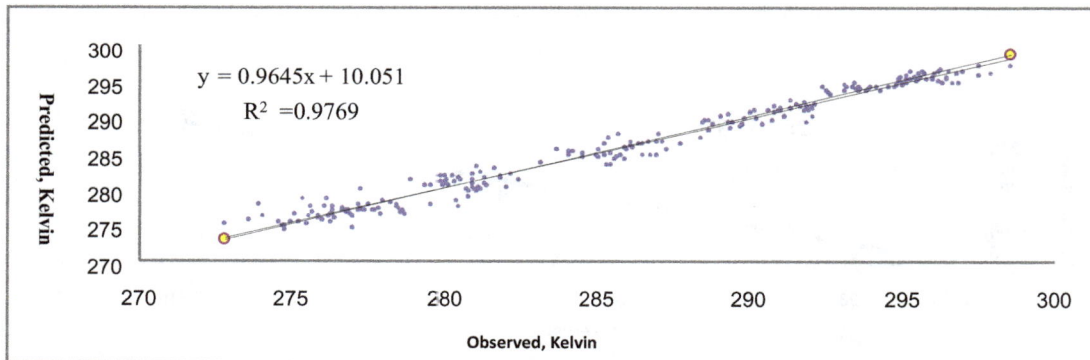

Figure 9. Scatter diagram for downscaling Method 6 for average monthly minimum temperature (1989-2005)

Briefly speaking, the results indicate that the multiple linear regression of stepwise type for downscaling GCM data in our study area is superior to linear, second and third order regression equations used in Methods 1 through 4.

3.3 Forecasts

Using all downscaling Methods already discussed, monthly and yearly minimum and maximum temperatures of the study area were predicted for period 1950 to 2100. A 15 year average of temperature data for 1950 to 2100 is typically shown in Figure 10 where the average temperature is calculated as follows:

$$Average \quad Temp = \frac{\overline{T}\,min + \overline{T}\,max}{2} \tag{14}$$

Since suitable regression equations were selected in calibration period for downscaling purpose, we have to use the same downscaling equations for future study to ascertain identical relationship between GCM and local temperature data. All downscaling Methods shown in this Figure indicate an overall average 4.91 degrees centigrade increase from present period or control run (1995-2009) to last 15 year period (2085-2099) in future for the study area. Separate data analysis indicated that temperature increase from present period to last 15 year period was 4.82 for minimum and 5.42 degrees for maximum temperature using Method 5 downscaling. The increase, however, depends on downscaling method and the GCM model type used in the study area (Table 12).

Table 12. Temperature increase using different downscaling methods from present (1995-2009) to future (2085-2099) in the study area. Temperature unit is in degrees Kelvin

Methods	Tmin Present	Tmin Future	Tmin increase	Tmax Present	Tmax Future	Tmax increase	Average Increase
Method 1	286.9154	290.8326	3.91721	298.1605	305.212	7.051527	5.484368
Method 2	289.172	293.0478	3.875824	297.6666	303.3017	5.635078	4.755451
Method 3	287.6803	291.4034	3.723104	297.2718	302.6302	5.35842	4.540762
Method 4	285.6941	290.0068	4.312736	300.0185	304.5322	4.513699	4.413217
Method 5	285.8574	290.674	4.816637	299.8011	305.2247	5.423676	5.120157
Method 6	285.8635	290.7341	4.8706	300.0366	305.4611	5.424597	5.147599
Average	286.8638	291.1165	4.252685	298.8258	304.3937	5.567833	4.910259

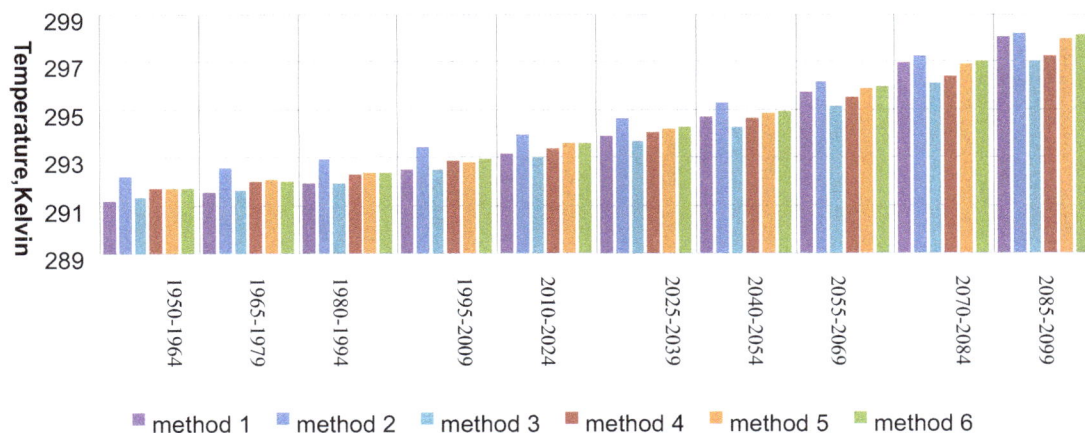

Figure 10. 15 year average of mean temperature change in the study area (1950-2099)

4. Conclusion

The significant finding of the study is that selection criteria (i.e., MSE or SS), type of regression equations for downscaling, type of variable retrieved from GCM models(in this case, minimum or maximum temperature) all can affect the type of GCM model selection in south west of Iran. Among all the regression equations used in this study, multiple linear regression of stepwise type proved to have the best fit. GCM temperature data downscaled with this regression equation was then used to obtain temperature from 1950 to 2100. The average monthly temperature for control run (1995-2009) was 292.83 and for future period (2085-2099) was 297.95 degrees Kelvin showing temperature increase of 5.12 degrees for the next 90 years.

References

Aghajanzadeh, J. (2010). *Evaluation of Climate Change Models and determination of a suitable model to forecast temperature in Fars Province, Iran.* A thesis submitted to graduate school of Water Resource Department in partial fullfilment of Master of Science Degree, Azad University, Shiraz, Iran.

Beldring S, Roald, L. A., Engen-Skaugen, T., & Forland, E. J. (2006). *Climate Change Impacts on Hydrological Processes in Norway 2071-2100.* Norwegian Water Resources and Energy Directorate.

Chen, Z., Grasby, S. E., & Osadetz, K. G. (2003). Relation between climate variability and ground water levels in upper carbonate aquifer, southern Manitoba, Canada. *Journal of Hydrology, 290*(1-2), 43-62. http://dx.doi.org/10.1016/j.jhydrol.2003.11.029

David, W., Piercea, L., Barnetta, T. P., Benjamin, D., Santerb, B., & Glecklerb, P. J. (2009). *Selecting global climate models for regional climate change studies.* Division of Climate, Atmospheric Sciences, and Physical Oceanography, Scripps Institution of Oceanography.

Fowler, H. J., Blenkisop, S., & Tebaldi, C. (2007). Linking climate change medeling to impacts studies. Recent advances in downscaling techniques for hydrological modeling. *International Journal and Climatology, 27,* 1547-1578.

Hamlet, A. F., Salathé, E. P., & Carrasco, P. (2009). *Statistical Downscaling Techniques for Global Climate Model Simulations of Temperature and Precipitation with Application to Water Resources Planning Studies* (Draft). Retrieved from http.//cses.washington.edu/cig/res/ia/waccia.shtml

Hoar, T., & Nychka, D. (2008). *Statistical downscaling of the Comunnity Climate System Model (CCSM), Monthly temperature and precipitation projections.* IMAGE/NCA.

Jones, P. D., Hewitson, B. C., Main, J., & Wilks, D. S. (1980). Statistical downscaling of general circulation model output. A comparison of methods. *Water Resource Research, 34,* 2995-3008.

Misra, V., & Kanamitsu, M. (2003). *Anomaly Nesting. A Methodology to Downscale Seasonal Climate Simulations from the AGCM. Center for Ocean-Land-Atmosphere Studies,* Institute of Global Environment and Society, Inc.

Wilby, R. L., Whitehead, P. G., Wade, A. J., Butterfield, D., Davis, R. J., & Watts, G. (2006). Integrated modeling of climate change impacts on water resources and quality in a lowland catchment. River Kennet, UK. *Journal of Hydrology, 330*(1-2), 204-220. http://dx.doi.org/10.1016/j.jhydrol.2006.04.033

14

A Greenhouse Gas Emissions Inventory and Ecological Footprint Analysis of Metro Vancouver Residents' Air Travel

Ruth Legg[1], Jennie Moore[1], Meidad Kissinger[2] & William Rees[1]

[1] School of Community and Regional Planning, University of British Columbia, Vancouver, BC, Canada

[2] Department of Geography and Environmental Development, Ben-Gurion University of the Negev, Beer-Sheva, Israel

Correspondence: Jennie Moore, School of Community and Regional Planning, University of British Columbia, Vancouver, BC, Canada. E-mail: jennielynnmoore@hotmail.com

Abstract

Ecological Footprint Analysis (EFA) at the city or regional scale does not typically include air travel due to a lack of readily available data. However, knowing the "load" placed on nature by various lifestyle choices, including air travel, is essential if we hope to enable society to live sustainably within ecological limits. This paper provides methods for including air travel in urban EFA, in a manner that is accessible to those that are interested in the complexities of urban sustainability. Our goal is to use the case of the Vancouver Metropolitan region to illustrate two methods in such a way that they can be replicated or adapted for use in other cities and regions. We found that the greenhouse gas emissions of air travel by Metro Vancouver residents for 2006 is between 1,191,070 and 1,402,420 tonnes of carbon dioxide equivalent (tCO$_2$e). The resulting ecological footprint is between 287,030 and 337,980 global hectares (gha), or between 0.136 and 0.160 gha/capita. The dedicated carbon sink required to neutralize the carbon dioxide emissions from Metro Vancouver residents' air travel alone is equivalent to twice the land area of the region (283,183 hectares).

Keywords: air travel, climate change, ecological footprint analysis, greenhouse gas emissions, methods, Metro Vancouver, peak oil, regional planning

1. Introduction

Demand for air travel and its ecological impacts should be scrutinized if society is to respond effectively to climate change, (Union of Concerned Scientists, 1992; Rees, 2008). According to the International Civil Aviation Organization (ICAO), global air passenger travel has been increasing at an annual rate of approximately 7% over the past decade with concomitant increases in greenhouse gas emissions (ICAO, 2011). In 2010, the ICAO established environmental objectives to limit greenhouse gas emissions from air travel and committed to developing a carbon dioxide emissions standard for aircraft (Benjamin, 2013). The ICAO also established a global annual fuel efficiency target of 2% per year to 2020 (Benjamin, 2013). Nevertheless, helping local governments and their citizens to understand their role in driving demand for air travel and the 'load' placed on nature by various lifestyle choices, including air travel, is also essential to enable society to live sustainably within ecological limits.

A greenhouse gas emissions inventory (GHG) measures the quantity of emissions associated with particular activities. Data from GHG emissions inventories can be used to inform ecological footprint analyses. An ecological footprint (EF) measures the amount of ecologically productive land and sea area required to supply the demand of a given population at its current level of affluence and technology (Rees, 1992; Wackernagel & Rees, 1996). Combining the two approaches can provide important feedback about demand on nature's services relative to global waste sink capacities.

Globally, cities and regions provide important leverage points for policy/programs designed to help people live within ecological limits. Ecological footprint analysis (EFA) can help local and regional governments communicate the ecological impacts of lifestyle choices and compare an urban populations' demands for biocapacity with locally and globally available supplies. In particular, understanding how these demands are distributed among commodity and waste flows through a city's or region's economy can inform strategic decision-making and provide a framework for setting goals, collecting data, and monitoring progress. EFA can also illuminate local/regional anomalies by enabling comparisons with other regions. While the method is still being refined, EFA is increasingly being applied in sustainability programs at the city and regional level (City of Vancouver, 2011; Kuzyk, 2011; Graymore, 2010; Wilson & Grant, 2009; McManus & Grant, 2006; Muniz & Galindo, 2005; James & Desai, 2003; Lewan & Simmons, 2001).

One major source of GHG emissions traceable to affluent urban populations is air travel. To date, while various researchers acknowledge its importance (e.g., Barrett, Vallack, Jones, & Haq, 2002; Wilson & Anielski, 2005), only a few out of dozens of published urban EFA studies have measured the air travel component, and these mostly use national data or economically derived input-output data with limited transparency of method and data sources (e.g., Collins, Flynn, Wiedmann, & Barrett, 2006; Aall & Norland, 2005; Xu & San Martin, 2010).

The purpose of this paper is to demonstrate a method to include air travel in urban EFA. Specifically, we estimate the GHG emissions and EF of air travel by residents of Metro Vancouver, Canada, over a one-year period (2006).

1.1 Air Travel, Climate Change and Ecological Limits

Canadians travel now more than ever. The total number of passengers through Vancouver's International Airport (also recognized by its call letters YVR) increased by 63% from 9,935,285 in 1992 to 16,177,438 in 2009 (Vancouver International Airport Website, 2010). Airport marketing and environmental department staff estimate that approximately 29% of this volume can be attributed to travel by Metro Vancouver residents (Frisby, personal communication, December 1, 2010). As a sector, Transportation contributed 37% of the growth in Canada's GHG emissions from 1990 to 2008 (National Inventory Report, 2010). While the impact of aviation's emissions is relatively minor within the entire transportation sector (4% of emissions in 2006), air travel's impact with regard to climate change is significant for two reasons. First, aircraft emissions affect the climate system differently than other forms of transportation because they are usually generated at high altitudes. Here, according to the Intergovernmental Panel on Climate Change (IPCC) they catalyze chemical reactions that enhance radiative forcing (IPCC, 2001a, 2001b). According to the David Suzuki Foundation (DSF), the net impact of aircraft emissions on climate change is 2-4 times that of comparable carbon dioxide emissions at lower altitudes (DSF, 2010). As a result, the per passenger kilometer impact of air travel is greater than that of other forms of transportation, including busses and trains (DSF, 2010). Second, airfares do not reflect either market or ecological costs. For example, over the five-year period from 2002/03 to 2006/07, the Canadian Government provided direct subsidies, grants and contributions to air travel totaling over $240M (Transport Canada, 2006) which compensated for rising fuel costs. This inflates demand for air travel and, with it, attendant GHG emissions and their contribution to the unaccounted costs of climate change.

1.2 Peak Oil and Globalization

Petroleum is needed to power air travel but there is a widening gap between supply and potential demand for oil globally (Hall & Day, 2009). Since 2010, approximately 1,500 commercial flights have utilized biofuels (Hupe, 2013); nevertheless, it is unknown whether alternatives to fossil fuel could sustain the current rate of air travel. The National Aeronautics and Space Administration (NASA) indicates that alternatives to fossil fuel "have good potential but presently appear to be better suited for use in ground transportation" (Daggett, 2006, p. 1).

There may soon be significant supply/price problems. Even the International Energy Agency (IEA) has publicly recognized that the global rate of conventional oil extraction peaked in 2006 and is now in decline (IEA, 2010). Recent enthusiasm over potentially abundant supplies of 'tight' oil (e.g., shale-oil in the US) overlooks the low energy-returns-on-energy-invested (EROEI) and rapid depletion rates of shale oil discoveries. In sum, declining fossil fuel reserves may well soon constrain future air travel.

Limited resource availability and waste absorption capacity highlight the importance of knowing the extent to which air travel contributes to urban populations' GHG emissions and ecological footprints. These data will enable more informed decisions about air travel relative to other modes of transportation and competing demands for petroleum.

1.3 Metro Vancouver Background

Located in the southwest corner of British Columbia, Canada near the United States (US) border, Metro Vancouver is home to approximately 2.3 million people and spans an area of 283,183 hectares (Metro Vancouver, 2006). It comprises 22 municipalities, including the City of Vancouver, and includes both urban and peri-urban environments (Metro Vancouver, 2011). The region contains some of Canada's most fertile agricultural land; the delta of one of Canada's largest rivers, the Fraser River; forested mountains; and coastal shores with several commercial fisheries. The region's main economic activities include: business services, tourism, manufacturing, and agriculture (BC Stats, 2010). Professional, technical and other services, such as public administration, retail trade and construction, comprise the majority of business employment (BC Stats, 2010).

Although recognized in the literature for its commitment to advancing sustainability through regional plans and initiatives (Pivo, 1996; Holden, 2006; Newman & Jennings, 2008; Wheeler & Beatley, 2009), Metro Vancouver is fairly typical of a high-consuming, first-world, urban region (Rees, 2009; Berelowitz, 2005). In 2006, the average income was $40,252 Cdn, with a labour-force participation rate of 67%. Household median income was $69,688, and the majority of households, 65%, owned their own home despite exceedingly high housing costs averaging $520,937 Cdn. Fifty-seven percent of the region's population is between 25 and 64 years of age and 62% of this group have some type of post-secondary education credential (BC Stats, 2010).

Regional transportation services are managed by the Greater Vancouver Transportation Authority (GVTA), also known as TransLink. This is a sister agency to Metro Vancouver that is operated jointly by Metro and the Province of British Columbia. Of particular significance to this study is the Vancouver International Airport, also known by its airport code, YVR (Note 1). The airport is situated within the Fraser River Delta, and also within the Pacific Flyway for migrating birds. It serves as Canada's air travel gateway to the Pacific and represents an important travel link for both Canadian and international business and tourist travelers.

2. Method

2.1 Ecological Footprint

Many studies have estimated the CO_2 emissions of air travel, and have described various methods for apportioning greenhouse gas emissions to air travel. Wood, Bows and Anderson (2010), for example, categorize the range of aviation greenhouse gas-apportioning methods by two filters—the emissions-allocation approach (involving local producers or consumers) and the emissions accounting method. In the former method, the "producer" variation involves estimates based on plane landings and take-offs; the "consumer" variation generates estimates that are based on passenger and freight data. The more top-down emissions-allocations approach employs national data accounts (e.g., such as those used by the Global Footprint Network to develop the National Footprint Accounts) combined with input-output multipliers. Wood et al. (2010) present a general framework but do not actually describe how to calculate GHG emissions. Our study describes two methods in detail, showing the step-by-step calculations, in order to make the methods accessible to planning practitioners and others at the community level interested in incorporating air travel in EFA.

While some report that the primary strength of EFA is as a communication tool (Weidman & Barrett 2010), it also has potential as a policy tool even at the local (municipal) level, particularly given growing interest in "one-planet living". The latter concept uses EFA as a primary metric (e.g., City of Cardiff, 2012; City of Vancouver, 2011; Desai, 2008). That said, three criticisms of EFA are relevant to this study. First, some analysts question the method's focus on GHGs, specifically carbon dioxide emissions to express waste (Fiala, 2008; Nijkamp, Rossi, & Vindign, 2004; Ayers, 2000; Van den Bergh & Verbruggen, 1999). Second, some researchers are concerned that EFA lacks transparency and that the required data are unavailable (Wilson & Grant, 2009; Aall & Norland, 2005). Third, critics argue that local governments lack the capacity to do EFA and that this undermines its relevance to planning practitioners (Curry, Maguire, Simmons, & Lewis, 2011; Wilson & Grant, 2009; Aall & Norland, 2005).

Let us take these points in order. We justify the focus on carbon dioxide emissions in EFA on several grounds: first, carbon dioxide emissions are the single largest waste product by weight in developed countries and increasingly in emerging economies; second, carbon dioxide is part of the natural carbon cycle and, because it is assimilated by growing plants, it can readily be represented by a corresponding exclusive (Note 2) area of productive ecosystems (e.g., carbon sink forest). Most other industrial wastes are not translatable into the eco-footprint metric. Third, carbon dioxide is the primary anthropogenic greenhouse gas and must be reduced to avoid excessive global warming. To date, the only practical means of re-absorbing atmospheric CO_2 is through natural or intentionally-created, dedicated carbon sink ecosystems and EFA provides an accurate estimate of the required ecosystem area. Finally, inclusion of carbon emissions in EFA is useful in policy discussions involving the public because an area-based carbon footprint is a graphically explicit link "between human consumption and [demand for] ecosystem services" (Weidman & Barrett, 2010, p. 1649). (N.B. Other tools are available to quantify and assess the impacts of other forms of waste, many of which are already being used by local governments.)

The second and third criticisms are not so much faults with EFA *per se* as they are comments on society's prevailing understanding of humankind-ecosystem interactions (which are indeed complicated and somewhat opaque) and with our corresponding perceptions of data requirements for sustainability planning. To the extent that EFA can contribute to developing that understanding and to balancing human demand with the supplies of ecosystems services, municipal data collection should be adapted to facilitate the analysis. In fact, using locally derived data contributes to methodological transparency and confidence because much of the data required to generate an EFA can be, or is already being, used for other policy and planning purposes in the community (e.g., waste management). Where local databases are not yet sufficiently sophisticated for EFA it is possible to adapt national production and consumption data to reflect local energy and material consumption—local data bases and EFA are still co-evolving (Wackernagel, 2009).

The objection that local governments often lack analytic capacity to do EFA may be true but, again, speaks more to the need for new forms of training for municipal planners, engineers and technicians than to a weakness in EFA. As society comes to recognize the reduction of energy and material throughput as key to sustainability, skills in such related methods as material flows analysis, life cycle analysis and eco-footprint analysis will become essential to municipal planning and management.

In the present study, we rely on publicly available local data and describe all steps in detail.

2.2 Air Travel by Metro Vancouver Residents

To estimate the GHG emissions and ecological footprint of air travel by Metro Vancouver residents, we collected data from several sources and organized them into three flight categories: international, domestic (national), and commuter (intra-provincial). All data are for the 2006 calendar year, January to December as this was the year with the most widely available data and coincides with Canada's national census. Seven steps in data collection and analysis are outlined below with notes on differences in data quality among flight categories.

2.2.1 Passenger Movements

Data from our two main sources, the Vancouver International Airport (YVR) and Statistics Canada's air travel survey, generated different estimates of passenger movements (Note 3). We therefore present a range of GHG emissions and corresponding carbon eco-footprint estimates.

2.2.2 Estimates Based Mainly on YVR Data (Approach "A")

This study required data on the number of passengers on outbound flights originating in YVR sorted by destination. For international flights to the United States (US), we retrieved data from Statistics Canada (2006a) on the number of passengers flying from Canada to the US by origin and destination. We then grouped data on flights originating in YVR into eleven different destination regions as defined in the International Travel Survey (2008), "Canadian Resident Trips Abroad". We recorded the most frequented airport in each of the eleven regions. For all other international flights, YVR (2010) provided data on the number of international outbound flights by destination. We classified these data by destination into five regions (Africa, Asia-Pacific, Europe, Latin America-Caribbean and the Middle East) and recorded the most frequented airport for each region. For domestic and commuter flights, YVR provided data on the number of outbound flights by destination (YVR, 2010).

2.2.3 Distance Traveled

To estimate passenger miles traveled we determined the Great Circle Distance (Note 4) between YVR and the most frequented destination airports in each region (identified from the International Travel Survey 2008), "Canadian Resident Trips Abroad", using the World Airport Codes (2010) web tool. This step is the same for all categories of travel: international, domestic and commuter.

2.2.4 Aircraft Information

We compiled data on the aircraft type, passenger capacity, average load factor and mileage for flights to each destination. YVR (2010) provided information on typical aircraft employed. We used an average load factor of 80% seat occupancy as cited in the National Greenhouse Gas Inventory Report (Environment Canada, 2010) (Note 5). Sources for passenger capacity and mileage differ by category. For international and domestic flights, we used Aircraft Charter World (Air Broker Center International, 2009) to determine passenger capacity for each aircraft and the Research and Innovative Technology Administration (RITA) Bureau of Transportation Statistics (2010) to determine the mileage (passenger miles per gallon) for each aircraft. Where we had data on the number of flights only we estimated passenger numbers on outbound flights by using the 80% average load factor (Environment Canada, 2010). While some data sources used imperial measurements, we converted our final calculations into metric.

For commuter flights, we used OAG Aviation (2010) and Airliners.net (2010) to determine passenger capacity for each aircraft, then converted number of flights to number of people on outbound flights by multiplying flights by passenger capacity—with an 80% average load factor (Environment Canada, 2010). We used the RITA Bureau of Transportation Statistics (2010) estimates of mileage (passenger miles per gallon) for most of the commuter aircraft. For mileage data not available on RITA, we used data from the Transportation Safety Board of Canada (2010); Airliners.net (2010); UK Department of Transport (2010); and All Experts Encyclopedia (2010) on gallons per hour fuel consumption rates. To convert gallons per hour to gallons per passenger, we consulted Kayak Flight Finder (2010) to find hours per one-way flight, multiplied gallons per hour by number of hours per one-way flight to find gallons of fuel consumed, then divided gallons of fuel consumed per flight by estimated passenger load to find gallons of fuel consumed per passenger.

2.2.5 Fuel Consumption

We calculated the total fuel consumed from outbound air travel originating in YVR in gallons, and then converted to Liters (L), using the standard conversion of 3.78541178 L per gallon (Quinn, 1996) (Note 6).

For international and domestic flights, we calculated gallons of fuel consumed per passenger by dividing distance (miles) by mileage (passenger miles per gallon). Second, we calculated total fuel consumption by multiplying the number of people on outbound flights from YVR by gallons of fuel consumed per passenger. We followed the calculations above for most of the commuter flights, but where we calculated fuel consumed per passenger (instead of passenger miles per gallon), we multiplied gallons of fuel consumed per passenger by number of people on outbound flights to find total fuel consumed.

2.2.6 Scale for Residents

Marketing research undertaken by YVR reveals that approximately 29% of passengers boarding flights are Metro Vancouver residents (YVR, 2010). We therefore estimated total fuel consumption by Metro residents by doubling the aggregate fuel consumption on all outbound flights to account for return flights, and multiplying by 0.29 to scale for flights by residents. This step is the same for all categories. (For Approach B, see section 2.8 below.)

2.2.7 Greenhouse Gas Emissions

We then calculated the GHG emissions (CO_2, CH_4 and N_2O), and the CO_2 equivalent (CO_2e) for non-CO_2 emissions, using emission factors from Environment Canada's National Greenhouse Gas Inventory Report (Environment Canada, 2010) and CO_2e conversion factors obtained from Natural Resources Canada (Leblanc personal communication November 9, 2010). All units were converted from grams (g) to tonnes (t) by dividing by 10^6. For international and domestic flights, we used emission factors, in grams per L for aviation turbo fuel, to complete the following equations: L fuel burned $\times$ 2,534 = g CO_2 emitted; L fuel burned $\times$ 0.080 = g CH_4 emitted; L fuel burned $\times$ 0.23 = g N_2O emitted.

We then used conversion factors for CH_4 and N_2O to calculate the CO_2e for CH_4 and N_2O through the following equations: CH_4 emitted $\times$ 21 = CO_2e; N_2O emitted $\times$ 310 = CO_2e (Note 7).

For commuter flights, we followed the steps above for all aircraft using aviation turbo fuel. For aircraft using aviation gasoline, we used the following emission factors (conversion factors same as above): L fuel burned $\times$ 2342 = g CO_2 emitted; L fuel burned $\times$ 2.2 = g CH_4 emitted; L fuel burned $\times$ 0.23 = g N_2O emitted.

2.2.8 Ecological Footprint

Last, we calculated the ecological footprint of all air travel by Metro Vancouver residents. Since only carbon dioxide is sequestered by ecosystems, our estimate of the ecological footprint of air travel is based on only the carbon dioxide emissions. We follow the protocol established by the Global Footprint Network (Ewing et al., 2008) for calculating the "Energy Land" meaning the dedicated carbon sink lands required to sequester carbon dioxide emissions, based on annual average global terrestrial sequestration capacity (Ewing et al., 2009). A simplified equation is:

$$EFc = (tCO_{2\text{Air Travel}}(1-S_{ocean}))/Yc \ \times \ EQF \qquad (1)$$

Where EFc stands for carbon sequestration land, tCO_2 is the weight of carbon dioxide emissions resulting from fuel combustion associated with Metro Vancouver residents' air travel, S_{ocean} represents the proportion of carbon dioxide sequestered by the world's oceans, and Yc is the annual rate of carbon sequestered per hectare of forest land (assuming world average yields). EQF is the equivalence factor prescribed by the Global Footprint Network (Ewing et al., 2009) used to convert global average yield for forest land to global average yield for all land types.

Specifically, since approximately 26% of carbon emissions are absorbed by the oceans, (IPCC, 2001c; Wackernagel & Monfreda, 2004; Scotti, Bondavalli, & Bodini, 2009; Kitzes & Wermer, 2006); the S_{Ocean} is equal to 0.26 and we subtract this proportion from the total emissions associated with Metro air travelers. We then convert the remaining emissions into the equivalent area of carbon sink forest, assuming a global average assimilation rate of 3.7 tones carbon dioxide (1.0 tone carbon) per hectare of forest (Scotti et al., 2009; Kitzes & Wermer, 2006). Finally, this forest area was multiplied by an equivalence factor of 1.24 (EQF) to represent the carbon sink in gha (reflecting the fact that forests are more productive than world average productive land (Ewing et al., 2009; Kitzes & Wermer, 2006).

2.2.9 Statistics Canada Air Travel Survey (Approach "B")

Based on YVR data, in approach A, we assumed that 29% of the people flying from YVR are Metro Vancouver residents and applied this ratio to flights to all destinations (international, domestic and commuter). This may be problematic. We therefore offer an alternative to the 29% blanket scaling method using data from a Statistics Canada travel survey on the travel patterns of BC residents classified by destination (Statistics Canada, 2006b). Aside from scaling, all other steps remain the same. We apply this alternative method to international and domestic flights, but not to commuting flights for lack of data. (Here we fall back on the original 29% scaling method.)

One shortcoming of this approach is that the Statistics Canada survey focused on the whole province, not just Metro Vancouver. We therefore estimated GHG emissions per BC resident and multiplied this by the Metro Vancouver population to obtain the required sub-sample. If anything, this probably underestimates actual emissions. Over 60% of BC residents live in Metro Vancouver, and socioeconomic data suggest that their share of overall BC travel is probably higher than the provincial average.

2.3 Limitations and Assumptions

As explained above, the study was constrained by cases of poor data which forced the use of informed estimates and by the need for certain methodological assumptions.

Two important data gaps were due largely to systemic and time/resource limitations respectively. First, YVR could provide data only on flights to first destinations. The lack of information on connecting flights could be a significant omission. Second, we were unable to account for Metro Vancouver residents who may drive to airports in close proximity to YVR such as Bellingham, Washington or Abbotsford, BC. Not including connecting flights or flights from nearby airports results in conservative estimates of Metro residents total air travel emissions.

One key assumption, that Metro Vancouver residents account for 29% of total outbound flights to all destinations, introduced the possibility of significant error and inspired an alternative estimate based on a Statistics Canada travel survey that included final destinations. Thus, our initial estimate using only YVR data treats a trip to Europe through Toronto as a flight to Toronto while the Statistics Canada data enabled us to document the entire journey to Europe. Developing this alternative approach enabled us to provide a plausible range of GHG emissions by Metro Residents.

Other emissions that also contribute to climate change, such as water vapor, are not quantified. Further, our estimates do not control for a variety of other factors that affect fuel consumption: type and age of engine(s) used in the airplane; weight; weight distribution; flight path factors such routing and distance flown; take-off and landing requirements of airports; and environmental factors such as wind speed and direction, temperature, altitude and air pressure. Keep in mind too that load factors are continually shifting as newer planes with improved fuel economy are introduced.

3. Results

We found that the GHG emissions from air travel by Metro Vancouver residents in 2006 were between 1,191,070 and 1,402,400 tCO_2e including 1,157,380 to 1,362,810 tonnes of carbon dioxide. By comparison, total GHG emissions from on-road transportation within the region were 5,386,785 tCO_2e, of which 4,505,287 tCO_2e was from light duty vehicles (Metro Vancouver, 2007) (Note 8). Therefore, GHG emissions from air travel represent an additional 26% to 31% of GHG emissions that could be attributed to personal travel by Metro Vancouver residents, or 22% to 26% considering the on-road transportation sector as a whole. While Metro Vancouver does not include air travel emissions beyond those from takeoff and landing in its GHG emissions inventory (Metro Vancouver, 2007), it does identify collaboration with senior government agencies (e.g. Environment Canada and Transport Canada) as important to effective emissions management. Such collaboration on managing GHG emissions from air travel may become necessary to fulfill Metro Vancouver's goal of becoming a sustainable region (Metro Vancouver, 2005).

We found that the ecological footprint of air travel by Metro Vancouver residents for 2006 is between 287,030 and 337 980 gha (Table 1). This is the area of productive terrestrial ecosystems (carbon sink forest) that would be required on a continuous basis to assimilate the typical annual CO_2 emissions of Metro Vancouver residents' air travel and is between 99% and 116% the size of the region's geographic land area (290 462 ha). Nevertheless, this underestimates the total ecological impact because it does not account for other GHG emissions that contribute to climate change such as CH_4, N_2O and H_2O (See Table 1 for all GHGs—except water vapor—in CO_2 equivalents for both analytic approaches).

Table 1. Metro Vancouver's 2006 air travel greenhouse gas emissions and ecological footprint

Flight Type	CO_2e Approach "A"	CO_2e Approach "B"	CO_2 Approach "A"	CO_2 Approach "B"	EFA Approach "A"	EFA Approach "B"
	tonnes	tonnes	tonnes	tonnes	global hectares	global hectares
International	743,970	1,097,790	723,150	1,067,060	179,340	264,630
Domestic	393,650	251,180	382,630	244,150	94,890	60,550
Commuter	53,450	53,448	51,600	51,600	12,800	12,800
Total	1,191,070	1,402,420	1,157,380	1,362,810	287,030	337,980

Table 1 reveals that for Approach A, international air travel accounts for by far the biggest portion of emissions among the three categories (international, domestic, commuter), comprising 63% of the total air travel footprint. Domestic travel takes second place, at 33% of the total. Commuter travel is a fairly small portion of the total, at just 4%. For Approach B, international travel accounts for an even larger share at 78% of the total travel footprint. Domestic travel trails in second place at 18%, and commuter emissions are negligible at less than 1%.

The table also shows that analysis using the Statistics Canada survey data (Approach 'B') produces an EF 15% larger than that derived from YVR's more limited data.

Figure 1 (a and b) breaks down Metro Vancouver air travel GHG emissions by specific destinations for both calculation procedures and shows there are considerable differences between them. For example, according to YVR data (Approach A) approximately 43% of GHG emissions were associated with flights to Asia. The Statistics Canada survey data produced a corresponding number of only 21%. Similarly Approach A suggests that flights to Europe account of 15% of GHG emissions while Approach B associates 23% of GHG emissions to European destinations.

Figure 1a. Metro Vancouver air travel GHG emissions by destination, Approach "A"

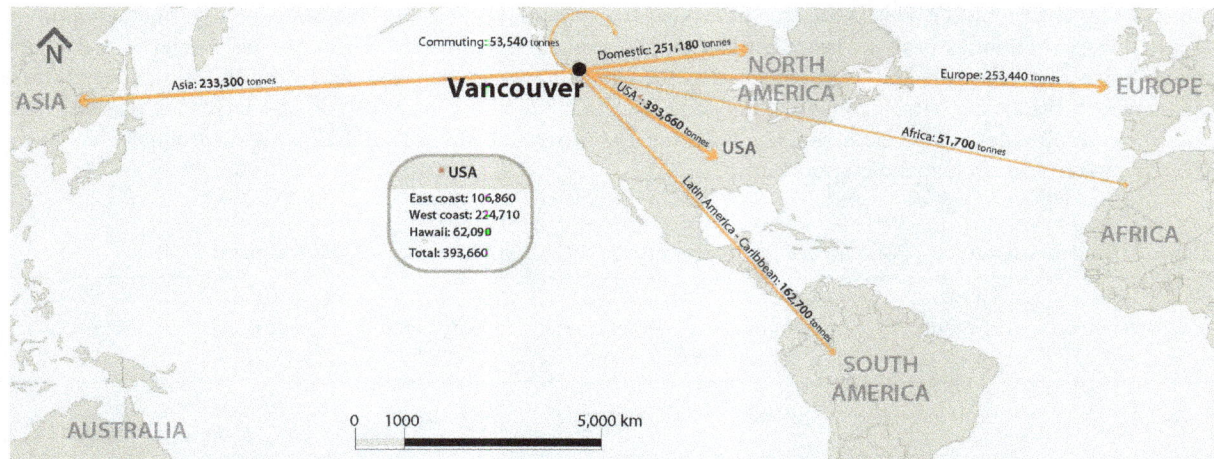

Figure 1b. Metro Vancouver air travel GHG emissions by destination, Approach "B"

4. Discussion

Through the use of two different methods described above, we have demonstrated how to estimate the GHG emissions and ecological footprint of air travel by residents of Metro Vancouver for a typical one-year period (2006). The main finding is that in 2006, the residents of Metro Vancouver relied on a productive land area larger than the region itself to meet their carbon sink demands from air travel.

A major strength of eco-footprinting is that it enables graphic comparisons of human demand for biocapacity with available supply. Our results show that, even if it were fully covered in healthy growing forest, the entire Metro Vancouver region would provide insufficient biocapacity to assimilate the carbon emissions from residents' air travel alone (and we would still have to account for all other carbon emissions). Metro Vancouver is clearly running a massive ecological deficit (Note 9) of which demand for carbon assimilation capacity is a major component. Such results underscore the extent to which wealthy high-income regions off-load their biophysical demands onto other regions and the global commons, the unspoken assumption being that there is surplus capacity 'elsewhere' to cover these regional deficits. The steady accumulation of carbon dioxide in the atmosphere shows this not to be the case. It follows that reducing the demand for air travel by Metro Vancouver residents or otherwise finding ways to mitigate air travel's ecological impacts could become increasingly important goals for a would-be sustainable region, given finite ecological resources and waste assimilative capacities and a lack of suitable technology to substitute for fossil fuels.

Although GHG emissions from air travel fall outside the jurisdiction of authority for air quality management granted to Metro Vancouver by the Province of British Columbia, it is useful to understand how this activity contributes to overall GHG emissions. For example Air travel by Metro Vancouver residents accounts for an additional 22% to 26% of GHG emissions, above what is already counted for on-road transportation in the regional emission inventory. Comparing the carbon eco-footprint of air travel by Metro Vancouver residents with that of other sectors could help to identify the most fruitful areas for emissions-reductions policy as well as areas for further research. In any case, the data should be available to Metro residents to improve their understanding of the need to reduce their personal ecological footprints and to underscore the importance of possible lifestyle changes.

Beyond illustrating the air travel implications for our case study region of Metro Vancouver, we have met our primary objective of developing methods for calculating the ecological footprint of air travel that are accessible to researchers, including urban planners, in other cities and regions. The confluence of climate change, ecological limits, peak oil and globalization affects cities and regions in complex ways. This paper underscores that air travel is an important impact factor that must be accounted for in consideration of regional sustainability. Limits to nature's capacity to absorb carbon dioxide emissions may well combine with declining fossil fuel reserves and may constrain opportunities for non-essential air travel in the future.

Acknowledgements

We appreciate the feedback provided by anonymous reviewers of this paper. We would like to thank the efforts of Cornelia Sussmann, PhD Candidate, School of Community and Regional Planning, University of British

Columbia for preliminary research into how to tackle this research project. We also wish to acknowledge the financial contributions of the Canadian Social Sciences and Humanities Research Council, the Government of British Columbia Pacific Century Graduate Scholarship, and the Pacific Institute for Climate Solutions for financial contributions in support of this research.

References

Aall, C., & Norland, I. (2005). The use of the ecological footprint in local politics and administration: Results and implications from Norway. *Local Environment, 10*, 159-172. http://dx.doi.org/10.1080/1354983052000330752

Air Broker Center International. (2009). *Aircraft Charter World*. Retrieved 2 November, 2010 from http://www.aircraft-charter-world.com

Airliners.net. (2010). *Aircraft Stats*. Retrieved 3 November, 2010 from http://www.airliners.net/aircraft-data/stats/

All Experts Encyclopedia. (2010). *SH6 Mileage*. Retrieved 14 November, 2010 from http://www.associatepublisher.com/

Barrett, J., Vallack, H., Jones, A., & Haq, G. (2002). *A material flow analysis and ecological footprint of York: Technical report*. Stockholm: Stockholm Environment Institute.

BC Stats. (2010). *Greater Vancouver Regional District: Community Facts* (BC Stats: 250-387-0327). Victoria BC: BC Government Ministry of Citizens Services. Retrieved 24 August, 2010 from http://www.bcstats.gov.bc.ca/data/dd/facsheet/cf170.pdf

Benjamin, R. (2013). Assessing our environmental progress. *The International Civil Aviation Organization Journal, 68*(2). Retrieved 10 September, 2013 from http://www.icao.int/publications/Pages/ICAO-Journal.aspx?year=2013&lang=en

Berelowitz, L. (2005). *Dream city: Vancouver and the global imagination*. Vancouver: Douglas and McIntyre.

Chambers, N., Simmons, C., & Wackernagel, M. (2001). *Sharing nature's interest: Ecological footprints as an indicator of sustainability*. London: Earthscan Publications.

City of Cardiff. (2012). *Cardiff one planet city. Draft*. Cardiff, UK: City of Cardiff. Retrieved 16 November, 2012 from http://www.cardiff.gov.uk/template/oneplanet/idea1/Oneplanetliving_eng.pdf

City of Vancouver. (2011). *Greenest city 2020 action plan*. Vancouver: City of Vancouver. Retrieved 18 April, 2012 from http://vancouver.ca/ctyclerk/cclerk/20110712/documents/rr1.pdf

Collins, A., Flynn, A., Wiedmann, T., & Barrett, J. (2008). The environmental impacts of consumption at a subnational level–the ecological footprint of Cardiff. *Journal of Industrial Ecology, 10*, 9-23. http://dx.doi.org/10.1162/jiec.2006.10.3.9

Curry, R,. Maguire, C., Simmons, C., & Lewis, K. (2011). The use of material flow analysis and the ecological footprint in regional policy-making: Application and insights from Northern Ireland. *Local Environment, 16*, 165-179. http://dx.doi.org/10.1080/13549839.2010.549118

Daggett, D., Hadaller, O., Hendricks, R., & Walther, R. (2006). *Alternative fuels and their potential impact on aviation*. Prepared for the National Aeronautics and Space Administration (NASA) 25th Congress of the International Council of the Aeronautical Sciences (ICAS).

David Suzuki Foundation (DSF). (2010). *Air travel and climate change*. Vancouver, BC: David Suzuki Foundation. Retrieved 4 December, 2010 from http://www.davidsuzuki.org/issues/climate-change/science/climate-change-basics/air-travel-and-climate-change/

Desai, P. (2008). Creating low carbon communities: One planet living solutions. *Globalizations, 5*, 67-71. http://dx.doi.org/10.1080/14747730701587462

Environment Canada. (2010). *National inventory report: Greenhouse gas sources and sinks in Canada*. Retrieved 2 November, 2010 from http://www.ec.gc.ca/Publications/default.asp?lang=En&xml=492D914C-2EAB-47AB-A045-C62B2CDAC C29

Ewing, B., Goldfinger, S., Oursler, A., Reed, A., Moore, D., & Wackernagel, M. (2009). *The ecological footprint atlas*. Oakland, CA: Global Footprint Network.

Ewing, B., Reed, A., Rizk, S. M., Galli, A., Wackernagel, M., & Kitzes J. (2008). *Calculation methodology for the national footprint accounts*. Oakland, CA: Global Footprint Network.

Fiala, N. (2008). Measuring sustainability: Why the ecological footprint is bad economics and bad environmental science. *Ecological Economics, 67*, 519-525. http://dx.doi.org/10.1016/j.ecolecon.2008.07.023

Global Footprint Network. (2012). *Footprint for Cities*. Retrieved April 5, 2012 from http://www.footprintnetwork.org/en/index.php/GFN/page/footprint_for_cities/

Graymore, M., Sipe, N., & Rickson, R. (2010). Sustaining human carrying capacity: a tool for regional sustainability assessment. *Ecological Economics, 69*, 459-468. http://dx.doi.org/10.1016/j.ecolecon.2009.08.016

Hall, C. A. S., & Day, J. W. (2009). Revisiting the limits to growth after peak oil. *American Scientist, 97*, 230-237. http://dx.doi.org/10.1511/2009.78.230

Holden, E. (2004). Ecological footprints and sustainable urban form. *Journal of Housing and the Built Environment, 19*, 91-109. http://dx.doi.org/10.1023/B:JOHO.0000017708.98013.cb

Holden, M. (2006). Urban indicators and the integrative ideals of cities. *Cities, 23*, 170-83. http://dx.doi.org/10.1016/j.cities.2006.03.001

Hupe, J. (2013). Driving progress through action on aviation and environment. *The International Civil Aviation Organization Journal, 68*(2). Retrieved 10 September, 2013 from http://www.icao.int/publications/Pages/ICAO-Journal.aspx?year=2013&lang=en

Intergovernmental Panel on Climate Change (IPCC). (2001a). *Aviation and the global atmosphere*. Retrieved 25 February, 2011 from http://www.grida.no/publications/other/ipcc_sr/?src=/climate/ipcc/aviation/index.htm

Intergovernmental Panel on Climate Change (IPCC). (2001b). *IPCC third assessment report: Climate change 2001: Working group I: The scientific basis. Chapter 3: The carbon cycle and the climate system*. Prentice, IC, lead author. Retrieved 11 July, 2011 from http://www.grida.no/publications/other/ipcc_tar/

Intergovernmental Panel on Climate Change (IPCC). (2001c). *Climate change: The scientific basis. Cambridge, UK: IPCC.International Travel Survey* (2008). Canadian Resident Trips Abroad. Retrieved 2 November, 2010 from http://www.nesstar.com/

International Civil Aviation Organization (ICAO). (2011). *Annual Report of the Council*. Retrieved 16 September, 2013 from http://www.icao.int/publications/pages/publication.aspx?docnum=9975

International Energy Agency (IEA). (2010). *World energy outlook*. Retrieved 3 December, 2010 from http://www.worldenergyoutlook.org/

James, N., & Desai, P. (2003). *One planet living in the thames gateway: A WWF-UK one million sustainable homes campaign report*. Surrey, UK: World Wide Fund for Nature.

Kayak Flight Finder. (2010). *Flight Search*. Retrieved 3 December, 2010 from http://www.kayak.co.uk/flights

Kitzes, J., & Wermer, P. (2006). *Technical memorandum: The carbon conversion factor in ecological footprint accounts: calculations and sources of variability*. Oakland, CA: Global Footprint Network.

Kuzyk, L. (2011). Ecological and carbon footprint by consumption and income in GIS: Down to a census village scale. *Local Environment, 16*, 871-886. http://dx.doi.org/10.1080/13549839.2011.615303

Lewan, L., & Simmons, C. (2001). *The use of ecological footprint and biocapacity analyses as sustainability indicators for sub-national geographic areas: a recommended way forward*. EUROCITIES, Ambiente. Italy: European Common Indicators Project.

McManus, P., & Haughton, G. (2006). Planning with ecological footprints: A sympathetic critique of theory and practice. *Environment and Urbanization, 18*, 113-127. http://dx.doi.org/10.1177/0956247806063963

Metro Vancouver. (2005). *Air quality management plan: Clean air, breath easy*. Burnaby, BC: Greater Vancouver Regional District.

Metro Vancouver. (2006). *Metro Vancouver's Generalized Land Use by Municipality*. Retrieved June 3, 2011 from http://www.metrovancouver.org/about/statistics/Pages/KeyFacts.aspx

Metro Vancouver. (2007). *Lower Fraser Valley air emissions inventory and forecast and backcast*. Burnaby, BC: Greater Vancouver Regional District.

Muniz, I., & Galindo, A. (2005). Urban form and the ecological footprint of commuting: The case of Barcelona.

Ecological Economics, 55, 499-514. http://dx.doi.org/10.1016/j.ecolecon.2004.12.008

Newman, P., & Jennings, I. (2008). *Cities as sustainable ecosystems: Principles and practice.* Washington, DC: Island Press.

Nijkamp, P., Rossi, E., & Vindign, G. (2004). Ecological footprints in plural: A meta-analytic comparison of empirical results. *Regional Studies, 38*, 747-765. http://dx.doi.org/10.1080/0034340042000265241

OAG Aviation. (2010). *Aircraft Statistics.* Retrieved 29 November, 2010 from http://www.oag.com/northamerica/airlineandairport/aircraftstatistics.asp/

Pivo, G. (1996). Toward sustainable urbanization on mainstreet Cascadia. *Cities, 13*, 339-54. http://dx.doi.org/10.1016/0264-2751(96)00021-2

Quinn, T. (1996). *Engineering conversion factors. University of Minnesota.* Retrieved 15 February, 2011 from http://www.cmrr.umn.edu/~strupp/units.html

Rees, W. E. (2006). Ecological footprints and biocapacity: essential elements in sustainability assessment. In J. Dewulf, & H. Van Langenhove (Eds.), *Renewables-based technology: Sustainability assessment.* Chichester, UK: John Wiley and Sons. http://dx.doi.org/10.1002/0470022442.ch9

Rees, W. E. (2008). Confounding integrity: humanity as dissipative structure. In L. Westra, K. Bosselmann, & R. Westra (Eds.), *Reconciling human existence with ecological integrity.* London: Earthscan.

Rees, W. E. (2009). The ecological crisis and self-delusion: implications for the building sector. *Building Research and Information, 37*, 300-311. http://dx.doi.org/10.1080/09613210902781470

Rees, W. E. (2013). Ecological footprint. In S. Levin (Ed.), *Encyclopedia of biodiversity* (2nd ed.). London: Academic Press.

Research and Innovative Technology Administration (RITA). (2010). *Transportation Statistics.* Retrieved 23 November, 2010 from www.transtats.bts.gov

Scotti, M., Bondavalli, C., & Bodini, A. (2009). Ecological footprint as a tool for local sustainability: The municipality of Piacenza (Italy) as a case study. *Environmental Impact Assessment Review, 29*, 39-50. http://dx.doi.org/10.1016/j.eiar.2008.07.001

Statistics Canada. (2006a). *Air Passenger Origin and Destination.* Canada-United States Report. Catalogue no. 51-205X. Table 3-21. Retrieved 2 November, 2010 from http://www.statcan.gc.ca/bsolc/olc-cel/olc-cel?catno=51-205-X&lang=eng

Statistics Canada. (2006b). *International Travel.* Catalogue no. 66-201-X.

Transport Canada. (2006). *Government Spending on Transportation: 2006 Annual Report.* Retrieved 3 December, 2010 from http://www.tc.gc.ca/eng/policy/anre-menu.htm

Transportation Safety Board of Canada. (2010). *DH3 Mileage.* Retrieved 4 December, 2010 from http://www.tsb.gc.ca/

UK Department of Transport. (2010). *Mileage for PAG Piper.* Retrieved 27 November, 2010 from http://www.aaib.gov.uk/

Union of Concerned Scientists. (1992). *World Scientists' Warning to Humanity.* Retrieved 28 November, 2010 from http://www.ucsusa.org/about/1992-world-scientists.html

Van den Bergh, J., & Verbruggen, H. (1999). Spatial sustainability, trade and indicators: An evaluation of the ecological footprint. *Ecological Economics, 29*, 61-72. http://dx.doi.org/10.1016/S0921-8009(99)00032-4

Vancouver International Airport Official (YVR) Website. (2010). *YVR Passengers (Enplaned + Deplaned) 1992-2009.* Retrieved 17 November, 2010 from: http://www.yvr.ca/en/about/facts-stats.aspx

Wackernagel, M., & Monfreda, C. (2004). Ecological footprint and energy. *Encyclopedia of Energy.* Vol 2. http://dx.doi.org/10.1016/B0-12-176480-X/00120-0

Wackernagel, M., & Rees, W. E. (1996). *Our ecological footprint: Reducing human impact on the earth.* Gabriola Island, BC: New Society Publishers.

Wackernagel, M., & Rees, W. E. (1996). *Our ecological footprint: Reducing human impact on the earth.* Gabriola Island, BC: New Society Publishers.

Wackernagel, M. (2009). Methodological advances in footprint analysis. *Ecological Economics, 68*, 1925-1927. http://dx.doi.org/10.1016/j.ecolecon.2009.03.012

Weidmann, T., Minx, J., Barrett, J., Vanner, R., & Ekins, P. (2006). Sustainable consumption and production–development of an evidence base. Project Reference SCP001-Resource Flows. Department of Environment, Food and Rural Affairs (DEFRA). Retrieved from http://www.resource-accounting.org.uk/uploads/Reports/scp001.pdf

Wheeler, S., & Beatley, T. (Eds.). (2009). *The sustainable urban development reader* (2nd ed.). London: Routledge, Taylor and Francis Group.

Wilson, J., & Grant, J. (2009). Calculating ecological footprints at the municipal level: What is a reasonable approach for Canada? *Local Environment, 14*, 963-979. http://dx.doi.org/10.1080/13549830903244433

Wilson, M., & Anielski, M. (2005). *Ecological footprints of Canadian municipalities and regions. Ottawa: Federation of Canadian Municipalities*. Retrieved 28 May, 2012 from http://www.fcm.ca/Documents/reports/Ecological_Footprints_of_Canadian_Municipalities_and_Regions_E N.pdf

World Airport Codes. (2010). *Search distance between airports*. Retrieved 13 November, 2010 from http://www.world-airport-codes.com/

Xu, S., & San Martin, I. (2010). *Ecological footprint for the twin cities: Impacts of consumption in the 7-county metro area*. Minnesota: Metropolitan Design Centre, College of Design, University of Minnesota.

Notes

Note 1. YVR is not part of the GVTA.

Note 2. Exclusive because carbon sinks cannot generally be exploited for other purposes. For example, a commercial forest being harvested for wood/fibre is a *source* of carbon emissions, not a sink.

Note 3. YVR only provided data on number of flights and types of aircraft. All other information has come from industry references and other sources. The assumed 80% load factor was confirmed through correspondence with YVR to be a reasonable average to use in our calculation.

Note 4. Great Circle Distance measures the shortest route between two points on a sphere's surface (i.e. the Earth).

Note 5. There was not publicly available data on the *actual* number of passengers.

Note 6. An alternative to calculating total fuel consumed is calculating total amount of fuel loaded on outbound aircraft. In this study, we did not use total fuel loaded onto outbound aircraft because that data was unavailable.

Note 7. Here we refer to GHG equivalence only, not EF equivalence. EF is reserved for impacts that can be expressed in terms of ecosystem area (Methane cannot be expressed as such).

Note 8. Emission inventory year is 2005.

Note 9. Consumption of biocapacity exceeds available 'income' in the form of local supply.

15

Spatial Evolution of Phosphorus Fractionation in the Sediments of Rhumel River in the Northeast Algeria

Sarah Azzouz[1], Smaine Chellat[2], Chahrazed Boukhalfa[1] & Abdeltif Amrane[3]

[1] Chemistry Department, University Constantine 1, Constantine, Algeria

[2] Earth Science Department, University Kasdi Merbah, Ouargla, Algeria

[3] Ecole Nationale Supérieure de Chimie de Rennes, Université Rennes1, CNRS, UMR 6226, Avenue du Général Leclerc, CS 50837, 35708 Rennes Cedex 7, France

Correspondence: Chahrazed Boukhalfa, Chemistry Department, University Constantine 1, Constantine, Algeria.
E-mail: chahrazed_boukhalfa@yahoo.com

Abstract

The objective of the present study is the characterization of the spatial evolution of phosphorus forms in sediments of Rhumel River located in northeast Algeria during winter conditions. Sediments samples were collected along the river in Constantine city during the year 2012. The samples were subjected to physicochemical characterization and metals analysis. Phosphorus was fractionated by sequential extractions procedure in exchangeable, oxyhydroxides bound; calcium bound; organic and residual fractions.

The distribution of the different forms of phosphorus in the sediments appears to be influenced by the physicochemical characteristics, which depend on the sampling location. Phosphorus speciation along the river is characterized by the predominance of inorganic phosphorus forms. The exchangeable fraction is the lowest. Phosphorus concentration in this fraction does not exceed 20 mg/kg. The fraction bound to calcium is the most important in retaining inorganic phosphorus with concentrations varying from 328 to 490 mg/kg. Phosphorus bound to oxyhydroxides represents an average of 172 mg/kg. Along the river, the contribution of the different fractions in the phosphorus retention follows the order: exchangeable < bound to oxyhydroxides ~ organic < bound to calcium < residual. As estimated by the sum of exchangeable, bound to oxyhydroxides and bound to organic matter, an average of about 28% of the total phosphorus can become bioavailable. The predominant fraction in the Rhumel sediments changes from residual at upstream Constantine city to bound to calcium at downstream from it.

Keywords: phosphorus, fractionation, mobility, sediment, Rhumel

1. Introduction

Phosphorus is an essential element in the functioning of aquatic ecosystems, it is considered as one of the major nutrients required by primary producers (Liu et al., 2012). However, it is also identified as a key nutrient responsible for eutrophication of aquatic environments which has become a serious environmental problem. Phosphorus is naturally present in the aquatic environment. It has various natural sources including leaching from rocks, drainage of forests and soil erosion. During the last century, the amount of phosphorus in freshwater has been greatly increased and amplified by human influence through industrial, agricultural and domestic activities. Sediments play an important role in the phosphorus cycle; they can adsorb large quantities of it and can also release it into the overlying water column when the concentration in water decreases and/or under the conditions of strong water dynamic or change of redox potential (Yang & He, 2010). The nature of the chemical and physical links of phosphorus with sediments is the most important factor that governs its release. The mechanisms involved can be of chemical or biological nature or a combination of both (Slomp, Van Raaphorst, Malschaert, Kok, & Sandee, 1993). Generally, phosphorus in sediments can be adsorbed by Fe, Al and Mn oxihydroxides, tied in organic substances and bound to calcium (Balzer, 1986). The mobilization of phosphorus can be affected by many factors such as temperature, dissolved oxygen, pH and the nature of the sediments (Hasnaoui et al., 2001). Under anoxic conditions, the release of the chemically bound phosphorus is due to reduction of iron oxides (Sallade & Sims, 1997), mineralization of organic matter (Golterman, 1995) and acidification of sediments (Golterman, 1998).

The main objective of the present work is the evaluation of phosphorus mobility in the sediments of Rhumel River which traverses Constantine city in eastern Algeria. In our knowledge no such study has been undertaken in the area.

2. Material and Methods

2.1 Study Site

Rhumel River is located in the northeastern Algeria (Figure 1). It originates from the northwestern Bellaa in Setif. It traverses the high plains of Constantine, with an orientation southwest - northeast until Constantine city. Then, it suddenly changes the direction and turns to the right and flows obliquely towards the northwest (Mébarki, 1984), it confluences with the Oued Endja around Sidi Merouene in Mila town. The main tributary of the river is Oued Boumerzoug which drains industrial and urban zones.

The climate of the area is a semiarid type; characterized by wet winters and dry and hot summers. The quality of the river water is characterized by neutral to alkaline pH and high electrical conductivity.

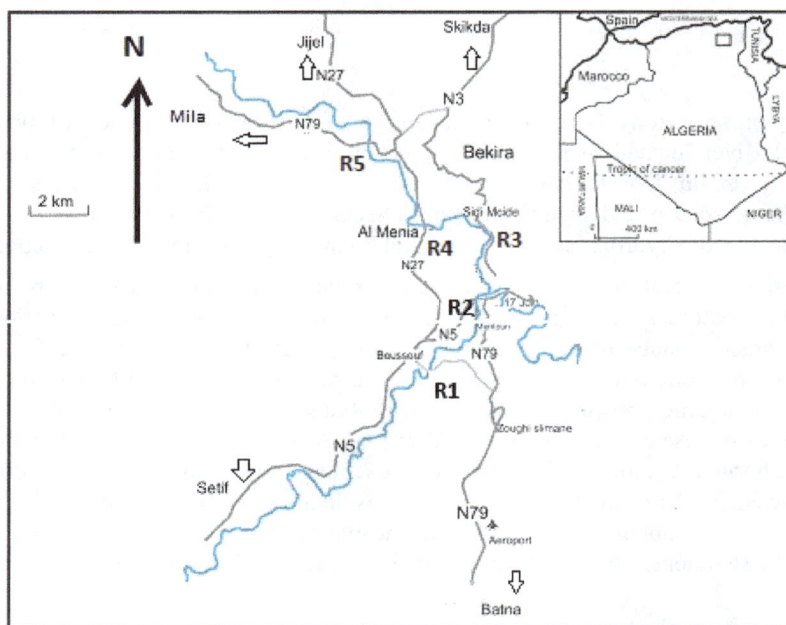

Figure 1. Localization of the Rhumel River and the sampling stations

2.2 Samples Collection and Pretreatment

The studied sediments were collected at five stations along the river (Figure 1, Table 1) in January 2012. Samples were placed in plastic bags and transported to the laboratory where they were dried at 40 °C then ground and sieved using a 0.215 mm sieve and conserved in polyethylene bottles until use.

Table1. Localization of the sampling stations

Location	Latitude	Longitude
R1	36°19'37.15"N	6°35'37.65"E
R2	36°20'53.45"N	6°36'49.61"E
R3	36°22'22.80"N	6°36'41.47"E
R4	36°22'22.90"N	6°35'32.63"E
R5	36°23'57.79"N	6°34'15.50"E

2.3 Physicochemical Characterization

Measurements of pH and electrical conductivity were performed in suspensions formed with distilled water. Organic matter was determined by loss on ignition at 550 °C. The total phosphorus was extracted with HCl (3.5 M) after calcination. Phosphorus was measured in extracts by UV-visible spectrophotometry using the method of Murphy and Riley (1962). In this method, orthophosphate ions react with molybdate to form a yellow phosphomolybdic complex. Ascorbic acid specifically reduced the phosphomolybdic complex to give a blue color. The absorbance was measured at 700 nm with a spectrophotometer Shimadzu UV-1650PC. The metals were determined after calcination and acid digestion by flame atomic absorption using an atomic absorption spectrometer Varian AA140.

2.4 Fractionation of Phosphorus in Sediments

The different forms of phosphorus were extracted using the fractionation procedure described by Hieltjes and Lijklema (1980). The target phases and the reagents used are illustrated in Table 2. Phosphorus in all extracts was determined by the method described above. All results are average values of triplicate determinations.

Table 2. Sequential extractions procedure

Target fraction	Extraction reagent
Exchangeable	NH$_4$Cl (1 M) 2 h shaking
Bound to oxyhydroxides	NaOH (1 M) 16 h shaking
Bound to calcium	HCl (0.5 M) 16 h shaking
Organic	Calcination at 550 °C 3 h, HCl (1 M) 16h shaking

3. Results and Discussion

3.1 Physicochemical Characterization

The physicochemical results are presented in Table 3. The sampled sediments have an alkaline pH reflecting the dominance of limestone and clay and the buffering capacity associated with these sedimentary materials (Nassali, Ben bouih, & Srhiri, 2002). At the first sampling station (R1), the lowest pH and the highest electrical conductivity are observed, showing the effect of the industrial zone located upstream. Generally, high values of electrical conductivity of the sediments are due to the enrichment by monovalent and divalent ions (Nassali et al., 2002). The low water contents reflect a low fluidity of these sediments (Abdallaoui, Derraz, Bhenabdallah, & Lek, 1998). The important organic matter contents ranging from 4% to 6% are probably due to the degradation of dead cells of the fauna and flora within the River and leaching of surrounding soils (Abdellaoui, 1998).

Calcium is the most abundant element in the studied sediments. The measured concentrations of this metal ranged from 134.48 g/kg to 182.35 g/kg. The Rhumel sediments are quite rich in iron and aluminum. The concentrations of the two metals vary between 15 g/kg – 20 g/kg and 13 g/kg – 18 g/kg respectively. Generally, the metals concentrations in the Rhumel sediments follow the order Mn < Al < Fe < Ca. Along the river, only calcium and manganese show a linear correlation in their spatial evolution (R: 0.86). Concentrations of total phosphorus vary from one site to another. The higher contents are observed at the two stations R1 and R4 located downstream the industrial zone and Constantine city respectively. Along the river, phosphorus is correlated with organic matter. The spatial evolution of its total concentration shows a decrease downstream. Phosphorus concentrations found in this study are similar to those measured in Oued D'Kor (1287 mg/kg) and Oued Beht (1343 mg/kg) in Morocco (Abdallaoui, 1998).

Table 3. Physicochemical analysis of sediments of Rhumel River

Sample	R1	R2	R3	R4	R5
pH	7.75	8.34	8.11	8.00	8.01
Electrical Conductivity(μs/cm)	1703	482	488	786	792
Water content (%)	3.25	2.61	2.05	1.31	0.89
Loss on ignition (%)	8.76	6.84	7.23	7.48	5.23
Total phosphorus (mg/kg)	1568	1140.29	1337.74	1547.98	1115.13
Ca (g/kg)	146.01	182.35	134.48	143.77	141.86
Fe (g/kg)	15.66	16.97	17.95	19.61	20.77
Al (g/kg)	18.43	15.66	12.95	14.58	13.70
Mn (mg/kg)	324.50	367.34	245.28	261.04	234.38

3.2 Fractionation of Phosphorus in Sediments

The used sequential fractionation scheme (Hieltjes & Lijklema, 1980) allowed us to distinguish five fractions: soluble phosphorus; phosphorus bound to iron, aluminum and manganese oxyhydroxides; phosphorus bound to calcium; organic phosphorus and residual fraction. The last fraction is calculated as the difference between total phosphorus and the sum of the four other fractions.

3.2.1 Spatial Evolution of Soluble Fraction

The exchangeable fraction represents mineral fraction adsorbed on exchange sites (Hieltjes & Lijklema, 1980), directly assimilated by algae. It is the most available fraction. In the Rhumel sediments (Figure 2), the soluble phosphorus does not exceed 20 mg/kg. The highest value is observed at the sampling station R4 and can be attributed to the settlement of finer particles due to slower water flow (Yang, He, Lin, & Stoffella, 2010). Low values of soluble phosphorus concentration have also been found in other studies (Salvia-Castellvi, Scholer, &Hoffmann, 2002; Taoufik & Dafir, 2002; Taoufik, Kemmou, Loukili Idrissi, & Dafir, 2004; Kemmou, Dafir, Wartiti, & Taoufik, 2006).

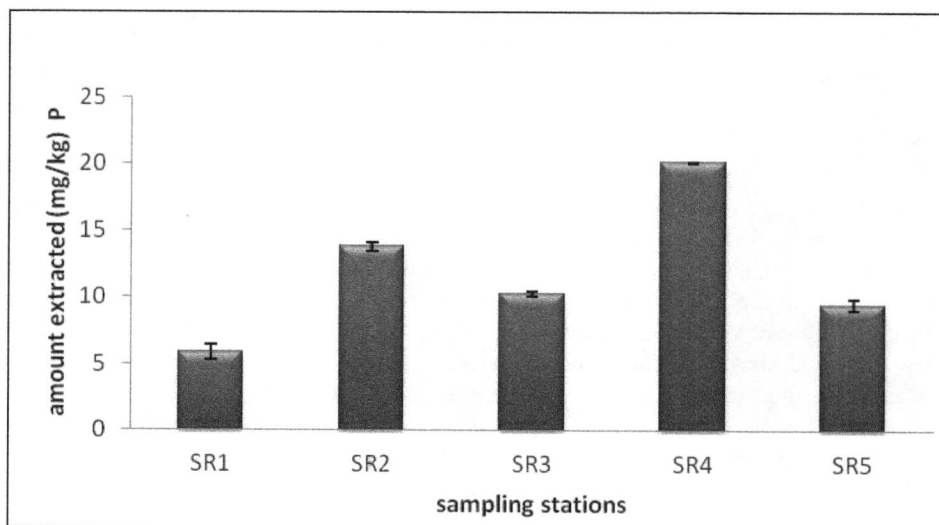

Figure 2. Spatial evolution of the exchangeable fraction

3.2.2 Spatial Evolution of the Fraction Bound to Oxyhydroxides

In the sediments, phosphorus is frequently associated with Fe, Al and Mn oxides and hydroxides (Pardo, Lopez-Sanchez, & Rauret, 2003). This fraction plays an important role in phosphorus exchange at the sediment-water interface (Kemmou et al., 2006). It is easily mobilized and is responsible for an increase of eutrophication (Zhou, Gibson, & Zhu, 2001). In the Rhumel sediments, concentrations of phosphorus extracted by NaOH ranged from 130 mg/kg to 221 mg/kg (Figure 3). The spatial evolution of this fraction shows that phosphorus is more closely related to Fe than Al and Mn. Consequently, anoxic conditions mediated by bacteria result in the release of sorbed phosphorus from iron oxyhydroxides. The Fe/P ratio of 2 has been regarded as a threshold of phosphorus saturation in soil or sediments (Blomqvist, Gunnars, & Elmgren, 2004). Elsewhere, the molar ratio P/(Fe+Al) has been considered as a better indicator of the potential availability of phosphorus in river sediments (Nair, Portier, Graetz, & Walker, 2004). In the present study, Fe/P ratios are above 2; the highest calculated value concern the sediments collected downstream from Constantine city. The calculated P/(Fe+Al) molar ratios vary around 0.05 implying the importance of phosphorus immobilization along the river.

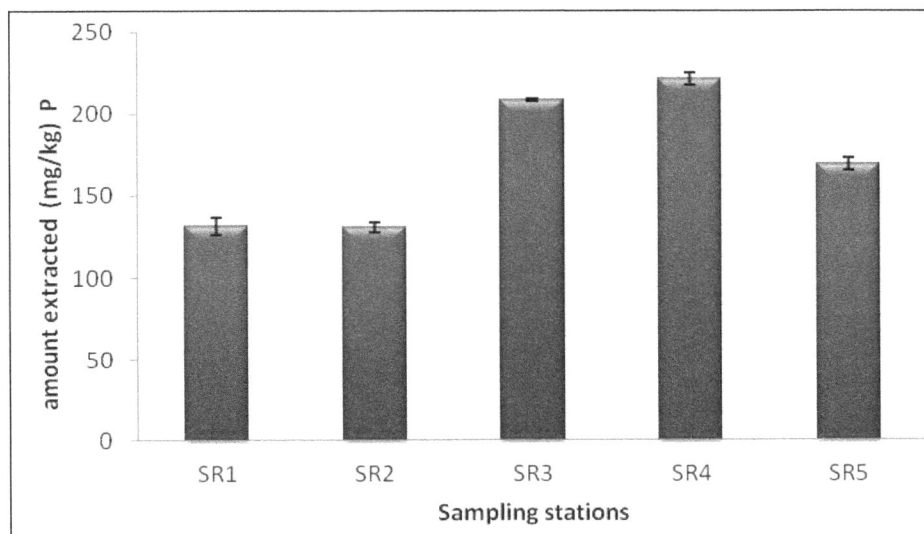

Figure 3. Spatial evolution of oxyhydroxides bound fraction

3.2.3 Spatial Evolution of the Fraction Bound to Calcium

This fraction is sensitive to low pH. It is assumed to be composed mainly of calcium bound phosphorus as apatite such as $Ca_5(PO_4)_3(OH, F, Cl)$ and phosphorus bound to calcium carbonate. The first fraction is highly insoluble and redox-insensitive; it can only be attacked by strong acids (Smolders et al., 2006). The Rhumel sediments are characterized by high concentrations of phosphorus bound to calcium varying from 328.35 mg/kg to 490.79 mg/kg (Figure 4). This is related to the significant calcium contents (Table 2). According to Golterman (1995), when sediments are acidified a part of the phosphorus bound to calcium carbonate might be solubilized. In the present study, pH does not vary significantly. Consequently, this fraction is the most uniform along the river. Generally, phosphorus bound to calcium is considered as the main route of permanent storage of phosphorus in sediments and soils (Gonsiorczyk, Casper, & Koschel, 1998). This fraction is released from sediments with difficulty. Consequently, it is not easily used by algae (Kozerski & Kleeberg, 1998; Kaiserli, Voutsa, & Samara, 2002).

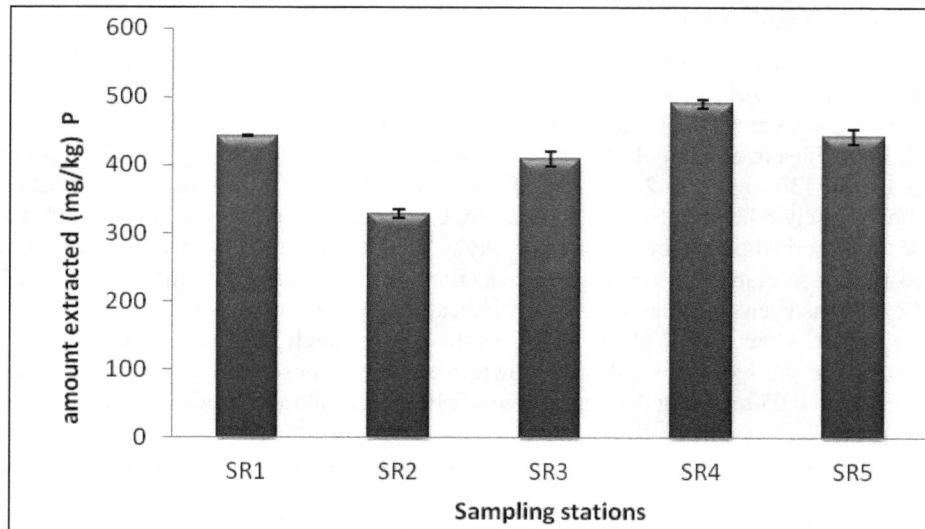

Figure 4. Spatial evolution of calcium bound fraction

3.2.4 Spatial Evolution of Organic Fraction

In the studied sediments, concentrations of organic phosphorus are lower than those of inorganic phosphorus (Figure 5). The spatial distribution of this fraction is generally similar to that of the organic matter. At the sampling stations located in the urban area (R2, R3, R4), a marked increase in organic phosphorus is observed. An increase in the organic matter amount of the sediment leads to an increase in the amount of associated phosphorus. It has been suggested that organic phosphorus in sediments is predominantly associated with humic material by complexation and chelation reactions involving metallic cations (Garcia & de Iorio, 2003). The complexes of organic matter with iron can also adsorb phosphorus (Kemmou et al., 2006). Under anoxic conditions, the organic fraction can become bioavailable after sediments mineralization.

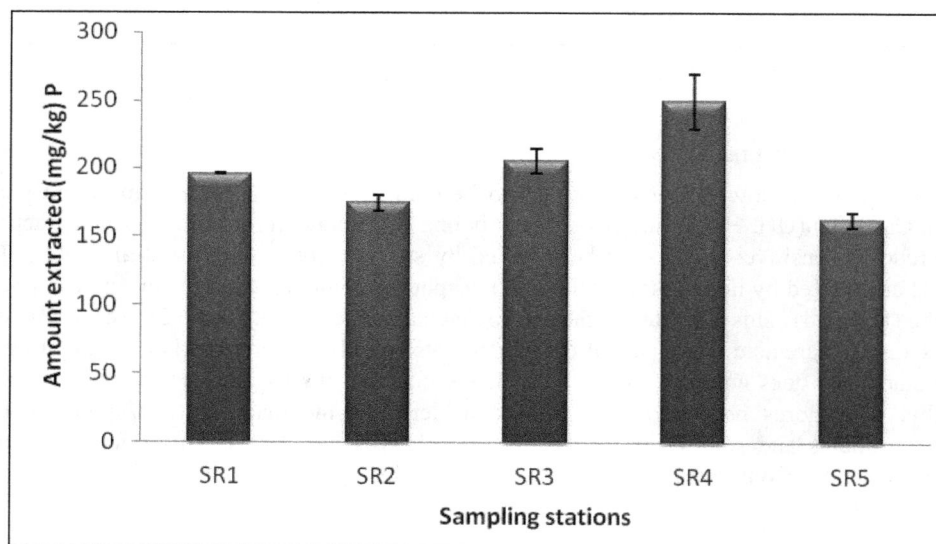

Figure 5. Spatial evolution of organic fraction

3.2.5 Distribution of Phosphorus in the Sediments

The speciation of sedimentary phosphorus in Rhumel River (Figure 6), show that it is mostly in inorganic forms. Along the river, the exchangeable form is the lowest compared to the other fractions. Phosphorus availability in the Rhumel sediments appeared to be related to phosphorus sorption by oxyhydroxides and complexation with organic material. Upstream the confluence with Boumerzoug tributary (R1, R2), the contribution of the organic fraction is more important than the one of the oxyhydroxides. However at downstream, the two fractions are closer. The fraction related to calcium is the most important part of the inorganic phosphorus. It has been suggested that the high phosphorus contents of this fraction could be also explained by the fact that a part of phosphorus extracted with NaOH is readsorbed on calcium (De Groot, & Golterman, 1990). In addition, a part of the organic fraction can be solubilized by acid extraction resulting in an overestimation of the phosphorus amount extracted during this step. The contribution of the residual fraction decreases along the river. In this fraction, phosphorus can be associated to crystalline iron oxides, silicates (Buffle, de Vitre, Perret, & Leppard, 1989) and crystalline aluminum-silicate species (Jonsson, 1997). The predominant phosphorus fraction in the Rhumel sediments changes from residual at upstream of Constantine city to bound to calcium downstream from it.

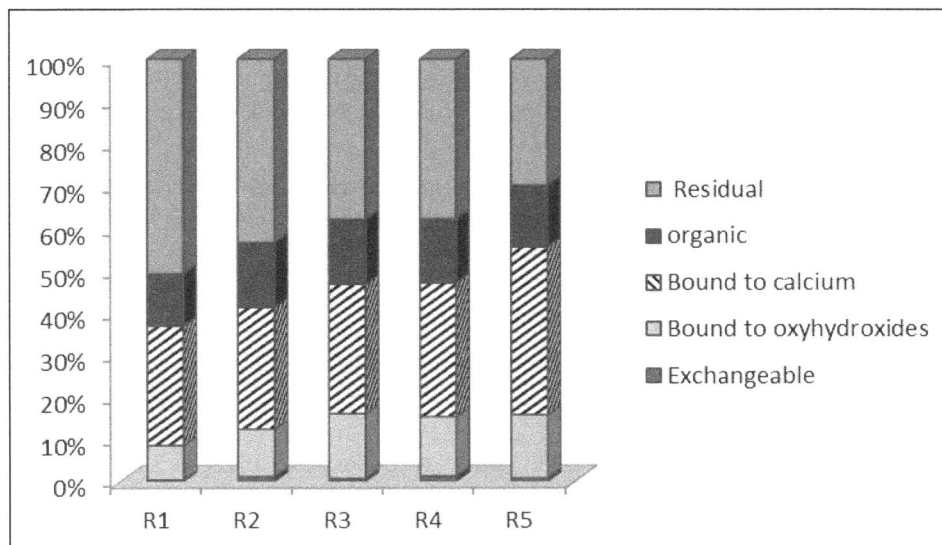

Figure 6. Spatial evolution of phosphorus distribution in Rhumel sediments

4. Conclusion

Sedimentary phosphorus in Rhumel River is mainly inorganic. The fraction directly available is the lowest. The two fractions, residual and bound to calcium considered as permanents, are the most important. As estimated by the sum of exchangeable, bound to oxyhydroxides and bound to organic matter, about 28% of the total phosphorus in Rhumel sediments can become bioavailable.

Acknowledgements

Authors thanks Mrs Oughebbi Laurence for her assistance in the analyses of metals.

References

Abdallaoui, A. (1998). Contribution à l'étude du phosphore et des métaux lourds contenusdans les sédimentset de leur influence sur les phénomènes d'eutrophisation et de la pollution. Cas du bassin versant de l'oued Beht et de la retenue de barrage El Kansera. PHD Thesis, Marroco. Retrieved from http://toubkal.imist.ma/handle/123456789/7171

Abdallaoui, A., Derraz, M., Bhenabdallah, M. Z., & Lek, S. (1998). Contribution à l'étude de la relation entre les Différentesformes du phosphore dans les sédiments d'une retenue de barrage eutrophe en climat méditerranéen (El Kansera, Maroc). *Revue des sciences de l'eau, 11*, 101-116. http://dx.doi.org/10.7202/705299ar

Balzer, W. (1986). Forms of phosphorus and its accumulation in coastal sediments of Kieler Bucht. *Ophelia, 26*, 19-35. http://dx.doi.org/10.1080/00785326.1986.10421976

Blomqvist, S., Gunnars, A., & Elmgren, R. (2004). Why the limiting nutrient differs between temperate coastal seas and freshwater lakes: a matter of salt. *Limnol. Oceanogr, 49*, 2236-2241. http://dx.doi.org/10.4319/lo.2004.49.6.2236

Buffle, J., De Vitre, R. R., Perret, D., & Leppard, G. G. (1989). Physico-chemical characteristics of a colloidal iron phosphate species formed at the oxic–anoxic interface of a eutrophic lake. *Geochimica et Cosmochimica Acta, 53*, 399-408. http://dx.doi.org/10.1016/0016-7037(89)90391-8

De Groot, C. J., & Golterman, H. L. (1990). Sequential fractionation of sediment phosphate. *Hydrobiologia, 192*, 143-148. http://dx.doi.org/10.1007/BF00006010

Garcia, A. R., & de Iorio, A. F. (2003). Phosphorus distribution in sediments of Morales Stream (tributary of the Matanza-Riachuelo River, Argentina). The influence of organic point source contamination. *Hydrobiologia, 492*, 129-138. http://dx.doi.org/10.1023/A:1024874030418

Golterman, H. L. (1995). The labyrinth of nutrient cycles and buffers in wetlands: results based on research in theCamargue (southern France). *Hydrobiologia, 315*, 39-58. http://dx.doi.org/10.1007/BF00028629

Golterman, H. L. (1998). Presence of and phosphate release from polyphosphates or phytate phosphate in lake sediments. *Hydrobiologia, 364*, 99-104. http://dx.doi.org/10.1023/A:1003212908511

Gonsiorczyk, T., Casper, P., & Koschel, R. (1998). Phosphorus binding forms in the sediment of an oligotrophic and an eutrophic hardwater lake of the Baltic district (Germany). *Water Science and Technology, 37*, 51-58. http://dx.doi.org/10.1016/S0273-1223(98)00055-9

Hasnaoui, M., Kassila, J., Loudiki, M., Droussi, M., Balvay, G., & Barrouin, G. (2001). Relargage du phosphore à l'interface eau-sédiment dans des étangs de pisciculture de la station Deroua (Béni Mellal, Maroc). *Revue des sciences de l'eau, 14*, 307-322. http://dx.doi.org/10.7202/705422ar

Hieltjes, A. H. M., & Lijklema, L. (1980). Fractionation of inorganic phosphate in calcareous sediments. *J. Env.Qual., 9*, 405-407. http://dx.doi.org/10.2134/jeq1980.93405x

Jonsson, A. (1997). Fe and Al sedimentation and their importance as carriers for P, N and C in a large humic lake in northern Sweden. *Water Air Soil Pollut, 99*, 283-295. http://dx.doi.org/10.1007/BF02406868

Kaiserli, A., Voutsa, D., & Samara, C. (2002). Phosphorus fractionation in lake sediments-Lakes Volvi and koronia N. Greece. *Chemosphere, 46*, 1147-1155. http://dx.doi.org/10.1016/S0045-6535(01)00242-9

Kemmou, S., Dafir, J. E., Wartiti, M., & Taoufik, M. (2006).Variations saisonnières et mobilité potentielle du phosphore sédimentaire de la retenue de barrage Al Massira (Maroc). *Water Qual. Res. J. Canada, 41*, 427-436.

Kozerski, H. P., & Kleeberg, A. (1998). The sediments and the benthic pelagic exchange in the shallow lake Muggelsee. *Int Rev Hydrobiol, 83*, 77-112. http://dx.doi.org/10.1002/iroh.19980830109

Liu, L., Zhang, Y., Efting, A., Barrow, T., Qian, B., & Fang, Z. (2012). Modeling bioavailable phosphorus via other phosphorus fractions in sediment cores from Jiulongkou Lake, China. *Environ Earth Sci., 65*, 945-956. http://dx.doi.org/10.1007/s12665-011-1295-2

Mébarki, A. (1984). *Ressources en eau et aménagement en Algérie. Le bassin du Kébir-Rhumel.* Alger, Office des Publications Universitaires.

Murphy, J., & Riley, J. (1962). A modified single method solution method for the determination of phosphates in natural water. *Anal. Chim. Acta, 27*, 31-36. http://dx.doi.org/10.1016/S0003-2670(00)88444-5

Nair, V. D., Portier, K. M., Graetz, D. A., & Walker, M. L. (2004). An environmental threshold for degree of phosphorus saturation in sandy soils. *J. Environ. Qual., 33*, 107-113. http://dx.doi.org/10.2134/jeq2004.1070

Nassali, H., Ben bouih, H., & Srhiri. A. (2002). Influence des eaux usées sur la dégradation de la qualité des eaux du lac Fouarate au maroc. *Proceedings of International Symposium on Environmental Pollution Control and Waste Management. Tunis (EPCOWM),* 3-14. Retrieved from http://www.geocities.jp/epcowmjp/EPCOWM2002/Volume1.htm

Pardo, P., Lopez-Sanchez, J. F., & Rauret, G. (2003).Relationships between phosphorus fractionation and major components in sediments using the SMT harmonized extraction procedure. *Analytical and Bioanalytical Chemistry, 376*, 248-254.

Sallade, Y. E., & Sims, J. T. (1997). Phosphorus transformations in the sediments of Delawareþs Agricultural Drainageways: Effect of reducing conditions on phosphorus release. *J. Envir. Qual., 26*, 1579-1588. http://dx.doi.org/10.2134/jeq1997.00472425002600060018x

Salvia-Castellvi, M., Scholer, C., & Hoffmann, L. (2002). Comparaison de différents protocoles de spéciation séquentielle du phosphore dans des sédiments de rivière. *Revue des sciences de l'eau, 15*, 223-233. http://dx.doi.org/10.7202/705448ar

Slomp, C. P., Van Raaphorst, W., Malschaert, J. F. P., Kok, A., & Sandee, A. J. J. (1993). The effect of deposition of organic matter on phosphorus dynamics in experimental marine sediment systems. *Hydrobiologia, 253*, 83-98. http://dx.doi.org/10.1007/BF00050724

Smolders, A. J. P., Lamers, L. P. M., Lucassen, E., Van Der Velde, G., & Roelofs, J. G. M. (2006). Internal eutrophication: how it works and what to do about it, a review. *Chem. Ecol., 22*(2), 93-111. http://dx.doi.org/10.1080/02757540600579730

Taoufik, M., & Dafir, J. E. (2002). Comportement du phosphore dans le sédiment des barrages de la partie aval du bassin versant d'Oum Rabiaa (Maroc). *Revue des sciences de l'eau, 15*, 235-249. http://dx.doi.org/10.7202/705449ar

Taoufik, M., Kemmou, S., Loukili Idrissi, L., & Dafir, J. E. (2004). Comparaison de deux méthodes de spéciation du phosphore dans des sédiments de la partie aval du basin Oum Rabiaa (Maroc). *Water Qual. Res. J. Canada, 39*, 50-56.

Yang, Y. G., He, Z. L., Lin, Y., & Stoffella, P. J. (2010). Phosphorus availability in sediments from a tidal river receiving runoff water from agricultural fields. *Agricultural Water Management, 97*, 1722-1730. http://dx.doi.org/10.1016/j.agwat.2010.06.003

Zhou, Q., Gibson, C. E., & Zhu, Y. (2001). Evaluation of phosphorus bioavailability in sediments of three contrasting lakes in china and UK. *Chemosphere, 42*, 221-225. http://dx.doi.org/10.1016/S0045-6535(00)00129-6

16

Spatial Interpolation of Two-Wavelengths Bio-Optical Models to Estimate the Concentration of Chlorophyll-*a* in A Tropical Aquatic System

Igor Ogashawara[1], Enner Herenio de Alcântara[2], Pétala Bianchi Augusto-Silva[1], Claudio Clemente Faria Barbosa[1] & José Luiz Stech[1]

[1] Remote Sensing Division, National Institute for Space Research, São José dos Campos, Brazil

[2] Cartography Department, São Paulo State University, Presidente Prudente, Brazil

Correspondence: Igor Ogashawara, Remote Sensing Division, National Institute for Space Research, São José dos Campos, SP 12227-010, Brazil. E-mail: igoroga@gmail.com

Abstract

Bio-optical models have been used to estimate and map the concentration of chlorophyll-a (chl-*a*) in aquatic systems. Bio-optical models' algorithms try to infer the concentration of optically active components in water from their inherent optical properties (IOPs). We proposed the use of two single wavelengths to retrieve chl-*a* concentration in a tropical aquatic system. The results were compared to *in situ* measurements of chl-*a*. To spatialized the results of the bio optical modeling, we tested the spatial interpolation following two methods: (1) ordinary kriging technique was used to spatialize the calculated values of estimated chl-*a* for each model; (2) ordinary kriging was used to spatialize the estimated R_{rs} from 470nm and 700nm then the two wavelengths models were calculated by a map algebra using the spatialized R_{rs}. We generated four different spatial interpolations of chl-*a* concentration. They were compared to the spatialized reference based on the *in situ* chl-*a* collected in the reservoir of Itumbiara–Goiás in the same period. The comparison was performed through the "Spatial Language for Algebraic Geoprocessing" (LEGAL) implemented at SPRING software. Results showed a better accuracy for the procedure using the spatialization of R_{rs} and map algebra of them. Thus the spatializaton of proximal remote sensing measurements in order to retrieve the optically active components in water should be performed through the interpolation of the R_{rs}.

Keywords: reflectance, bio-optical models, chlorophyll-*a*, interpolation, water quality

1. Introduction

Chlorophyll-*a* (chl-*a*) has been use as a parameter to calculate the trophic state index (TSIs) of an aquatic system. It also has been use as an indicator of photoautotrophic biomass which could be related to primary productivity (Huot et al., 2007). Blooms of chl-*a* are visible symptoms of eutrophication process in aquatic system (Mudroch, 1999). It is also related to the increased growth of algae in which cyanobacteria is a troublesome group. It is the responsible to cause severe oxygen depletion which causes fish mortalities. It is also the responsible for the death of cattle and other animals from ingestion of cyanotoxins (Melack, 2000). Another consequence of algal blooms is the high concentrations of dissolved organic carbon (DOC). Phytoplankton blooms also interferes in the decrease of light penetration through the water column (Boyer, Kelble, Ortner, & Rudnick, 2009). This process could depress seagrass growth and productivity.

Chl-*a* monitoring is important for water quality measures. Mainly because it may vary temporally and spatially in an aquatic system. However traditional field sampling methods used to estimate Chl-*a* concentration are expensive, time consuming (Duan, Ma, Xu, Y. Zhang, & B. Zhang, 2010) and difficult to make for large studies areas (Gons, 1999). These observations have highlighted the importance of the monitoring of chl-*a* concentration. Remote sensing has been use as a tool to estimate chl-*a* concentration in case 1 and case 2 waters. This water classification varies according to the optical properties of the water. In case 1 waters the optical properties are determined primarily by phytoplankton and related constituents as dissolved organic matter (CDOM) and detritus. However, in case 2 waters the optical properties can be influenced by other constituents such as mineral particles, CDOM, or microbubbles, whose concentrations do not covary with the phytoplankton concentration

(Mobley, Stramski, Bisset, & Boss, 2003).

1.1 Bio-Optical Models

Bio-optical models describe the interaction between water's optically active constituents and electromagnetic radiation. The application of these algorithms to case 2 waters still a challenge due to the presence of chl-a's non-covarying constituents (S. Mishra & D. Mishra, 2012). Empiricals and semi-analytical algorithms have been developed to estimate chl-*a* in case 2 waters (Moses, Gitelson, Berdnikov, Saprygin, & Povazhnyi, 2012; Duan et al., 2010; S. Mishra & D. Mishra, 2012). Gilerson et al. (2010) and S. Mishra and D. Mishra (2012) developed a two band algorithm for inland waters using red-near infrared (NIR) bands. Both studies used the bands 7 and 9 from the MEdium Resolution Imaging Spectrometer (MERIS) from the European Space Agency (ESA).

These two bands were centered on the following wavelengths: 665 and 708.75 nm (ESA, 2006). The 665nm wavelength represents a wide spectral absorption peak which is generally assigned to the absorption by chl-*a* pigment. The reflectance peak centered at 700 nm which is maximally sensitive to the variations in chl-*a* concentration in water (S. Mishra & D. Mishra, 2012). The parameters of both algorithms were empirically set by comparing radiometric data and *in situ* measured data from a particular geographical location under a particular seasonal regime. Therefore they only represent a specific region on Earth which has a unique environmental dynamic. These only demonstrate the need for studies in tropical regions since both models were developed in aquatic systems placed in latitudes higher than 30°.

Nevertheless there is no model developed for tropical case 2 waters, which is a distinct environment with a unique primary production dynamic. In order to analyze a tropical aquatic system, we adapted two models (Gilerson et al., 2010; S. Mishra & D. Mishra, 2012) using single wavelengths from proximal remote sensing. To spatialize the results of bio-optical modeling we tested two approaches of spatial interpolation using ordinary kriging algorithms.

The purpose of our study was to evaluate the accuracy of the two-wavelengths algorithms to estimate chl-*a* concentration in a tropical aquatic system. It was also a goal to analyze two procedures to interpolate the bio-optical modeling results: (1) by the spatialization of the estimated reflectance for 470 and 700 nm wavelengths followed by an map algebra of them; and (2) by the spatialization of the calculated model values for the same two wavelengths.

1.2 Bio-Optical Modeling

Empirical and semi-analytical bio-optical models have been developed to estimate water constituents concentrations. Empirical models are based on calculation of statistical relation between the water constituent concentration and water leaving radiance or reflectance (Dekker, 1993). Semi-analytical models describe the reflectance from Inherent Optical Properties (IOPs) and water constituent concentration, by means of the radiative transfer equation.

To estimate the remote sensing reflectance (R_{rs}) from each wavelength (Gordon, Brown, & Jacobs, 1975; Gordon et al., 1988) proposed a relation where the R_{rs} is calculated just beneath the water surface and can be expressed as:

$$R_{rs}(\lambda, 0^-) = \gamma \frac{b_b(\lambda)}{a(\lambda) + b_b(\lambda)} \tag{1}$$

Where a(λ) and b$_b$(λ) are the total spectral absorption and backscattering coefficients. γ is a proportionality constant which depends on the geometry of the field of light emerging from the water body. It can be estimate by 2 different ways: γ= f/Q or γ=f. Where f is the anisotropic factor of the downwelling light field (Kirk, 1994) and Q is the geometrical factor (Gons, 1999).

To determinate the f value, it was previous found that it was a function of the solar elevation angle (Kirk, 1994). It was reasonably well expressed as a linear function of μ_0–the mean cosine of the zenith angle–and can be estimated as:

$$f = 0.975 - 0.629\mu_0 \tag{2}$$

The value of μ_0 was calculated according to the sampling time, locations (latitude and longitude) and solar zenith angle (Martin & McCutcheon, 1999; Rees, 2001). For the Q value it was proposed (Gons, 1999) an empirical equation for turbid inland waters under different solar elevation angles, as expressed:

$$Q = \frac{2.38}{\mu_0} \tag{3}$$

A factor of 0.544 (Austin, 1980) was proposed to relate the radiance just above the surface to radiance just beneath the surface. Therefore, R_{rs} just above the water surface can be determined as follows:

$$R_{rs}(\lambda, 0^+) = 0.544\gamma \frac{b_b(\lambda)}{a(\lambda) + b_b(\lambda)} \tag{4}$$

To calculate the $a(\lambda)$ it is possible to use the following relation:

$$a(\lambda) = a_W(\lambda) + a_{ph}(\lambda) + a_{CDOM}(\lambda) + a_{TR}(\lambda) \tag{5}$$

Where: $a_w(\lambda)$ is the absorption coefficient of pure water, and $a_{ph}(\lambda)$, $a_{tr}(\lambda)$ and $a_{CDOM}(\lambda)$ are specific absorption coefficients for phytoplankton, tripton and CDOM (Colored Dissolved Organic Matter), respectively.

The chl-a bio-optical models proposed by S Mishra and D. Mishra (2012) and Gilerson et al. (2010) use two bands from MERIS. Their models were given through a division of the reflectances. Equation 6 shows the bio-optical model developed by S. Mishra and D. Mishra (2012). Equation 7 shows the model developed by Gilerson et al. (2010).

$$C_{chl} \propto (R\lambda_{708} - R\lambda_{665}) \cdot (R\lambda_{708} + R\lambda_{665})^{-1} \tag{6}$$

And:

$$C_{chl} \propto [35,75[R_{\lambda 665}^{-1} \cdot R_{\lambda 708}] - 19.3]^{1.124} \tag{7}$$

Where: C_{chl} is the concentration of chl-a and $R\lambda_n$ is the reflectance of the MERIS bands centered in the "n" wavelength.

If we apply Eauqtion 4 to estimate the reflectance to be use in both models, the fact of using a reflectance division, nulls the factors (0.544 and γ). Therefore the R_{rs} for these two models can be estimate through the use of the coefficients of total absorption and backscattering as it is expressed in Equation 8.

$$R_{rs}(\lambda, 0^+) = \frac{b_b(\lambda)}{a(\lambda) + b_b(\lambda)} \tag{8}$$

Equation 8 was used to estimate the R_{rs} in 2 wavelengths. It was possible because we used it on Equations 6 and 7 which refer to the two-bands chl-a bio-optical models. This process adapted the models which used MERIS bands to single wavelengths.

Through these procedures it is possible to calculate the R_{rs} for the sampling points from the IOPs of the water constituents. It was also possible to calculate the bio-optical values for the same points. However there is no standard method to spatially interpolate these calculated values. This fact enhances the importance for our investigation about the spatialization of bio-optical models values for an aquatic system.

2. Method

2.1 Study Site

The study was conducted in Itumbiara Reservoir (18°25' S, 49°06' W), located in Brazilian West-Central between the States of Minas Gerais and Goiás (Figure 1). The region is typically a tropical grassland savanna, and the reservoir was formed by damming of Paranaíba River resulting in the flooding of its main tributaries, Corumbá and Araguari Rivers. The geomorphology of the basin resulted in a reservoir with a dendritic pattern covering an area of approximately 814 km² and a volume of 17.03×109 m³ (Alcantara et al., 2010). The surface of reservoir is at 520 meters above the sea level. The length of major axis is 30 km and the maximum width is about 15 km. The depth of reservoir ranges from 0.5 m to 78 m with mean depth of 32 m.

According to Köppen (1931) Climate Classification System (CCS) the climate on the Itumbiara Reservoir region is classified as "tropical savanna" with two well defined seasons, dry (May-October) and wet (December-April). Monthly precipitation ranges from 5 mm (dry season) to 250 mm (wet season). The air temperature is high during the wet season (24 °C to 26 °C) and a little lower in the dry season (20 °C. The relative humidity has a similar pattern, but with a small shift in the minimum value towards September (47%). The wind intensity is lower during the wet season (1.6 m s^{-1}) than in the dry season (3.3 m s^{-1}) (Curtarelli, 2012).

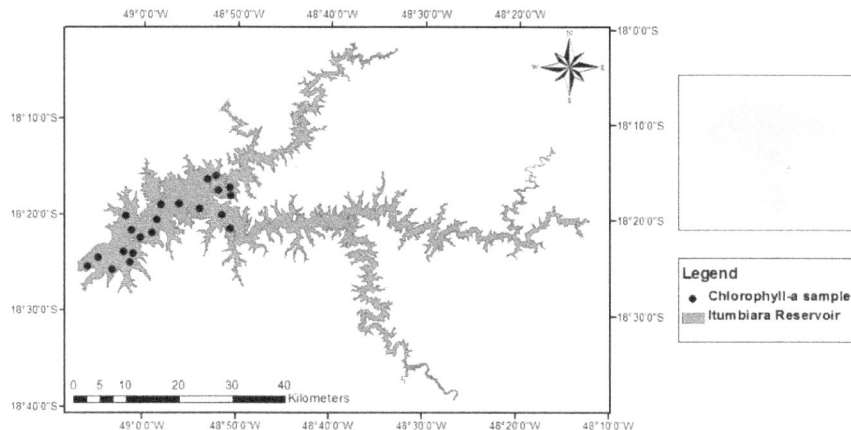

Figure 1. Localization of Itumbiara Reservoir and the distribution of the chl-*a* samples points

2.2 Data Collection

The proximal remote sensing data and water samples for laboratory analysis were collected from 12 to 13 of May, 2009, obtained between 10:00 and 14:00 local to provide representative daily readings (Nascimento, 2010). The samples were collected in 21 different locations in the Itumbiara's Reservoir (Figure 1). Absorptions coefficients were calculated from surface water samples were collected in 1 L Niskin bottles and immediately filtered onto 0.7 μm Whatman GF/F filters. The volume of water filtered were 250 ml. Particulate absorption coefficient, $ap(\lambda)$, and absorption coefficient of detrital matter, $ad(\lambda)$, were determined using standard quantitative filtration technique (QFT) as described in Mitchell, Kahru, Wieland, and Stramska (2003). A UV-2450 Shimadzu spectrophotometer with an integrating sphere was used to measure absorbance of the samples within a spectral range from 350 to 700 nm. For $a_{CDOM}(\lambda)$ the water samples analysis were filtered immediately after collection through 0.2 μm nucleopore membrane filters and read in a U-3010 Hitachi, spectrophotometer. Backscattering coefficients for 470 and 700nm were calculated using a Satlantic profiler (Satlantic, 2000) at 1m and were used to estimate $R_{rs.}$

Concentration of limnological parameters–concentration of chl-*a*, suspended organic and inorganic matter and dissolved organic and inorganic carbon–were obtained from analytical process in the laboratory. The collected water samples from the subsurface–approximately 10 cm from the surface–were at cool temperature until the delivery at the laboratory for the analysis without duplicates. The method of chlorophyll-a analysis (Nush, 1980) consisted in filtered the collected samples in the same day of the collect and then analyzed the filters through a spectrophotometer. Total Suspended Solids (TSS) were determined based on Wetzel and Likens (1991) Water physical parameters as water temperature, pH and turbidity were also measured in situ using a YSI Inc and Horiba multi-parameter sensors. Euphotic zone was estimated multiplying the Secchi Depth by 2.7 (Cole, 1994). The euphotic zone–or euthotic depth–is defined as the compartment where the solar radiation levels are at least 1% of the levels measured just below the free surface. It is in this depth where the primary production of lakes and reservoirs concentrates (Tundisi & Matsumura-Tundisi, 2008).

2.3 Bio-Optical Model Calculation and Validation

Using IOPs obtained from the profiler and QFT, we calculated the R_{rs} for the two wavelengths (470 nm and 700nm) through Equation 8. Then the two-wavelengths bio-optical models values were calculated based on Equations 6 and 7 for the estimated $R_{rs.}$

Although the two-band models (Eauqtions 6 and 7) used red-NIR bands, we tested wavelengths at 470 nm and 700 nm. We chose them because of the chl-*a* reflectance peak at 400–450 nm and 650–700 nm (Kirk, 2011). It was also considered departures from detritus form at 440 and 470nm associated with phytoplankton pigments made the absorption signals small (Ruddick, Gons, Rijkeboer, & Tilstone, 2001). At 700 nm the radiance spectrum correlated strongly with the chl-*a* concentration (Gitelson, 1992). It is the reflectance peak due to the minimum sum of absorption of phytoplankton, particulate and dissolved organic matter and pure water (Gitelson, 1992).

To validate the adapted algorithms we used a linear pearson correlation analysis. It was performed among the calculated two-wavelength bio-optical values and chl-*a* concentration. All of the values were normalized by minimum-maximum normalization. It was also calculated indicators that provide a quantitative estimate of the differences between model and the reference. Bias, Mean Square Error (MSE), Mean Absolute Error (MAE) and Root Mean Square Error (RMSE) were used evaluate the adapted models and were calculated according to Table 1.

Table 1. Error estimators and their formulas

Estimator	Formulas
Bias	$Bias = \dfrac{1}{n}\sum_{i=1}^{n}(y_i - x_i)$
MAE	$MAE = \dfrac{1}{n}\sum_{i=1}^{n}|y_i - x_i|$
MSE	$MSE = \dfrac{1}{n}\sum_{i=1}^{n}(y_i - x_i)^2$
RMSE	$RMSE = \sqrt{MSE}$

2.4 Spatial Interpolation

The spatial interpolation of the models were conducted using a geostatistic technique called ordinary kriging. It is a generic name for a family of generalized least-squares regressions algorithms (Goovaerts, 1997). The calculation of the Kriging weights is based upon the estimation of a semivariogram model. The semivariogram was fitted using the the Geographical Information System (GIS) SPRING, version 5.2, developed at the National Institute for Space Research (INPE) (Camara, Souza, Freitas, & Garrido, 1995). It was chosen since it was possible to apply several theoretical models (spherical, exponential, Gaussian, linear and power) using the weighted least square method.

The adapted models were interpolated by two different procedures: (1) kriging technique was used to spatialized the calculated values of chl-*a* for each model; (2) kriging was used to spatialized the estimated R_{rs} from 470nm and 700nm then the models were calculated by a map algebra using the spatialized R_{rs}.

In order to compare the results, the spatial interpolated data was normalized by the minimum and maximum values in a range from 0 to 1. The bio-optical models were applied by the Spatial Language for Algebraic Geoprocessing (LEGAL) for SPRING (Camara et al., 1995).

To evaluate the precision of these two procedures to accurately retrieve the chl-*a* concentration, kappa statistic (Cohen, 1960) was used. The reference data was use to calculate the statistic through the SPRING GIS, and the prediction error plots were examined.

The use of Kappa statistic is a way to measure the magnitude of agreement between estimations results and the reference - the in situ data (Cohen, 1960). Then, the kappa index varies between -1 to 1, within 1 indicating perfect agreement and ≤ 0 indicating agreement equivalent to chance. For the calculation of kappa statistic, the estimated chlorophyll-a concentration probabilities are classified as not correlated or correlated status (≤ 0 or 1). It was also used the Mean Error (ME) in order to realize a spatial analysis. ME is equal to zero if the predictions were completely unbiased, i.e. centered on the measurement values.

3. Results and Discussions

3.1 Bio-Optical Models Validation

The results from the validation of our adapted models are expressed in Table 2. Adaptation 1 used S. Mishra and D. Mishra (2012) structure and adaptation 2 used Gilerson et al. (2010) structure.

Table 2. Adapted models validation

Estimator	Adaptation 1	Adaptation 2
Bias	-0.314	-0.250
MAE	0.347	0.294
MSE	0.156	0.122
RMSE	0.395	0.349
R	0.571	0.590

3.2 Spatial Interpolation Results

The first method of spatial interpolation (calculated bio-optical models) to the estimated values of chl-*a* by the adapted bio-optical models and the reference are showed in Figure 2. Figure 2a shows the reference values spatialized by ordinary kriging to all the reservoir of Itumbiara. Figures 2b and 2c show the estimated concentration of chl-*a* from the adaptation of two-wavelengths bio-optical models presented. These results represented the first procedure using the calculated models values to estimate the chl-*a*.

Figure 3 shows the results from the second method of spatial interpolation (spatialization of R_{rs} and then map algebra of them). The reference (Figure 3a) and estimations of chl-*a* concentration by two-wavelenghts models (Figures 3b and 3c) were interpolated for the spatial distribution of the reservoir.

Both figures (Figure 2 and Figure 3) illustrate the distribution of chl-*a* in the Itumbiara Reservoir. It revealed 2 regions which have different dynamics. The first region (1) indentified is located near the entrance of Corumbá River. It was a region with low concentration of chl-*a*, probably due to its clean water and low input of nutrients. The second region (2) had a high concentration of chl-*a*. It was located near the entrance of Paranaíba River–in the border of the reservoir. The high concentration probably happened due to the fact that the river follows a meandric shape. This shape was the responsible for the accumulation of the organic matter in decomposition which were transported into the meander.

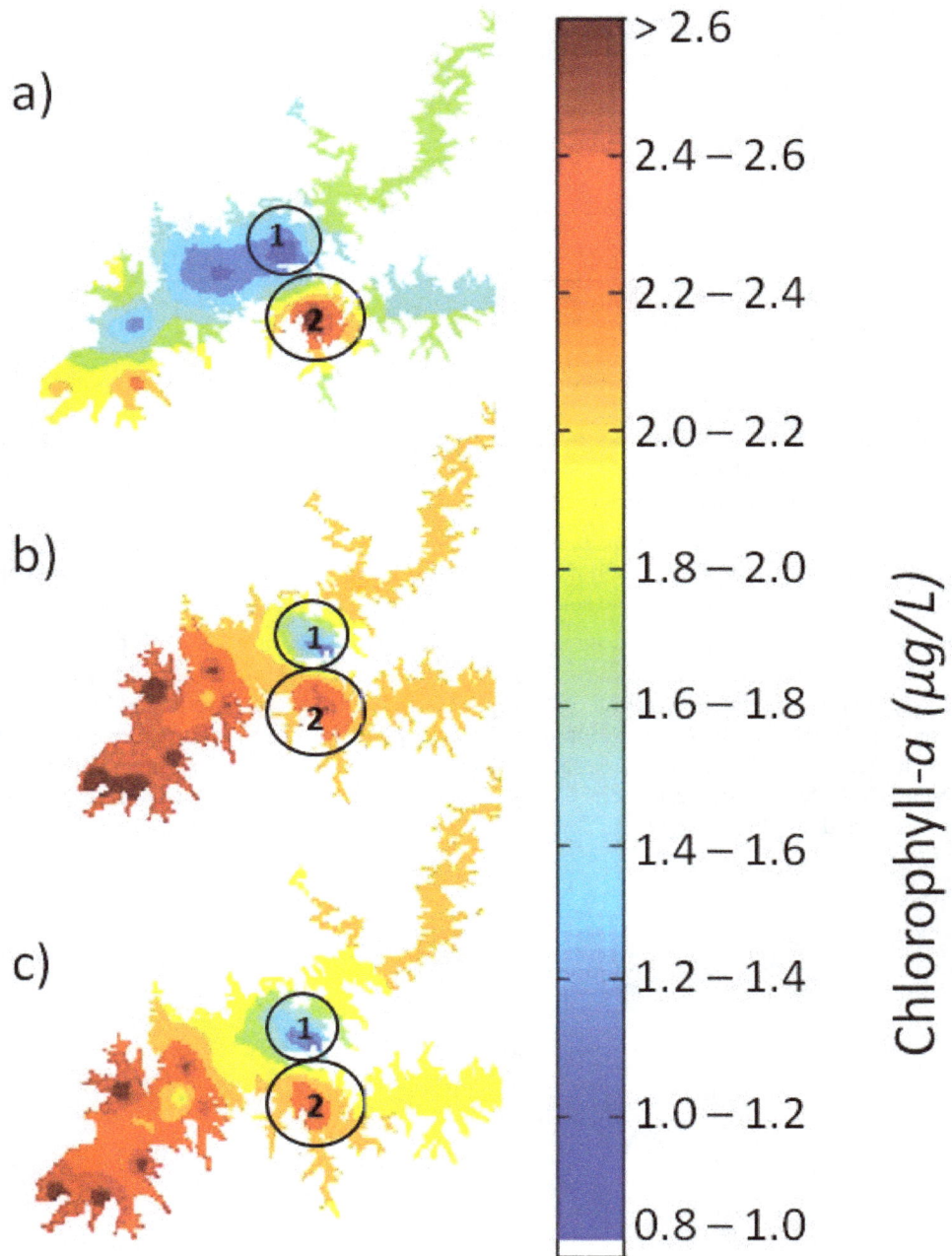

Figure 2. (a) Chl-*a* reference data. (b) Spatial interpolated chl-*a* concentration based on the spatialization of calculated adaptation 1 model values. (c) Spatial interpolatedf chl-*a* concentration based on the spatialization of calculated adaptation 2 model values

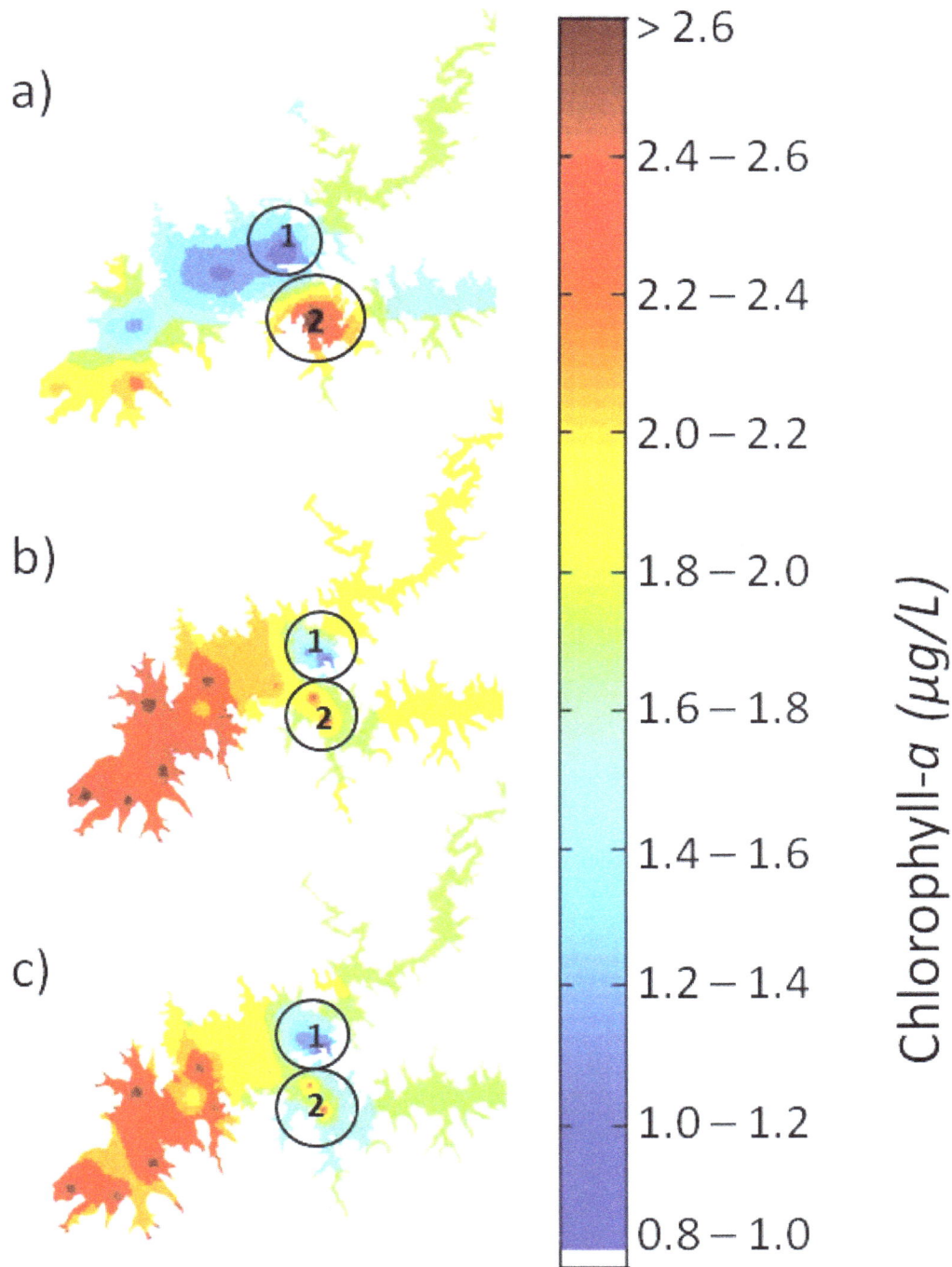

Figure 3. (a) Chl-*a* reference data. (b) Spatial interpolated chl-*a* concentration based on the spatialization of R_{rs} and map algebra of adaptation 1 equation. (c) Spatial interpolated chl-*a* concentration based on the spatialization of R_{rs} and map algebra of adaptation 2 equation

Kappa statistical result–used to compare the spatialized adapted models' estimations to the spatialized reference data–to the adaptation 1 was -0.0806 for the procedure based on the calculated model values. A kappa equals -0.0098 was found to the procedure based on the spatialization of R_{rs} and map algebra. To the second adaptation, the kappa indices were -0.0267 and 0.0861 for the calculated model values and the spatialization of R_{rs} and map algebra respectively.

Kappa statistical results suggest that the four proposed spatial interpolations were not correlated to the reference.

Furthermore they also suggest that the spatialization of R_{rs} and map algebra method for the estimation got the best kappa statistic when compared to the calculated model values method. These results revealed that the adaptation 2 was more appropriated for a tropical and deep reservoir. Probably due to the structure of the algorithm which is more accurate than the first adaptaion. This fact can be explained by the algorithms themselves. The second algorithm (Gilerson et al., 2010) was calibrated with data from previous works and from the literature. In the other hand, the first algorithm (S. Mishra & D. Mishra, 2012) was a normalized difference index proposal, so it just depends on the reflectance of the wavelength.

We also tried to understand the reason for this poor statistical performance which is probably related to the reflectance estimation. It could be affected by innumerous factors like the instability on the aquatic system surface. It is also possible that Gordon's algorithm (1988) is not useful for case 2 waters since it was developed for case 1 waters.

A previous study (Nixdorf, 1994) reported that the reflectance on the surface of a water body depends on physical agents. These agents can be related to the water column depth and the bottom type which are the responsible for the magnitude and spectral composition of the backscattering flux.

To analyze the spatial changes among the estimations and the reference we used a mean error (ME) approach. It was calculated through a simple spatial difference between the estimations and the reference using LEGAL. Figure 4 shows the mean error for the two adaptations models following the two spatial interpolations methods. The results from the first method (calculated models values) are represented in Figure 4a and 4b. The ME results for the second method (spatialization of R_{rs} and map algebra) are represented in Figures 4c and 4d.

As well as the kappa analysis, adaptation 2 got the best results for both methods. However the spatialization of R_{rs} and map algebra method had the best results for the braces of the reservoir. The calculated model values method had the best results to the main body of the reservoir. Furthermore both two adaptation and methods had registered a good accuracy in the region with a high concentration of chl-*a* (region 2 in Figures 2, 3, and 4). But they had a low accuracy in the middle of the reservoir (region 3 in Figure 4) which is the region that had the lowest levels of chl-*a* concentration.

The differences in performance at each region can be related to the instability of the aquatic system' surface. The instable areas can be correlated to reservoirs' bathymetry (Figure 5) since it influences the water column stability. Nixdorf (1994) established that stability of a water column is determined by its depth and the gradients of temperature, i.e. the differences in density. Stability is one of the various factors contributing to the stratification and mixing of an aquatic system. It is also a measure for physical forcing on phytoplankton communities.

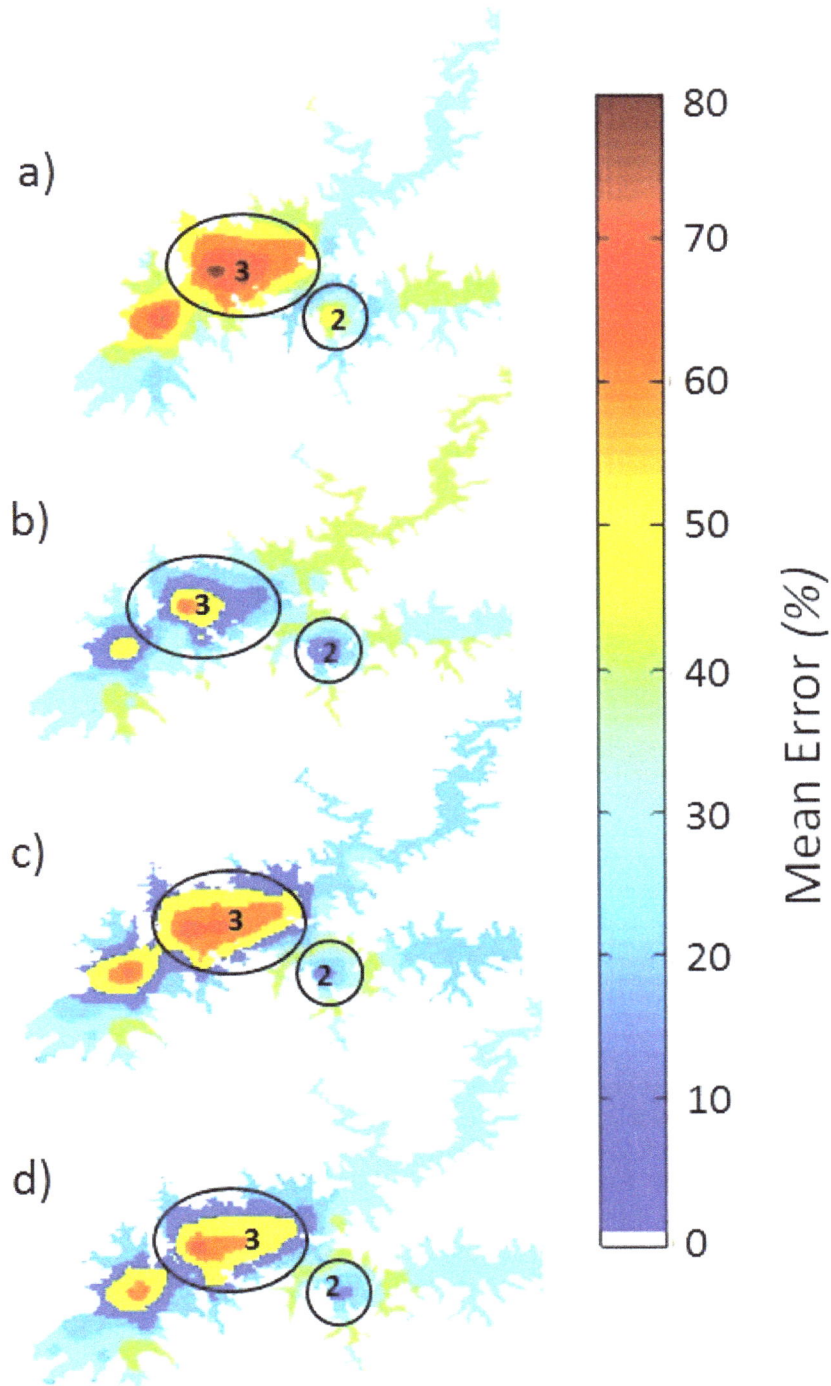

Figure 4. (a) Error of chl-*a* estimation from adaptation 1 by calculated model values method. (b) Error of chl-*a* estimation from adaptation 2 by calculated model values method. (c) Error of chl-*a* estimation from adaptation 1 by spatialization of R_{rs} and map algebra method. (d) Error of chl-*a* estimation from adaptation 2 by spatialization of R_{rs} and map algebra method

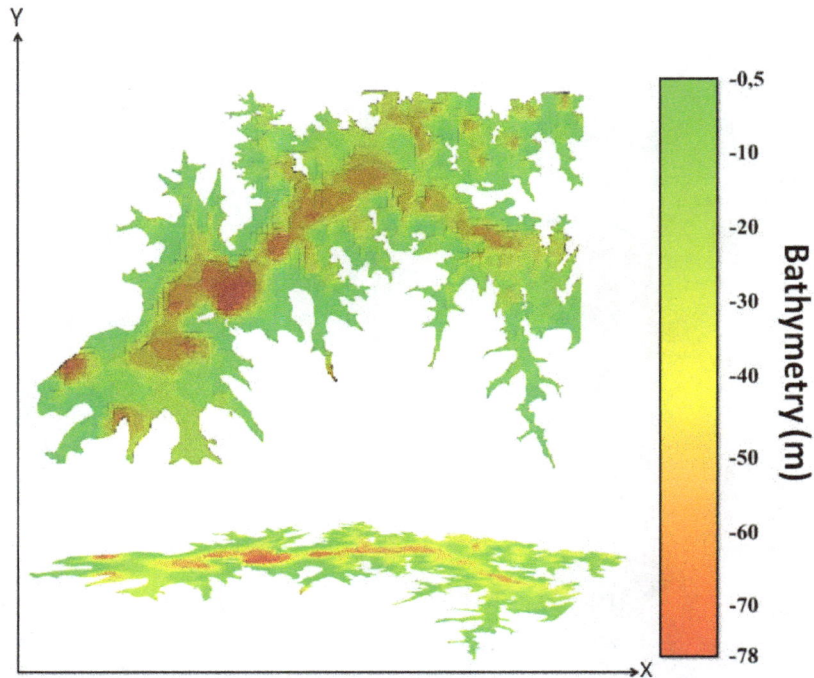

Figure 5. Bathymetry of Itumbiara's Reservoir in meters (adapted from Curtarelli, 2012)

Changes in the aquatic system surface can spoil the estimation of remote sensing reflectance. Besides the deepness of the reservoir, another important factor to the stability of the water column and to R_{rs} estimation is the wind intensity. For a water surface illuminated by irradiance propagating in a direction, it is necessary to compute the irradiance reflectance as a function of wind speed (Cox & Munk, 1954). A Monte Carlo simulation was developed to estimate the upwelling radiance varying the angles and wind speeds (Morel & Gentili, 1996). Through it the authors circumvented their results by using the results for a wind speed of 0. It was noticed that it could lead to significant error for large viewing angles and wind speeds.

Figure 6 shows the spatialization of the euphotic zone–or euphotic depth–for the reservoir. We noticed a dependence of chl-*a* to the highest values of the euphotic zone–regions 2 and 4 (in Figure 6). In region 1, the euphotic zone was shallow mainly due to the entrance of sediments from the Corumbá River. Euphotic zone can also be related to low concentration of chl-*a* in aquatic systems, since it also have a daily variation due to meteorological events as cold fronts, circulation periods and water column stability (Tundisi & Matsumura-Tundisi, 2008).

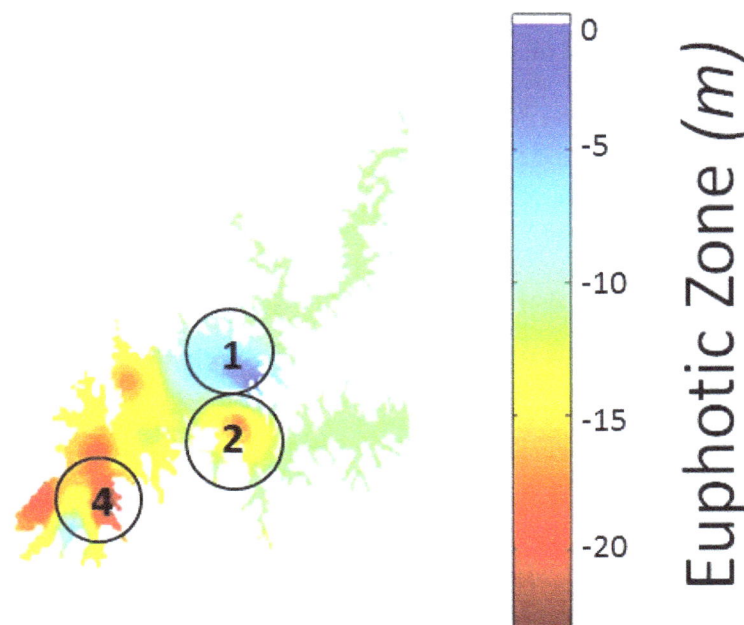

Figure 6. Euphotic Zone of Itumbiara's Reservoir in meters

These results demonstrate the importance for a parameterization of a bio-optical model for tropical complex aquatic systems. The lack of studies for tropical inland waters is important because of the complexity of the environment. They have different atmospheric dynamics, dissolved matter and depth when compared to conventional studies in shallow, small and controlled reservoirs or lakes in latitudes higher than 30°.

The tropical reservoir of Itumbiara is not optically complex as others Brazilian reservoirs. However, due to the instability on the aquatic system surface—due to wind intensity and the reservoir's morphology—we could not estimate the chl-a concentration in some deep regions of the reservoir. The opposite occurred to shallow regions. This implies that the highest errors of the adaptations and even the original models we have tested had a correlation to the most instable regions of the reservoir.

The spatialization of the reference values of chl-a concentration showed a dependence to the instability in the water column. It is related to the mixing process in aquatic systems which generally depends on thermal stratification and wind-induced circulation (Alcântara et al., 2011; Tundisi et al., 2004, 2010). This process can be accepted as a hydrological option for controlling phytoplankton blooms since it physically breaks the blooms.

In the other hand, to the radiometric data, the instability in the water column is totally miscorrelated to the accuracy of the retrieval of R_{rs}. The equations proposed to estimate the R_{rs} from aquatic systems were adequate to estimate the shallow-water reflectance and not deeper waters (Marioneta, Morel, & Gentili, 1994). The spectral shape of the surface reflectance was a function of the spatial variability of bottom albedo, the turbidity of the water and to depth (Burns, Taylor, & Sidhu, 2010).

4. Conclusion

A proximal remote sensing was used to calculate the bio-optical and R_{rs} for the water surface of Itumbiara's reservoir. Two approaches to interpolate the proximal remote sensing data were proposed and analyzed. The results showed that the spatialization of the reflectance had better results when compared to the spatialization of bio-optical model's values. The difference noted between the both approaches is the range used in the spatialization. Since the range of the R_{rs} values were smaller than the range of the bio-optical models values.

More extensive work needs to be done on aquatic systems from various geographic locations, depths and dissolved matters. However the use of proximal remote sensing will allow us to fully understand and quantify the limits of the potential universal applicability of the Remote Sensing products to tropical optically complex aquatic systems.

Acknowledgements

The authors want to thanks Renata Fernandes Figueira Nascimento for allowing us to use material and information regarding her Master Thesis for the Itumbiara Reservoir. The authors also want to thanks Marcelo Pedroso Curtarelli for allowing us to use the bathymetry from the Itumbiara Reservoir. Special thanks to FAPESP Project 2010/15075-8 and CNPq Project 471223/2011-5.

References

Alcântara, E. H., Bonnet, M. P., Assireu, A. T., Stech, J. L., Novo, E. M. L. M., & Lorenzzetti, J. A. (2010). On the water thermal response to the passage of cold fronts: initial results for Itumbiara reservoir (Brazil). *Hydrol. Earth Syst. Sci. Discuss., 7*, 9437-9465. http://dx.doi.org/10.5194/hessd-7-9437-2010

Alcântara, E. H., Novo, E. M. L. M., Barbosa, C. C. F., Bonnet, M. P., Stech, J. L., & Ometto, J. P. (2011). Environmental factors associated with long-term changes in chlorophyll-a concentration in the Amazon floodplain. *Biogeosciences Discuss., 8*, 3739-3770. http://dx.doi.org/10.5194/bgd-8-3739-2011

Austin, R. W. (1980). Gulf of Mexico, ocean-color surface-truth measurements. *Bound-Lay. Meteorol., 18*(3), 269-285. http://dx.doi.org/10.1007/BF00122024

Boyer, J. N., Kelble, C. R., Ortner, P. B., & Rudnick, D. T. (2009). Phytoplankton bloom status: Chlorophyll a biomass as an indicator of water quality condition in the southern estuaries of Florida, USA. *Ecological indicators, 9*, 56-67. http://dx.doi.org/10.1016/j.ecolind.2008.11.013

Burns, B. A., Taylor, J. R., & Sidhu, H. (2010). Uncertainties in Bathymetric Retrievals. *IOP Conf. Series: Earth and Environmental Science, 11*, 1-6. http://dx.doi.org/10.1088/1755-1315/11/1/012032

Camara, G., Souza, R. C. M., Freitas, U. M., & Garrido, J. (1995). SPRING: Integrating remote sensing and GIS by object-oriented data modelling. *Comput. Graph., 20*(3), 395-403. http://dx.doi.org/10.1016/0097-8493(96)00008-8

Cohen, J. (1960). A coefficient of agreement for nominal scales. *Educ. Psychol. Meas., 2*, 37-46. http://dx.doi.org/10.1177/001316446002000104

Cole, G. A. (1994). *Textbook of Limnology* (4th ed.). Illinois, USA: Waveland PressInc.

Cox, C., & Munk, W. (1954). Measurements of the roughness of the sea surface from photographs of the Sun's glitter. *J. Opt. Soc. Am., 44*, 838-850. http://dx.doi.org/10.1364/JOSA.44.000838

Curtarelli, M. P. (2012). *Estudo da influência de frentes frias sobre a circulação e os processos de estratificação e mistura no reservatório de Itumbiara (GO): um enfoque por modelagem hidrodinâmica e sensoriamento remoto.* (Master thesis, Instituto Nacional de Pesquisas Espaciais (INPE), São José dos Campos. (in Portuguese))

Dall'olmo, G., & Gitelson, A. A. (2005). Effect of bio-optical parameter variability on the remote estimation of chlorophyll-a concentration in turbid productive waters: Experimental results. *Appl. Optics, 44*, 412-422. http://dx.doi.org/10.1364/AO.44.000412

Dekker, A. G. (1993). *Detection of optical water quality parameters for eutrophic waters by high resolution remote sensing.* (PhD. thesis. p. 222,Vrije Universiteit, Amsterdam)

Duan, H., Ma, R., Xu, J., Zhang, Y. & Zhang, B. (2010). Comparison of different semi-empirical algorithms to estimate chlorophyll-a concentration in inland lake water. Environmental Monitoring and Assessment, *170*(1-4), 231-44. http://dx.doi.org/10.1007/s10661-009-1228-7

European Space Agency. (2006). *Envisat MERIS Product Handbook*; Issue 2.1, ESA: Paris, France. Retrieved December 24, 2012 from http://envisat.esa.int/pub/ESA_DOC/ENVISAT/MERIS/meris.ProductHandbook.2_1.pdf

Fougnie, B., Frouin, R., Lecomte, P., & Deschamps, P. Y. (1999). Reduction of Skylight Reflection Effects in the Above-Water Measurement of Diffuse Marine Reflectance. *Appl. Optics, 38*(18), 3844-3856. http://dx.doi.org/10.1364/AO.38.003844

Gilerson, Al. A., Gitelson, An. A., Zhou, J., Gurlin, D., Moses, W., Ioannou, I., & Ahmed, S. A. (2010). Algorithms for remote estimation of chlorophyll-a in coastal and inland waters using red and near infrared bands. *Opt. Express, 18*(23-24), 109-124. http://dx.doi.org/10.1364/OE.18.024109

Gitelson, A. A. (1992). The peak near 700 nm on radiance spectra of algae and water: relationships of its magnitude and position with chlorophyll concentration. *Int. J. Remote Sens., 13*(17), 3367-3373.

http://dx.doi.org/10.1080/01431169208904125

Gitelson, A. A., Gritz, U., & Merzlyak, M. N. (2003). Relationships between leaf chlorophyll content and spectral reflectance and algorithms for non-destructive chlorophyll assessment in higher plant leaves. *J. Plant Physiol., 160*, 271-282. http://dx.doi.org/10.1078/0176-1617-00887

Gons, H. J. (1999). Optical teledetection of chlorophyll a in turbid inland waters. *Environ. Sci. Technol., 33*, 1127-1132. http://dx.doi.org/10.1021/es9809657

Gons, H. J., Auer, M. T., & Effler, S. W. (2008). MERIS satellite chlorophyll mapping of oligotrophic and eutrophic waters in the Laurentian great Lakes. *Remote Sens. Environ., 112*, 4098-4106. http://dx.doi.org/10.1016/j.rse.2007.06.029

Goovaerts, P. (1997). *Geostatistics for Natural Resources Evaluation*. New York, USA: Oxford University Press.

Gordon, H. R., & Morel, A. Y. (1983). *Remote Assessment of Ocean Color for Interpretation of Satellite Visible Imagery: A Review*. New York, United States: Springer-Verlag.

Gordon, H. R., Brown, O. B., & Jacobs, M. M. (1975). Computed relationship between the inherent and apparent optical properties of a flat homogeneous ocean. *Appl. Optics, 14*, 417-427. http://dx.doi.org/10.1364/AO.14.000417

Gordon, H. R., Brown, O. B., Evans, R. H., Brown, J. W., Smith, R. C., Baker, K. S., & Clark, D. K. (1988). A semianalytic radiance model of ocean color. *J. Geophys. Res., 93*, 10909-10924. http://dx.doi.org/10.1029/JD093iD09p10909

Huot, Y., Babin, M., Bruyant, F., Grob, C., Twardowski, M. S., & Claustre, H. (2007). Does chlorophyll a provide the best index of phytoplankton biomass for primary productivity studies? *Biogeosciences Discuss., 4*, 707-745. http://dx.doi.org/10.5194/bgd-4-707-2007

Kirk, J. T. O. (1994). *Light & photosynthesis in aquatic ecosystems* (2nd ed.). Melbourne, Australia: Cambridge University Press.

Kirk, J. T. O. (2011). *Light & photosynthesis in aquatic ecosystems* (3rd ed.). Melbourne, Australia: Cambridge University Press.

Köppen, W. (1931). *Grundriss der Klimakund*. Berlin: Walter de Gruyter.

Le, C., Li, Y., Zha, Y., Sun, D., Huang, C., & Lu, H. (2009). A four-band semi-analytical model for estimating chlorophyll a in highly turbid lakes: the case of Taihu Lake, China. *Remote Sens. Environ., 113*, 1175-1182. http://dx.doi.org/10.1016/j.rse.2009.02.005

Maritorena, S., Morel, A., & Gentili, B. (1994). Diffuse reflectance of oceanic shallow waters: Influence of water depth and bottom albedo. *Limnol. Oceanogr., 39*(7), 1689-1703.

Martin, J. L., & McCutcheon, S. C. (1999). *Hydrodynamics and Transport for Water Quality Modeling*. Boca Raton, USA: Lewis Publishers.

Melack, J. (Ed.) (2000). *Planning and Management of Lakes and Reservoirs, An Integrated Approach to Eutrophication–A training module*. Tech. Publ. Ser. 12. Shiga, Japan: UNEP International Environmental Technology Centre.

Mishra, S., & Mishra, D. R. (2012). Normalized difference chlorophyll index: A novel model for remote estimation of chlorophyll-a concentration in turbid productive waters. *Remote Sens. Environ., 117*, 394-406. http://dx.doi.org/10.1016/j.rse.2011.10.016

Mitchell, B. G., Kahru, M., Wieland, J., & Stramska, M. (2003). Ocean optics protocols for satellite ocean color sensor validation, rev. 4, in vol. IV. In J. L. Mueller, G. S. Fargion, & C. R. McClain (Eds.), *Inherent optical properties: Instruments, characterizations, field measurements and data analysis protocols* (pp. 39-64) (TM-2003-211621/Rev4-Vol.IV (NASA, 2003)).

Mobley, C. D., Stramski, D., Bisset, W. P., & Boss, E. (2003). Optical Modeling of Ocean Water: Is the Case1 - Case 2 Classification still useful? *Oceanography, 17*(2), 60-67. http://dx.doi.org/10.5670/oceanog.2004.48#sthash.h8tXHlX3.dpuf

Morel, A., & Gentili, B. (1996). Diffuse reflectance of oceanic waters. III. Implication of bidirectionality for the remote sensing problem. *Appl. Optics, 35*, 4850-4862. http://dx.doi.org/10.1364/AO.35.004850.

Moses, W. J., Gitelson, A. A., Berdnikov, S., Saprygin, V., & Povazhnyi, V. (2012). Operational MERIS-based NIR-red algorithms for estimating chlorophyll-a concentrations in coastal waters–The Azov Sea case study.

Remote Sens. Environ., 121, 118-124. http://dx.doi.org/10.1016/j.rse.2012.01.024

Mudroch, A. (Ed.) (1999). *Planning and Management of Lakes and Reservoirs, An Integrated Approach to Eutrophication.* Tech. Publ. Ser. 11. Shiga, Japan: UNEP International Environmental Technology Centre.

Nascimento, R. F. F. (2010). *Utilização de dados MERIS e in situ para a caracterização bio-óptica do reservatório de Itumbiara, GO.* (Master thesis, Instituto Nacional de Pesquisas Espaciais (INPE), São José dos Campos (in Portuguese)).

Nixdorf, B. (1994). Polymixis of a shallow lake (GroBer Miiggelsee, Berlin) and its influence on seasonal phytoplankton dynamics. *Hydrobiologia, 275/276*, 173-186. http://dx.doi.org/10.1007/BF00026709

Nush, E. A. (1980). Comparison of different methods for chlorophyll and phaeopigment determination. *Arch Hydrobiol Beih Ergebn Limnol, 14*, 14-36.

Rees, W. G. (2001). *Physical principles of remote sensing.* Cambridge, England: Cambridge University Press.

Ruddick, K. G., Gons, H. J., Rijkeboer, M., & Tilstone, G. (2001). Optical remote sensing of chlorophyll a in case 2 waters by use of an adaptive two-band algorithm with optimal error properties. *Appl. Optics, 40*(21), 3575-3585. http://dx.doi.org/10.1364/AO.40.003575

Satlantic. (2000). *SeawiFS Proling Multichanel Radiometer User's Manual* SPMR/SMSR 041, Issue/Rev. 1/1.

Tundisi, J. G., Matsumura-Tundisi, T., Arantes Jr, J. D., Tundisi, J. E., Manzini, N. F., & Ducrot, R. (2004). The response of Carlos Botelho (Lobo, Broa) reservoir to the passage of cold fronts as reflected by physical, chemical, and biological variables. *Braz. J. Biol., 64*(1), 177-186. http://dx.doi.org/10.1590/S1519-69842004000100020.

Tundisi, J. G., Matsumura-Tundisi, T., Pereira, K. C., Luzia, A. P., Passerini, M. D., Chiba, W. A. C., ... Sebastien, N. Y. (2010). Cold fronts and reservoir limnology: an integrated approach towards the ecological dynamics of freshwater ecosystems. *Braz. J. Biol., 70*(3), 815-824. http://dx.doi.org/10.1590/S1519-69842010000400012

Tundisi, J. G., & Matsumura-Tundisi, T. (2008). *Limnologia.* São Paulo, Brasil: Oficina de Textos (in Portuguese).

Wetzel, R. G., & Likins, G. E. (1991). *Limnological analyses* (2nd ed.). New York, USA: Springer.

Permissions

The contributors of this book come from diverse backgrounds, making this book a truly international effort. This book will bring forth new frontiers with its revolutionizing research information and detailed analysis of the nascent developments around the world.

We would like to thank all the contributing authors for lending their expertise to make the book truly unique. They have played a crucial role in the development of this book. Without their invaluable contributions this book wouldn't have been possible. They have made vital efforts to compile up to date information on the varied aspects of this subject to make this book a valuable addition to the collection of many professionals and students.

This book was conceptualized with the vision of imparting up-to-date information and advanced data in this field. To ensure the same, a matchless editorial board was set up. Every individual on the board went through rigorous rounds of assessment to prove their worth. After which they invested a large part of their time researching and compiling the most relevant data for our readers.

The editorial board has been involved in producing this book since its inception. They have spent rigorous hours researching and exploring the diverse topics which have resulted in the successful publishing of this book. They have passed on their knowledge of decades through this book. To expedite this challenging task, the publisher supported the team at every step. A small team of assistant editors was also appointed to further simplify the editing procedure and attain best results for the readers.

Apart from the editorial board, the designing team has also invested a significant amount of their time in understanding the subject and creating the most relevant covers. They scrutinized every image to scout for the most suitable representation of the subject and create an appropriate cover for the book.

The publishing team has been an ardent support to the editorial, designing and production team. Their endless efforts to recruit the best for this project, has resulted in the accomplishment of this book. They are a veteran in the field of academics and their pool of knowledge is as vast as their experience in printing. Their expertise and guidance has proved useful at every step. Their uncompromising quality standards have made this book an exceptional effort. Their encouragement from time to time has been an inspiration for everyone.

The publisher and the editorial board hope that this book will prove to be a valuable piece of knowledge for researchers, students, practitioners and scholars across the globe.

List of Contributors

Matobola J. Mihale
Department of Analytical and Environmental Chemistry, Vrije Universiteit Brussel (VUB), Pleinlaan, Brussels, Belgium
Department of Physical Sciences, Open University of Tanzania (OUT), Dar es Salaam, Tanzania

Kim Croes
Department of Analytical and Environmental Chemistry, Vrije Universiteit Brussel (VUB), Pleinlaan, Brussels, Belgium

Clavery Tungaraza
Department of Physical Sciences, Sokoine University of Agriculture (SUA), Chuo Kikuu Morogoro, Tanzania

Willy Baeyens
Department of Analytical and Environmental Chemistry, Vrije Universiteit Brussel (VUB), Pleinlaan, Brussels, Belgium

Kersten Van Langenhove
Department of Analytical and Environmental Chemistry, Vrije Universiteit Brussel (VUB), Pleinlaan, Brussels, Belgium

Leonid Tartakovsky
Faculty of Mechanical Engineering, Technion–Israel Institute of Technology, Haifa, Israel

Marcel Gutman
Faculty of Mechanical Engineering, Technion–Israel Institute of Technology, Haifa, Israel

Doron Popescu
Faculty of Mechanical Engineering, Technion–Israel Institute of Technology, Haifa, Israel

Michael Shapiro
Faculty of Mechanical Engineering, Technion–Israel Institute of Technology, Haifa, Israel

Chuanqi Fan
College of Economics & Management, Sichuan Agricultural University, Chengdu, China

Xiaojun Zheng
School of International Law, Southwest University of Political Science and Law, Chongqing, China

Ashar Hasairin
Biology Education Department, Faculty of Mathematic and Science, State University of Medan, Indonesia

Nursahara Pasaribu
North Sumatra University, Medan, Indonesia

Lisdar I. Sudirman
Bogor Agricultural Institute, Indonesia

Retno Widhiastuti
North Sumatra University, Medan, Indonesia

Hiromi Ikeura
Organization for the Strategic Coordination of Research and Intellectual Properties, Meiji University, Kanagawa, Japan

Nanako Narishima
School of Agriculture, Meiji University, Kanagawa, Japan

Masahiko Tamaki
School of Agriculture, Meiji University, Kanagawa, Japan

Edward B Ilgren
Formerly, University of Oxford, Faculty of Biological & Agricultural Sciences, Department of Pathology, Oxford, UK

Drew Van Orden
RJ Lee Group, Monroeville, PA, USA

Richard Lee
RJ Lee Group, Monroeville, PA, USA

Yumi Kamiya
Formely, Bryn Mawr College for Women, PA, USA

John A Hoskins
Formerly, Medical Research Council, Leicester, UK

Douglas B. Sims
College of Southern Nevada, Department of Physical Sciences, North Las Vegas, Nevada, USA

Peter S. Hooda
Centre for Earth and Environmental Sciences Research, Kingston University London, Kingston upon Thames, UK

Gavin K. Gillmore
Centre for Earth and Environmental Sciences Research, Kingston University London, Kingston upon Thames, UK

Morris E. Demitry
Utah Water Research Laboratory, USA

Michael J. McFarland
Utah Water Research Laboratory, USA

Kohei Makita
School of Veterinary Medicine, Rakuno Gakuen University, 582 Bunkyodai Midorimachi, Ebetsu, Japan

Kazuto Inoshita
School of Veterinary Medicine, Rakuno Gakuen University, 582 Bunkyodai Midorimachi, Ebetsu, Japan

Taishi Kayano
School of Veterinary Medicine, Rakuno Gakuen University, 582 Bunkyodai Midorimachi, Ebetsu, Japan

Kei Uenoyama
School of Veterinary Medicine, Rakuno Gakuen University, 582 Bunkyodai Midorimachi, Ebetsu, Japan

Katsuro Hagiwara
School of Veterinary Medicine, Rakuno Gakuen University, 582 Bunkyodai Midorimachi, Ebetsu, Japan

Mitsuhiko Asakawa
School of Veterinary Medicine, Rakuno Gakuen University, 582 Bunkyodai Midorimachi, Ebetsu, Japan

Kenta Ogawa
College of Agriculture, Food and Environmental Sciences, Rakuno Gakuen University, Bunkyodai Midorimachi, Ebetsu, Japan

Shin'ya Kawamura
Department of Human Sciences, Graduate School of Letters, Hokkaido University, Kita-ku, Sapporo, Japan

Jun Noda
School of Veterinary Medicine, Rakuno Gakuen University, 582 Bunkyodai Midorimachi, Ebetsu, Japan

Koichiro Sera
Cyclotron Research Center, Iwate Medical University, Takizawa, Iwate, Japan

Hitoshi Sasaki
College of Agriculture, Food and Environmental Sciences, Rakuno Gakuen University, Bunkyodai Midorimachi, Ebetsu, Japan

Nobutake Nakatani
College of Agriculture, Food and Environmental Sciences, Rakuno Gakuen University, Bunkyodai Midorimachi, Ebetsu, Japan

Hidetoshi Higuchi
School of Veterinary Medicine, Rakuno Gakuen University, 582 Bunkyodai Midorimachi, Ebetsu, Japan

Naohito Ishikawa
Action for Peace, Capability and Sustainability (APCAS), Colombo, Sri Lanka

Hidetomo Iwano
School of Veterinary Medicine, Rakuno Gakuen University, 582 Bunkyodai Midorimachi, Ebetsu, Japan

Yutaka Tamura
School of Veterinary Medicine, Rakuno Gakuen University, 582 Bunkyodai Midorimachi, Ebetsu, Japan

Katarzyna Sygit
Department of Physical Education and Health Promotion, University of Szczecin, Poland

Witold Kołłątaj
Department of Paediatric Endocrinology and Diabetology, Medical University of Lublin, Poland

Marian Sygit
Department of Physical Education and Health Promotion, University of Szczecin, Poland
Department of Physical Education and Health Education, University of Szczecin; Institute of Rural Health, Lublin, Poland

Barbara Kołłątaj
Department of Epidemiology, Medical University of Lublin, Poland

Ryszard Kolmer
Voivodship Sanitary and Epidemiological Station in Szczecin, Poland

Renata Opiela
Voivodship Sanitary and Epidemiological Station in Szczecin, Poland

Paweł Zienkiewicz
Department of Physical Education and Health Promotion, University of Szczecin, Poland

Mathieu Nsenga Kumwimba
Institute of Environment and Sustainable Development in Agriculture, Chinese Academy of Agricultural Sciences / Key Laboratory of Agro-Environment, Ministry of Agriculture, Beijing, China

Xibai Zeng
Institute of Environment and Sustainable Development in Agriculture, Chinese Academy of Agricultural Sciences / Key Laboratory of Agro-Environment, Ministry of Agriculture, Beijing, China

Lingyu Bai
Institute of Environment and Sustainable Development in Agriculture, Chinese Academy of Agricultural Sciences / Key Laboratory of Agro-Environment, Ministry of Agriculture, Beijing, China

Jinjin Wang
Institute of Environment and Sustainable Development in Agriculture, Chinese Academy of Agricultural Sciences / Key Laboratory of Agro-Environment, Ministry of Agriculture, Beijing, China

Teresa da Silva-Rosa
Vila Velha University, Center of Socioenvironmental and Urban Studies -- NEUS/UVV, Brazil

Michelle Bonatti
University Buenos Aires -UBA, Argentina

Andrea Vanini
Technical Office for Management and Environmental Education at the Fiocruz Atlantic Forest Campus, Brazil

Catia Zuffo
Federal University at Rondônia – UNIR, Brazil

Mohamad Aflatooni
Islamic Azad University, Shiraz Branch, Department of Water Resources, Shiraz, Iran

Jafar Aghajanzadeh
Islamic Azad University, Shiraz Branch, Department of Water Resources, Shiraz, Iran

Ruth Legg
School of Community and Regional Planning, University of British Columbia, Vancouver, BC, Canada

Jennie Moore
School of Community and Regional Planning, University of British Columbia, Vancouver, BC, Canada

Meidad Kissinger
Department of Geography and Environmental Development, Ben-Gurion University of the Negev, Beer-Sheva, Israel

William Rees
School of Community and Regional Planning, University of British Columbia, Vancouver, BC, Canada

Sarah Azzouz
Chemistry Department, University Constantine 1, Constantine, Algeria

Smaine Chellat
Earth Science Department, University Kasdi Merbah, Ouargla, Algeria

Chahrazed Boukhalfa
Chemistry Department, University Constantine 1, Constantine, Algeria

Abdeltif Amrane
Ecole Nationale Supérieure de Chimie de Rennes, Université Rennes1, CNRS, UMR 6226, Avenue du Général Leclerc, CS 50837, 35708 Rennes Cedex 7, France

Igor Ogashawara
Remote Sensing Division, National Institute for Space Research, São José dos Campos, Brazil

Enner Herenio de Alcântara
Cartography Department, São Paulo State University, Presidente Prudente, Brazil

Pétala Bianchi Augusto-Silva
Remote Sensing Division, National Institute for Space Research, São José dos Campos, Brazil

Claudio Clemente Faria Barbosa
Remote Sensing Division, National Institute for Space Research, São José dos Campos, Brazil

José Luiz Stech
Remote Sensing Division, National Institute for Space Research, São José dos Campos, Brazil

www.ingramcontent.com/pod-product-compliance
Lightning Source LLC
Chambersburg PA
CBHW080637200326
41458CB00013B/4659